Medical Applications of Colloids

Erratum

XPS Analysis of Biosystems and Biomaterials

Michel J. Genet, Christine C. Dupont-Gillain, and Paul G. Rouxhet

Unité de chimie des interfaces, Université catholique de Louvain,
Croix du Sud 2/18, B-1348, Louvain-la-Neuve, Belgium

E. Matijević (ed.) *Medical Applications of Colloids*, DOI: 10.1007/978-0-387-76921-9,
pp. 177–307, © Springer Science+Business Media, LLC 2008

DOI 10.1007/978-0-387-76921-9_6

The publisher regrets that in the online version of this book the metadata for the title, authors, and authors' affiliation for the chapter entitled "XPS Analysis of Biosystems and Biomaterials" by Michel J. Genet, Christine C. Dupont-Gillain, and Paul G. Rouxhet are incorrect. The .pdf file for the chapter is correct.

The online version of the original chapter can be found at
http://dx.doi.org/10.1007/978-0-387-76921-9_5

Egon Matijević
Editor

Medical Applications of Colloids

Editor
Egon Matijević
Department of Chemistry
Clarkson University
Potsdam, NY 13699, USA
matiegon@clarkson.edu

ISBN: 978-0-387-76920-2   e-ISBN: 978-0-387-76921-9
DOI: 10.1007/978-0-387-76921-9

Library of Congress Control Number: 2008924175

© 2008 Springer Science+Business Media, LLC
All rights reserved. This work may not be translated or copied in whole or in part without the written permission of the publisher (Springer Science+Business Media, LLC, 233 Spring Street, New York, NY 10013, USA), except for brief excerpts in connection with reviews or scholarly analysis. Use in connection with any form of information storage and retrieval, electronic adaptation, computer software, or by similar or dissimilar methodology now known or hereafter developed is forbidden.
The use in this publication of trade names, trademarks, service marks, and similar terms, even if they are not identified as such, is not to be taken as an expression of opinion as to whether or not they are subject to proprietary rights.

Cover illustration: Follicles of colloid in thyroid

Printed on acid-free paper

9 8 7 6 5 4 3 2 1

springer.com

Preface

The important role of finely dispersed matter and surfaces in medicine is not always fully understood and appreciated. Specifically, fine particles (solid or liquid) in the size range of several nanometers to several micrometers have a tremendous effect on our lives, because they can be beneficial or detrimental to our well-being. Such particles are present in living bodies as red blood cells or cholesterol crystals in the gall bladder. They are ubiquitous in the environment, where they can cause many diseases, such as asbestosis, silicosis, and black lung disease, but they are also used in diagnostic tests, drug delivery, and numerous other applications. More recently, evidence has become available that drug formulations with active components in a finely dispersed state may significantly affect their functionality. Furthermore, with miniaturization of medical instrumentation, the size of the components is necessarily reduced to colloid or even smaller range.

This volume is a collection of several chapters dealing with diverse topics of colloids and surfaces relevant to medical applications. Thus, Siiman describes the use of optical properties of uniform colloidal particles as probes in flow cytometry.

Giesche focuses on the preparations and properties of exceedingly uniform silica spheres for different uses, such as in chromatography. In modified forms, silica particles with incorporated dyes are employed in diagnostics and those combined with tiny magnetic entities in drug delivery.

Bosch deals with the role of fine particles in pharmaceutical and therapeutic applications, with special emphasis on the size of the dispersed bioactive materials. The latter can be controlled by mechanical means or by precipitation processes.

Ahmadi and McLaughlin describe biomedical applications of particle transport and deposition. Special attention is given to recent advances in the use of computational models for predicting the transport, dispersion, and deposition of particles in the human airway passages. These include airflow and particle transport in the nose, oral airways, lung bifurcation, and alveolar cavities. In addition, an overview of advances in blood flow simulations in various arteries is presented.

Genet et al. contribute a comprehensive chapter on X-ray photoelectron spectroscopy of biomaterials and biosystems. Thus, valuable information can be obtained on the composition and interfacial phenomena of microorganisms. Other uses involve detection of the modification of materials to biological exposure, as exemplified by titanium implants in reconstruction and many other cases.

In summary, this volume offers insight into a number of applications of colloid and surface science and techniques of essence in diverse medical applications.

Potsdam, New York, USA

Egon Matijević

Contents

Colloids as Light Scattering and Emission Markers for Analysis of Blood 1
Olavi Siiman

Medical and Technological Application of Monodispersed Colloidal Silica Particles 43
Herbert Giesche

Pharmaceutical Applications of Finely Dispersed Systems 69
H. William Bosch

Transport, Deposition, and Removal of Fine Particles: Biomedical Applications 95
Goodarz Ahmadi and John B. McLaughlin

XPS Analysis of Biosystems and Biomaterials 177
Michel J. Genet, Christine C. Dupont-Gillain, and Paul G. Rouxhet

Index 309

Contributors

Goodarz Ahmadi
Department of Mechanical and Aeronautical Engineering, Clarkson University, Potsdam, NY 13699, USA, ahmadi@clarkson.edu

H. William Bosch
Philadelphia Naval Business Center iCeutica, Inc. Philadelphia, PA 19112, USA, bill@iceutica.com

Christine C. Dupont-Gillain
Unité de chimie des interfaces, Université catholique de Louvain, Croix du Sud 2/18, B-1348, Louvain-la-Neuve, Belgium

Michel J. Genet
Unité de chimie des interfaces, Université catholique de Louvain, Croix du Sud 2/18, B-1348, Louvain-la-Neuve, Belgium

Herbert Giesche
School of Ceramic Engineering and Science, New York State College of Ceramics at Alfred University, Alfred, NY 14802-1296, USA, giesche@alfred.edu

John B. McLaughlin
Department of Chemical and Biomolecular Engineering, Clarkson University, Potsdam, NY 13699, USA, jmclau@clarkson.edu

Paul G. Rouxhet
Unité de chimie des interfaces, Université catholique de Louvain, Croix du Sud 2/18, B-1348, Louvain-la-Neuve, Belgium, rouxhet@cifa.ucl.ac.be

Olavi Siiman
5920 Cobblestone Court, Davie, FL 33331, USA, siiman@ix.netcom.com

Colloids as Light Scattering and Emission Markers for Analysis of Blood

Olavi Siiman

Introduction

Topics in this review are focused on the author's specific interests in particle probes for flow cytometry/cell sorting. Five different classes of bead probes for flow cytometric analyses are surveyed. Two of the five classes—light scatter and fluorescence emission/light scatter bead probes—have the longest history of use, and one, the fluorescent beads, is enjoying a renaissance as reagents in multiplex flow analyses. Surface plasmon resonance/light scatter probes have yet to find their niche; however, activity in this area has expanded lately. The relatively new field of luminescence emission/light scatter probes is experiencing growing pains as methods of stabilizing the high emission intensities of highly reactive, semiconductor nanoparticles are perfected. Finally, the future is wide open for expansion into the field of enhanced Raman/light scatter probes. This would add another dimension to flow cytometry, which has not experienced a major modification since the introduction of the first commercial instruments in the mid-1970s.

To give physical scientists an appreciation for the impact that flow cytometry has had in the biological sciences, it has been said that flow cytometry is to the cell biologist as nuclear magnetic resonance is to the organic chemist. Since flow cytometry/flow cytometers are not usually found in the toolbox of physical scientists, it is appropriate to start with a brief introduction to the instrumentation and technique.

Description of Flow Cytometry

Flow cytometry involves particles or cells flowing through a laser beam. Signals are produced and provide information about the particle. The process begins with an isotonic liquid called sheath fluid, which enters the flow chamber from the side and flows downward in the Coulter® EPICS™ Elite™ series flow cytometers that are used for analyses as well as for sorting cells, but flows upward in the Coulter® EPICS™ XL™ flow cytometer that is used solely for analyses. A particle/cell suspension is inserted into the flow chamber. The particles are directed downward/upward in a stream through the center of the sheath fluid and in the same direction as the flow of sheath fluid. The flow of sample suspension and sheath fluid through the flow cell is laminar and the fluids do not mix. The pressure of the sheath fluid against the suspension aligns the particles in single file by hydrodynamic focusing. Inside the flow chamber, particles are passed one by one through a very narrow cross-section beam of monochromatic light from a laser. The stream of sample particles flows through the laser beam, intersecting it at a right angle. Light is scattered from each particle in all directions, but is typically collected by a forward scatter detector and a side scatter detector, each converted to electronic pulses with voltages proportional to the intensity of light. The electronic pulses are amplified and processed digitally, so that the peak voltage is converted to a proportional channel number for a histogram or list mode data. The magnitude of the forward scatter pulse is roughly proportional to the size of the particle. For particles larger than 20 µm, a neutral density filter may be inserted in the light path to reduce the amount of light and the pulse height. Laser light deflected off internal structures or granules within a particle and surface irregularities supplies the majority of the side scatter light. Fluorescent dyes may also be bound within or on a particle. These dyes absorb the incident laser light energy and emit fluorescence of different colors. Emitted light of different colors are separated by filters and directed to photomultiplier tubes (PMTs), which act as sensors. The PMT converts the collected light to a voltage pulse. Greater than eight PMTs have been used to allow detection of as many colors simultaneously; however, more typically 4 or 6 colors are used simultaneously.

Furthermore, in the analysis of data, the correlated channel numbers for the parameters acquired for each particle that is analyzed are stored and used with any gates to create histograms. A single-parameter histogram is a graph of particle count on the y-axis and the measurement parameter on the x-axis. For each particle, the channel value from the analog-to-digital convertor is taken and the count is incremented in the histogram channel. Height above the axis represents channel count. All one-parameter histograms have 1,024 channels. A gate for a single-parameter histogram is a simple range of channels. A two-parameter histogram is a graph (similar to a topographical map) of cell count and two measurement parameters. The x- and y-axes are assigned to the parameters with channel 0,0 at the lower left. A computer takes the channel values of the parameters and increments the count at the intersection of the channels on the two-parameter histogram. On each axis of the two-parameter histograms 64 or 256 channels can be selected. Particle counts at the intersection of two channels are shown by dot density. A gate for a dual-parameter histogram can be rectilinear or amorphous. An amorphous gate (bitmap) is an area defined within a two-parameter histogram, which can have any shape. In the calculation, the channel values of the two-parameter measurements are taken and the coordinates are compared to the amorphous region coordinates. If the particle falls inside the region, the correlated measurements are included in any histograms gated on the region. A more detailed exposition[1] of various elements of flow cytometry can be found elsewhere.

Normal Light Scatter Bead/Nanoparticle Probes

Background

Polystyrene (PS) latex beads of uniform size and shape in the 0.2 to ~5.0 μm diameter range have found use in flow cytometric assays for cells in whole blood. The light scatter patterns of forward versus side scatter histograms of targeted biological cells, when the beads have been conjugated with specific binding agents such as monoclonal antibodies, are changed so that the position of the targeted cell population does not overlap with the position of any nontargeted, nonshifted set of cells. In general, biological assays for white blood cell (WBC) subclasses could be designed with a single specific bead targeting only one subset of WBCs, processed in series. If parallel assays were desired, multiple bead types targeting two or more subsets could be used simultaneously so long as the subsets represented mutually exclusive populations of cells.

Overviews[2,3] of recent developments in beads, including fluorescent ones, have been provided by The Latex Course, recently sponsored by Bangs Laboratories. Also, the most recent edition[1] of the "nuts-and-bolts" book of flow cytometry provides an update on literature in the area of bead probes. Furthermore, the International Federation of Clinical Chemistry (IFCC) Working Group on Nanotechnology, chaired by Larry J. Kricka (University of Pennsylvania Medical Center), has recently compiled literature[4] on microarrays and microchips, the fourth part of which will cover literature on protein, peptide, and antibody microarrays and microchips to the middle of 2003, and will be made available via the IFCC web page. The latter literature includes that of bead suspension arrays.

Early work in this field for flow cytometric applications was suggested by Mack Fulwyler. Seminal publications[5–8] were followed by the work of Thomas Russell and coworkers, which gave rise to a series of patents[9–13] describing the use of adsorbed antibodies on PS latex beads in bead assays for subpopulations of WBCs.

Our own work[14–19] provided more stable systems by using covalently conjugated antibodies on gelatin-, aminodextran-, and diaminopropane-coated PS latex or magnetic beads. Micrographs of CD4+ lymphocytes in a sample of whole blood from a normal donor as stained with 2.13 μm diameter PS latex beads, conjugated with CD4 antibody (IgG1, clone SFCI12T4D11) are shown in Fig. 1. The term "CD" refers to "Cluster Designation" adopted by the International Workshops on Human Leukocyte Differentiation Antigens. Targeted CD4+ lymphocytes are seen coated with a closest-packed layer of beads of 2.13 μm diameter, which appear to be physically pushed aside by the applied cover slip.

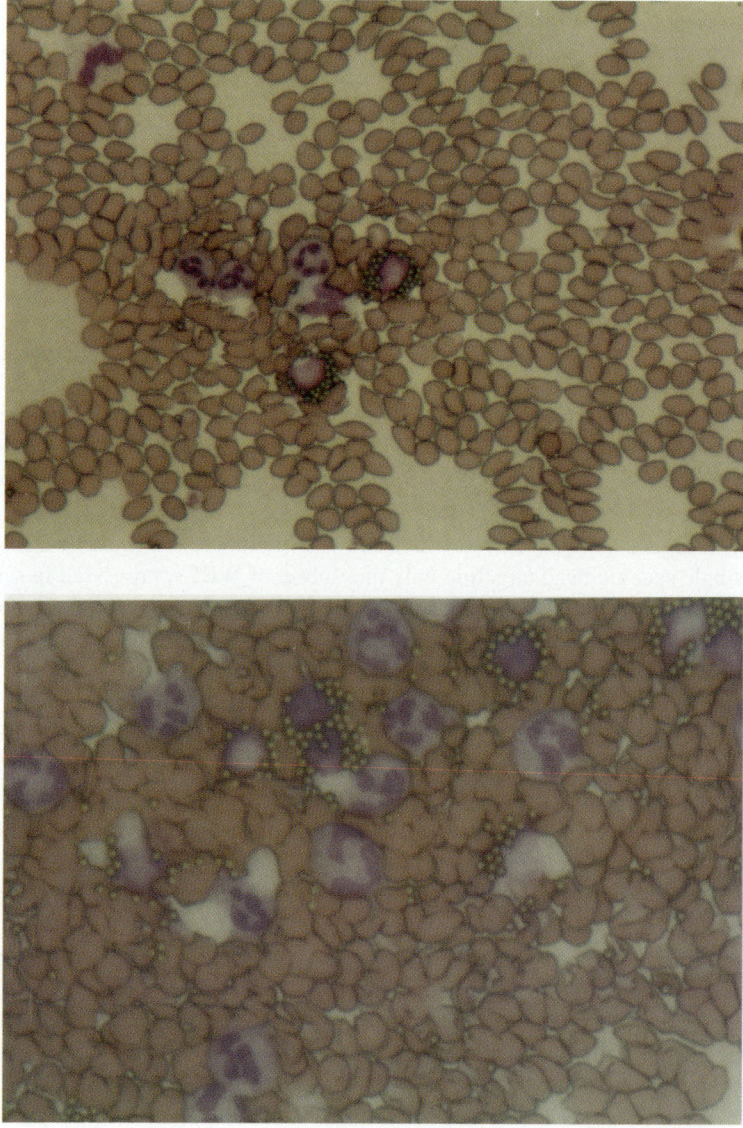

Fig. 1 Wright's stain smears (air dried and mounted) of EDTA stabilized whole blood mixed with CD4 antibody-conjugated PS latex beads of 2.13 μm diameter

CD3 Antibody–Beads

The following example taken from our unpublished work [O. Siiman and A. Burshteyn] on CD3 antibody–beads illustrates the strategies involved in obtaining optimum antibody–bead performance in bead–blood assays. The main factors affecting the bead assay performance with whole blood were identified as (1) bead size, (2) antibody avidity, (3) bead titer, and (4) bead–blood mixing time. Different sizes of PS latex beads in the 2.0–4.0 μm diameter were examined for their light scatter shift properties on targeted lymphocytes, typically of 8–12 μm diameter in whole blood. PS aldehyde/sulfate latex beads (Interfacial Dynamics Corp.) of mean diameters (coefficients of variation), 2.13 (2.7%), 2.4 (2.0%), 3.2 (7.0%), and 4.00 (3.3%), were used to prepare CD3 antibody conjugates(lots 1A to 4A and 1B to 4B) according to the covalent antibody coupling procedure described previously[19] for CD4 and CD8 antibody beads. In each preparation, the same lot of CD3 antibody (IgG1, clone UCHT1), activated at 15:1(A) and 7.5:1(B) IT:Ab (2-iminothiolane: antibody) ratios, was used. The Coulter® STKS2B™ hematology analyzer assay for CD3+ lymphocytes was performed with 40 μL of 4% w/v solid beads per 400 μL of whole blood. The bead–blood mixing times were 2 and 8 min. Lyse of red blood cells is automatically performed in the analyzer. The results for %CD3+ lymphocytes in one run are shown in Table 1.

The Coulter® STKS2B™ hematology analyzer histograms of DF 8 (median angle light scatter) versus Coulter volume shown in Fig. 2 for the above bead–blood mixtures suggest that the best results are obtained with 2.40 μm beads, since the angle at which the targeted population is shifted with larger 3.20 and 4.00 μm beads is increased too far so that it begins to overlap with the nonshifted granulocyte population of WBCs. The slightly larger 2.40 μm versus 2.13 μm beads offer some

Table 1 Effects of bead size, antibody activation, and bead–blood mixing time on percentage of CD3+ lymphocytes in whole blood of normal donor

Bead dia. (μm)	IDC bead lot no.	IT:Ab ratio	CD3 bead lot	Mixing time (min)	% CD3 Trial 1	Trial 2	Trial 3
2.13	540,3	15:1	1A	2	a.f.	a.f.	a.f.
				8	a.f.	86.18	a.f.
2.13	540,3	7.5:1	1B	2	a.f.	a.f.	a.f.
				8	83.18	83.11	84.87
2.4	561,1	15:1	2A	2	a.f.	a.f.	83.58
				8	86.54	86.77	87.36
2.4	561,1	7.5:1	2B	2	73.36	a.f.	a.f.
				8	87.41	85.69	86.79
3.2	542,2	15:1	3A	2	82.39	79.69	80.03
				8	85.19	84.54	86.22
3.2	542,2	7.5:1	3B	2	78.54	83.52	80.88
				8	82.79	83.17	83.54
4.00	506	15:1	4A	2	78.99	79.71	81.99
				8	82.28	81.99	82.11
4.00	506	7.5:1	4B	2	80.55	78.64	81.15
				8	83.41	82.21	85.35
Flow Comparator						86.5	

a.f. = algorithm failure.

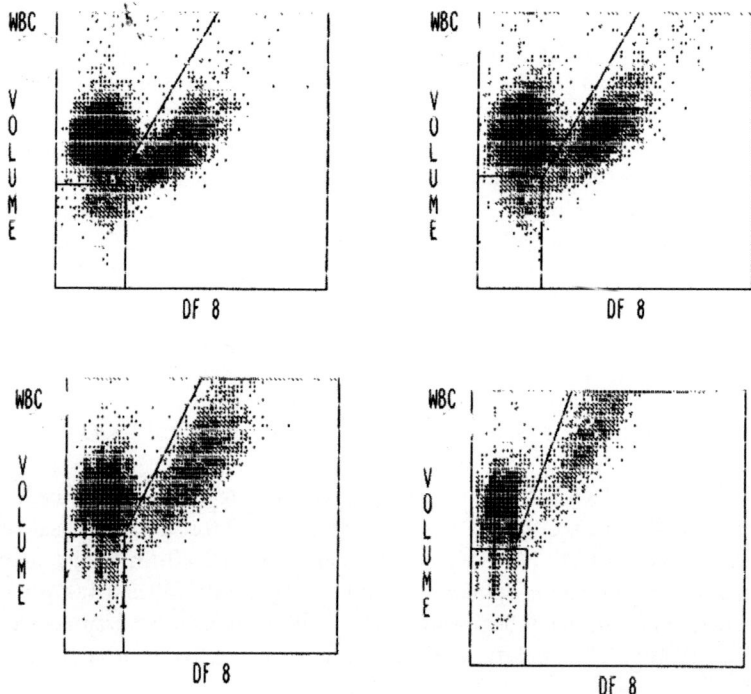

Fig. 2 Median angle light scatter (DF8) versus Coulter volume histograms for whole blood mixed with various sizes (2.13, 2.40, 3.20, and 4.00 μm diameter; left to right, top to bottom) of CD3 antibody-conjugated PS latex beads

advantage in their larger shift of targeted T lymphocytes away from nontargeted B cells in the scattergram. Further, the standard IT activation ratio of 15:1 rather than 7.5:1 gave results that consistently agreed better with the flow comparator (targeted cells in same blood sample labeled with CD3-FITC, FITC ≡ fluorescein isothiocyanate, and analyzed on a flow cytometer).

Antibody Lot Dependence

Duplicate preparations of CD3-beads (IDC lot 561,1) with each of the following CD3 antibody lots were carried out – lot 4573E1C7 to give CD3-bead lots 1A and 1B; lot 4574A1C8 to give CD3-bead lots 2A and 2B. About 40 μL of 4% w/v solid beads were mixed with 400 μL of whole blood from each of three donors for either 2 or 8 min, and tested in the Coulter® STKS2B™ hematology analyzer assay to give the results in Table 2.

Table 2 Effects of antibody lot and bead–blood mixing time on percentage of CD3+ lymphocytes in whole blood of normal donors

Mix time	Donor 30692		Donor 31246		Donor 49070	
	2 min	8 min	2 min	8 min	2 min	8 min
lot–1A	71.34	a.f./74.43	78.21	a.f./80.29	64.42	64.70/65.28
lot–1B	74.09	73.68/77.06	79.60	79.25/80.45	64.56	67.77/66.02
lot–2A	62.84	75.92/78.48	60.93	76.25/79.54	63.35	66.84/65.97
lot–2B	a.f.	72.26/77.06	a.f.	75.36/79.78	a.f.	68.32/64.14
Flow comp	74.8		80.6		65.6	

a.f. = algorithm failure.

%CD3 obtained with lot no. 1 antibody–beads was consistently closer to the %CD3 measured by flow cytometry. This was consistent with CD3/CD3-PE (PE ≡ R-phycoerythrin) competitive binding experiments performed by flow cytometry, which showed that lot no. 1 CD3 antibody had a higher affinity for its antigenic receptors on T-cells, $K_{assoc} = 1.9 \times 10^{11} \, M^{-1}$, than lot no. 2 CD3 antibody, which had $K_{assoc} = 1.1$–$4.5 \times 10^{10} \, M^{-1}$. The binding constant of CD3-PE for the same receptors was determined by flow cytometry to be 1.5–$2.5 \times 10^9 \, M^{-1}$. Our methods used to determine binding constants have been previously described.[20,21] Also, an 8 min bead–blood mixing time gave more consistent results than the 2 min mix. It is clear that the more avid the antibody in binding to its antigenic partner, the less mixing time is required to test collisions that stick.

Bead Titer Dependence

Various titers, 1–50 µL, of lot no. 1, 4.2% w/v solids, CD3 antibody beads were mixed in duplicate for 8 min with 400 µL of whole blood from three donors to establish the dependence of %CD3 (Table 3) in the Coulter® STKS2B™ hematology analyzer assay on bead titer.

Graphical representations of %CD3 from the Coulter® STKS2B™ hematology analyzer assays versus bead titers with whole blood yield curves that reach a plateau between 10 and 15 µL titer of 4% w/v solid beads per 400 µL of whole blood that were mixed to aspirate into the Coulter® STKS2B™ hematology analyzer. The optimum bead titer was selected as 25 µL of 4% w/v solid beads per 400 µL of whole blood from the consistency of %CD3 results without algorithm failure and agreement with the flow cytometry comparator. Finally, various mixing times of 1–8 min were then tested with lot no. 1 CD3 beads and whole blood from three donors. The 6 min mixing time was chosen as the shortest time that gave the most consistent %CD3 results that had low Cvs below 5% and showed good agreement with the flow cytometry comparator.

Table 3 Effect of bead titer on percentage of CD3+ lymphocytes in whole blood of three normal donors in duplicate

Bead titer (μL)	Donor 52616		Donor 53554		Donor 55048	
1	1.61	1.56	9.13	8.64	i.s.	i.s.
5	i.s.	i.s.	62.96	i.s.	73.84	69.34
10	61.09	63.99	65.90	65.09	73.05	79.32
15	64.92	71.04	67.60	63.03	74.77	74.61
20	69.57	70.97	69.77	69.25	76.15	74.07
25	68.32	66.75	66.77	66.44	76.15	78.36
30	72.98	71.72	69.46	a.f.	77.79	77.67
35	69.53	71.31	a.f.	a.f.	79.33	a.f.
40	69.63	a.f.	a.f.	a.f.	77.67	a.f.
50	71.73	a.f.	a.f.	a.f.	a.f.	a.f.
Flow comp.	71.1		69.7		78.5	

i.s. = insufficient shift; a.f. = algorithm failure.

Thus, lot no. 1 beads were then used in a statistical measurement run with two donors and eleven measurements per donor at bead titers of 10 μL and 25 μL of 4% w/v solid beads per 400 μL of whole blood, mixed together for 6 min. For one donor, the mean %CD3+ lymphocytes was 74.28(2.45) at a 10 μL titer and 76.98(1.40) at the 25 μL titer versus 77.1 for the flow comparator. The second donor gave 69.60(1.95) at the 10 μL titer and 74.50(1.54) at the 25 μL titer versus 76.6 for the flow comparator. The mean percentage CD3 should have an interassay precision within three standard deviations, and preferably, within two standard deviations. The accuracy of the Coulter® STKS2B™ hematology analyzer measurements as determined by a comparison with data from flow cytometry was within 5% of the comparator for the standard titer of 25 μL of 4% w/v solid beads per 400 μL of whole blood.

Bead-to-Cell Ratios for Targeted Cell Populations

For an average 75% CD3+ lymphocytes out of $4.8\text{--}13.6 \times 10^5$ lymphocytes per 400 μL of whole blood, we have $3.6\text{--}10.2 \times 10^5$ CD3+ cells. For 2.4 μm diameter beads at 4.2% w/v solids, we have 5.50×10^9 beads/mL, so that our standard titer of 25 μL of 4.2% w/v solid beads per 400 μL of whole blood on the Coulter® STKS2B™ hematology analyzer will contain 1.375×10^8 beads. The range of bead-to-cell ratio is then 135–382.

A theoretical limit for a monolayer coating of beads around lymphocytes can be calculated for comparison. For lymphocytes of 8–12 μm diameter size range, the surface area, $4\pi r^2$, is 201–452 μm². Thus, the maximum number of 2.4 μm diameter beads that can be packed around a single lymphocyte in a closest-packed lattice with an area per bead of $\sqrt{3}\, d^2/2$ will range from 40 to 91.

About 1–5 times the latter range was observed for a minimum bead titer of 15 µL of 4.2% w/v solid beads, or a bead-to-cell ratio from 81 to 229 was sufficient in almost every assay.

Magnetic Beads

Recent interest in aminodextran coatings for beads has been spurred by their low, minimal, nonspecific interactions with nontargeted cells and macromolecules for use as media in affinity chromatography, etc. The beads do not have to be pure PS latex ones. Refractive index changes in the beads were studied. PS–magnetite beads (polymer beads, ~1.0 µm diameter, containing ferrofluid size nanoparticles, ≤20 nm diameter) produced light scatter shifts in different directions, and manganese ferrite beads (~300 nm diameter) produced their own characteristic light scatter shift. The importance of a narrow size distribution and sufficient magnetism in the beads is demonstrated in Fig. 3, which compares the depletion characteristics of two types of CD45 (IgG1, clone KC56 (T-200)) antibody–magnetic beads. Raw PS–magnetite (23% magnetite, 0.9 µm average diameter, 0.3–3.0 µm range) particles were obtained commercially from Rhone-Poulenc (through distributor, Bangs Laboratories) and coated with aminodextran followed by conjugation of antibody as described[16,18] previously. The method of preparing manganese ferrite particles was disclosed in our patent.[22] The size distribution of the ferrite particles is shown in the histogram (292(83) nm diameter) in Fig. 4, which was constructed from 414 individual particles seen in the SEM (scanning electron microscope)

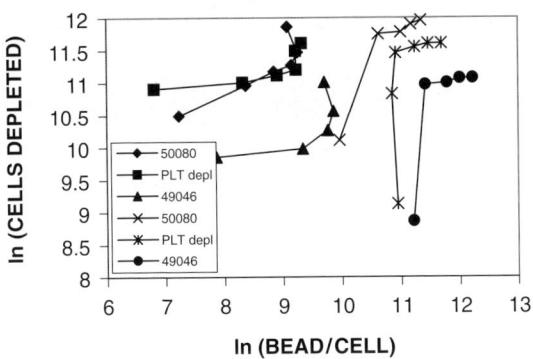

Fig. 3 Ln–ln plot of beads per cell versus number of cells depleted for depletion of WBCs from a leuko-rich sample (buffy-coat) of whole blood with CD45 antibody-conjugated magnetic beads. Three curves on left-hand side represent depletion runs with Rhone-Poulenc magnetic latex beads; on right-hand side are runs with manganese ferrite beads. Samples analyzed were blood from donors 50080 and 49046, and 49046 after depletion of 98% platelets with CD41 antibody-conjugated beads

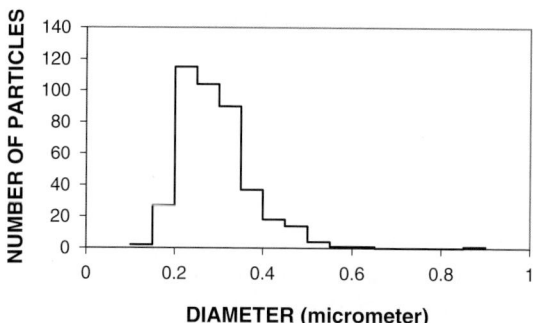

Fig. 4 Histogram of diameter of manganese ferrite nanoparticles versus number of particles with diameter within 0.05 μm intervals (data for 414 particles in photograph in Fig. 5)

Fig. 5 SEM photomicrograph of sample of manganese ferrite particles dried on aluminum stub. Length of white bar in mid-bottom section of photograph represents 1 μm

photomicrograph shown in Fig. 5. The SDP (size distribution processor) differential histogram (306(52) nm diameter) in Fig. 6, obtained from light scattering analysis in a Coulter® N4MD™ analyzer, gives similar data for a sample of manganese ferrite particles suspended in 1% aminodextran[23] (1×-Amdex) aqueous solution. Coating of the ferrite particles with gelatin or gelatin–aminodextran followed by conjugation of antibody has been previously[15,17] described.

The Rhone-Poulenc magnetic beads of wide dispersion in size and, thus, in magnetic content present problems in being able to deplete WBCs from whole blood. Out of the polydisperse distribution of PS–magnetite beads, the small magnetic

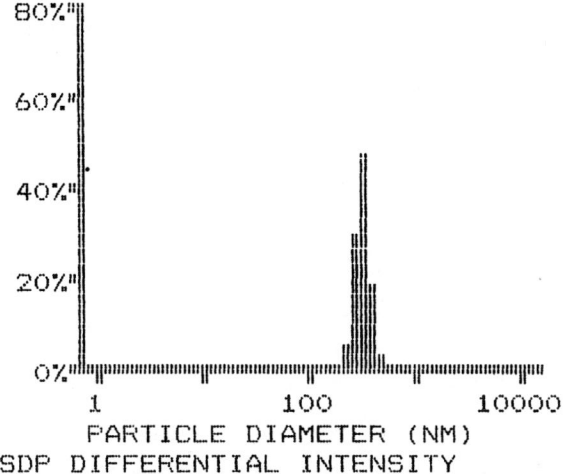

Fig. 6 Histogram of particle diameter in nanometers versus percentage of particles in diameter interval obtained from 90° light scattering analysis

beads of 200–300 nm or smaller in diameter tend to be more mobile (more kinetic energy) and occupy targeted antigenic sites on cells first. This makes it difficult, as shown in Fig. 3 on the left-hand side, to reach a plateau in the log–log plot of beads/cell versus number of cells depleted, and show complete removal of larger WBCs such as granulocytes that are targeted for depletion. There are other advantages in using the intermediate size, 50–500 nm diameter, ferrite particles rather than the recently described[24,25] 1–10 nm diameter manganese ferrite particles or particles larger than 500 nm in diameter. As light scattering probes, the smaller particles will not show any perturbation of scattering from targeted WBCs of about 3–15 μm diameter. Larger particles, which approach the size of targeted cells, may interfere in scattergrams as free excess beads that are not bound to the cell surface. Also, attaching heavy inorganic particles to cells will tend to bring about sedimentation of the targeted cells by gravity.

The intermediate size ferrite particles are sufficient to allow magnetic separation of targeted WBCs with standard, hand-held rare earth magnets. The depletion efficiency graphs for WBCs, neutrophils, red blood cells (RBCs), and platelets (PLTs) from whole blood are shown in Fig. 7 in the form of plots of bead-to-cell (B/C) ratios versus number of cells depleted. The monoclonal antibodies conjugated to the ferrite beads were anti-CD45 (IgG1, clone KC56 (T-200)), anti-CD16b (IgM, clone 1D3), anti-CD235a (glycophorin A, IgG1, clone 11E4B-7-6 (KC16)), and anti-CD41 (IgM, clone 69 (PLT-1)). The most efficient use of magnetic beads is reached at the end-point or plateau after the inflection point of each curve, where the B/C ratios were about 5,000, 1,000, 75, and 300, respectively.

In addition, the intermediate size ferrite-nanoparticle–monoclonal-antibody conjugates have been used as light scatter markers for immunophenotyping and enumeration of subsets of WBCs in whole blood. Enumeration of WBC subsets could

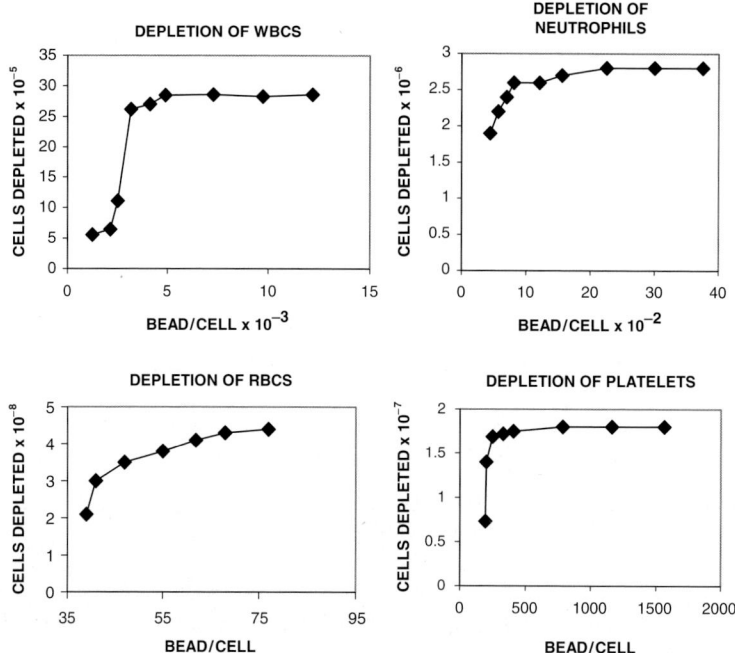

Fig. 7 Graphs of beads per cell versus number of cells depleted from whole blood with CD45, CD16b, CD235a, and CD41 antibody-conjugated manganese ferrite beads (left to right, top to bottom, respectively) as determined by flow cytometry and/or hematology(cell counter) instrumentation

be done either by magnetic bead depletion or by light scatter shifts. This was demonstrated in detail for neutrophils targeted by 1D3 antibody magnetic beads.[15,16] Figure 8 shows the stepwise effect of increasing 1D3 antibody–ferrite bead titers (0, 20, 50, 75, 100, and 200 μL of 0.25 w/v solids) on the position of neutrophils in a sample of leuko-rich, buffy-coat cells in the side versus forward scatter histograms obtained by flow cytometry. The progressive decrease in forward light scatter intensity with larger titers of beads reflects the rising absorption of incident laser light by the black ferrite beads on targeted neutrophils; while the increasing side light scatter intensity arises from the structural complexity of the cell–bead conjugate. Unlike the enormous relative size effects achieved from adding 2–4 μm diameter PS latex beads coated with CD3 or CD4/CD8 antibodies to target lymphocytes in whole blood, as summarized in the previous section, the 0.3 μm diameter ferrite beads will contribute little to alter the size of typical neutrophils in the 15–20 μm diameter range. For the 100 μL titer, the total analysis results were 26,604 lymphocyte counts, 1,045 monocyte counts, 17,449 unshifted granulocyte counts, and 88,145 shifted granulocyte counts. Thus, the percent neutrophils (=shifted granulocytes) in the total WBC count was calculated as 66.2%. Magnetic depletion data with the same magnetic beads and sample of blood were also obtained from flow cytometric

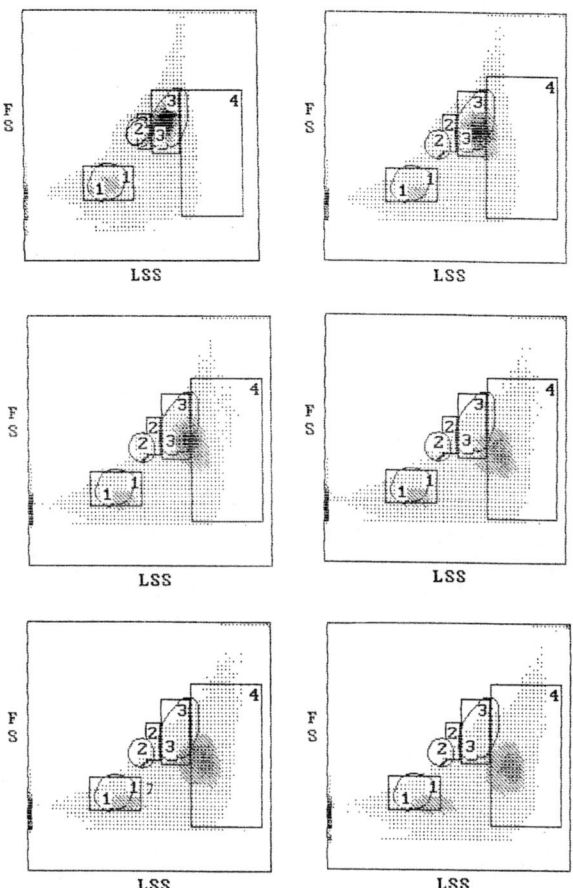

Fig. 8 Side(LSS) versus forward(FS) light scatter histograms showing the effect of increasing titers, 0–200 μL (as given in text, left to right, top to bottom, respectively), of 0.25% w/v solids CD16b (1D3) antibody-conjugated manganese ferrite nanoparticle suspension in 1 × PBS, 1% BSA buffer mixed with whole blood, and analyzed by flow cytometry

analyses of the control (undepleted whole blood sample) and supernatants after depletion, as stained with fluorescent antibody, 1D3-FITC. These latter results gave 66.8% neutrophils, which agreed well with the number obtained by the population shift method.

One product from the use of a cocktail of monoclonal antibody-conjugated magnetic beads [CD15 (IgM, clone 80H5) to remove granulocytes, CD235a (glycophorin A, IgG1, clone 11E4B-7-6 (KC16)) to remove RBCs, CD41 (IgM, clone 69 (PLT-1)) to remove platelets, CD16b (IgM, clone 1D3) to remove neutrophils, and CD14 (IgM, clone 116 (Mo2)) to remove monocytes] in negative depletion from the whole blood of a pool of normal donors is a suspension of WBCs to be used as standards or control cells[26] for process control in flow cytometry. The monoclonal antibodies referred to herein are identifying designations used by Beckman Coulter,

Inc. for monoclonal antibodies made by Coulter. Beckman Coulter, Inc. markets the control cells in lyophilized form as CYTO-TROL™ and IMMUNO-TROL™ control cells.

Another application of magnetic beads is in the removal of normal cells from a cellular sample so that enhanced throughput and precision can be obtained in tumor cell DNA analysis. The interfering cells include leukocytes, fibroblasts, and other normal cell types. Pan-leukocyte antibody (CD45)–manganese ferrite beads were used[27,28] to quickly remove the infiltrating leukocytes from hemorrhagic samples for tumor cell enrichment, followed by flow cytometric determination of the DNA ploidy and proliferation index of the tumor cells present in the sample. For example, in a mixed population of cells containing 95% peripheral blood leukocytes and 5% breast tumor cells, the leukocytes were depleted in one cycle of CD45-conjugated ferrite beads, thus increasing the percentage of tumor cells from 5% to 99.6% prior to analysis. The sample was then labeled with monoclonal anti-cytokeratin (CK) or anti-Epithelial Membrane Antigen (EMA, clone Mc5) and processed on the Coulter® DNA-Prep System for DNA staining. With gated flow cytometric analysis, the tumor (EMA+)/epithelial (CK+) cells were identified and the proliferative index was accurately assessed. Similar strategies, using preliminary magnetic bead depletion of cells such as WBCs, RBCs, and/or PLTs that are present in large numbers, can be applied to obtain enrichment of other cellular populations for rare event analyses by flow cytometry.

In large scale, magnetic bead depletion, nothing beats the heavy nickel beads in efficiency and speed of biological cell sorting. These were initially described with adsorbed antibody[29] and then with covalently conjugated antibody linked to the nickel surface by a new crosslinker, tris(3-mercaptopropyl)-N-glycylaminomethane.[30] This "spider" ligand, originally synthesized[31] to serve as a linker to polymerize alanine on a gold surface, is now one of the few examples, if not the only one, of a crosslinker designed to bridge a metal surface with a protein, in this case a monoclonal antibody. B/C ratios reached the 2–4 range in one run for depletion of CD8+ cells with nickel beads conjugated with CD8 antibody (IgG1, clone SFCI21Thy2D3), and 30–35 in another run. Excellent RBC-to-bead ratios (C/B) of 1.2 and 1.5 in one run, and 9–10 in another run were also obtained with the nickel beads conjugated with CD235a (KC16) antibody.

A start-up company, Coulter Cellular Therapies, later renamed Eligix, Inc., was born to use CD8 antibody– and other antibody–nickel and manganese ferrite beads to remove leukemic CD8 and other unwanted cells from bone marrow, peripheral blood stem cell, and donor leukocyte grafts used in transplant procedures. Typically, the patient's bone marrow is extracted and externally purged with magnetic beads, the patient undergoes radiation treatment, and autologous bone marrow transplantation follows. Furthermore, the selective depletion of CD8+ lymphocytes from donor lymphocyte infusions was used to achieve remission in patients relapsing after bone marrow transplantation, and decrease the incidence of graft versus host disease. The Eligix High Density Microparticle (HDM™) Cell Separation System was recently acquired from BioTransplant, Inc., Medford, MA, in the Chap. 11 process, by Miltenyi Biotec GmbH.

Surface Plasmon Resonance: Light Scatter Nanoparticle–Bead Probes

Background

More dramatic refractive index changes, brought about by the incorporation of metal nanoparticles as probes, give rise to larger light scatter changes that depend dramatically on the wavelength of excitation. Early work[32–35] with gold and silver nanoparticles was restricted to the use of available adsorbed antibody–gold nanoprobes for immunolabelling of cells. The necessity for excitation into surface plasmon resonance extinction bands of the nanoparticles for the enhancement of light scatter intensity was not realized by authors in the early work. Research in the areas of flow cytometry of gold colloid-labeled cells and surface plasmon resonance, especially in the then, newly discovered area of surface-enhanced Raman scattering in 1979 (vide infra) continued along parallel tracks into the mid-1990s without any lateral communication. Our own patented work on flow cytometry of gold- and silver-nanoparticle–aminodextran-coated PS latex beads only[36] and flow cytometry of antibody-conjugated beads and targeted WBCs[37] was later published.[38,39] Therein, we suggested the need to excite into surface plasmon resonance bands of gold or silver nanostructures and established the wavelength dependence of light scatter intensities from these beads by flow cytometry. The results for these macroscopic size nanoparticles of ∼20–200 nm diameter, which display the more discrete, collective modes of oscillation of the conduction electrons in a metal particle as opposed to the wide band absorption of bulk metals, could be explained by the principles of classical Mie scattering theory.[40–42] The light scattering results obtained by flow cytometry for silver- and gold-nanoparticle–PS beads are believed to represent the first quantitative verification of the predictions of Lorentz–Mie scattering theory for silver and gold structures in the intermediate, 50–200 nm diameter, size range.

Other work in this area has led to the formation of several start-up companies. Notably, nanobarcodes[43] was registered as a trademark by Michael Natan when he was chief technical officer at SurroMed, Inc., Mountain View, CA, out of which was spun Nanoplex Technologies, Inc. They propose[44] to use metallic microrods in simultaneous detection of numerous biological analytes. Similar goals were initiated by the Yguerabides[45] and Genicon Sciences,[46] San Diego, CA, for resonance light scattering from metal nanoparticle suspensions, particularly those of gold and silver as a vehicle for multiplex assays using light scatter/transmission from various sizes and types of metal nanoparticles. However, in the efforts of neither start-up company is there any indication that they would functionalize the nanoparticles with specific ligands such as antibody or other protein targeting various analytes. Here, the "spider" ligand, which was used as a crosslinker between nickel[30] or gold[37,39] particles and antibody, would be very useful.

Light Scatter Shifts

Common things to note for bead probes, in which beads are of the same size and similar refractive index properties, are that light scatter shifts will be in same direction for multiple bead types on the same target biological cell. Thus, assays would necessarily be carried out in series. For cellular assays, a threshold in the filling of receptor sites on the surface of cells must be reached before a homogeneous light scatter pattern is obtained for the targeted subpopulation of cells, which is distinguishable from the original cellular population without any beads attached to the surface. Although theoretically feasible, the use of beads of various sizes to perform separate biological cell assays faces some serious practical problems, especially if each separately targeted cell is not saturated with a close-packed array of beads of a distinguishable size. Here, a definite restriction is that the various targeted subpopulations of WBCs be mutually exclusive. This is a consequence of using a single detector to analyze all the light scattering patterns.

Since PS, Ag–PS, and Au–PS beads of the same size can be distinguished by their side scatter intensities when surface plasmon resonance bands are excited by incident laser radiation, the beads described in the first two sections do not require any additional detectors from the standard ones used in conventional flow cytometers for forward and side scatter. Parallel biological cell assays were demonstrated[39] by us for the first time by measuring side and median angle light scattering parameters for mixtures of whole blood with the following pairwise combinations of antibody-conjugated beads: CD4–PS, CD8–Au–PS; CD4–Au–PS, CD8–PS; CD4–PS, CD8–Ag–PS; and CD4–Ag–PS, CD8–PS. Three bead probes could, in principle, be used simultaneously to assay for three different subpopulations of cells. However, the results would not always provide a direct percentage for each target. Some strongly shifted populations (Ag–PS and/or Au–PS) and other shifted but not strongly shifted populations depending on excitation wavelength (488 or 633 nm) may include overlapping populations that will have to be subtracted out with nonoverlapped populations to obtain final results. Further, the size distribution of gold or silver nanoparticles on the PS surface will have to be made narrower and aggregation of beads will have to be minimized in order to achieve narrower distributions in the side scatter intensities (especially those of the gold–PS bead system) to allow a three or more bead mixture to be used simultaneously in successfully doing whole blood analyses.

The enhanced scattering excitation profiles for metal structures between 50 and 300 nm in diameter, even when they are very uniform in size and shape, are intrinsically broader[38] in wavelength than emission bands in q-dots™ (vide infra). Thus, the number of simultaneous probes will be limited. Also, the sensitivity of their excitation wavelength–scattering intensity characteristics to aggregation complicates their practical use in extensive multiplexing. Structural homogeneity of the gold colloid structures on PS beads was not as good as silver colloid on PS beads, as seen[38] from the wide distribution in side scatter intensity for the former. Tighter size distributions would be needed in the gold/silver nanoparticles to obtain narrower plasmon extinction bandwidths, and, in turn, narrower excitation profiles. Progress in this direction has been recently made with the preparation of size-tunable nanoparticle arrays by nanosphere lithography.[47–50]

Nevertheless, metal nanoparticles have some distinct advantages over organic fluorophores and luminescent q-dots™ as probes. Organic emitters are photobleachable, blink, and become optically saturated at high exciting intensities. Q-dots™ resist photobleaching longer than organic dyes, but do eventually succumb to bleaching and do blink. Metal nanoparticles are not photobleachable, but do blink. Single, gold nanoparticles have been shown[51] to emit Raman scattered photons in an intermittent fashion on the millisecond-to-second time scales. Further, the light scattering (Mie–Rayleigh and Raman) signals from metal nanoparticles are not subject to saturation.

Fluorescence Emission: Light Scatter Nanoparticle–Bead Probes

The bead probes to be described in next two sections will require additional fluorescence/luminescence emission detector(s). The number of these detectors depends on the number of different "colors" or wavelength regions of emission to be detected. A typical maximum that has currently been feasible with organic dye emitters has been about six. This may be augmented due to the observed narrow bandwidths (typically, 30–50 nm, and as low as 26 nm, reproducibly produced) of emission from semiconductor nanoparticles, which have been recently shown to allow 10 "colors."

Background

PS beads with embedded fluorescent dyes have a long history as calibration standards for flow cytometry. An overview of recent developments in beads, including fluorescent ones, has been provided by The Latex Course.[2,3] More recent uses in multiparametric analyses have proliferated as attested to by presentations at the three (2002, 2000, 1998) most recent ISAC (International Society for Analytical Cytology) meetings in San Diego, CA; Montpellier, France; and Colorado Springs, CO. Luminex Corp.,[52–56] Austin, TX, markets an array of fluorescent beads of different "colors"/emission wavelengths and varying concentrations of fluorescent dyes corresponding to different emission intensities. For example, multiplex fluorescent bead immunoassays for human cytokines have been described in several articles.[57–60] Multianalyte microsphere assays were covered in a recent workshop report.[61]

The geometric and electronic structures of the molecular, fluorescent dyes are generally perturbed only to a minor degree by the medium in which the dye is embedded. Thus, the emission intensity and position of the wavelength maximum can be considered to be uncoupled from the medium in which the molecular emitter resides.

Furthermore, monodisperse droplets and particles of 10–0.15 µm diameter have been produced[62] by electrified coaxial liquid jets. Moreover, a joint effort between an Alabama group and one in California at Beckman Coulter has produced bead chemical sensors for flow cytometric analyses,[63–66] using a method of doping plasticized poly(vinyl chloride) containing organic solvent with a variety of components including sodium- or potassium-selective ionophore, H^+-selective fluorescent chromoionophore, and cation-exchanger, followed by sonic stream casting of the composite beads and hardening them in an aqueous tank by allowing diffusion of organic solvent such as dichloromethane out of the initial soft beads into the aqueous phase. The process was started by Mack Fulwyler[5] and has been used at Coulter for many years to manufacture fluorescent dye–bead standards for flow cytometry. An example of a proposed addition[67] to the product line at Beckman Coulter, Inc. is the Flow-Cal 575 fluorospheres, a premixed suspension of PS fluorospheres (7 populations: 1 blank and 6 fluorescent populations obtained by dye loading polymer beads at six different dye concentrations; PE fluorescent channel specific; all populations in suspension are the same size) whose fluorescence intensity with 488.0 Ar+ laser excitation spans four decades, as shown in Fig. 9. With appropriate software, the fluorescence intensities are translated into user definable, dye equivalent units. An application for Flow-Cal fluorospheres to a CD38-PE biological assay by flow cytometry is displayed in Fig. 10, wherein the Flow-Cal 575 beads are run on the Coulter® EPICS™ XL™ flow cytometer at settings for the CD38 assay of an

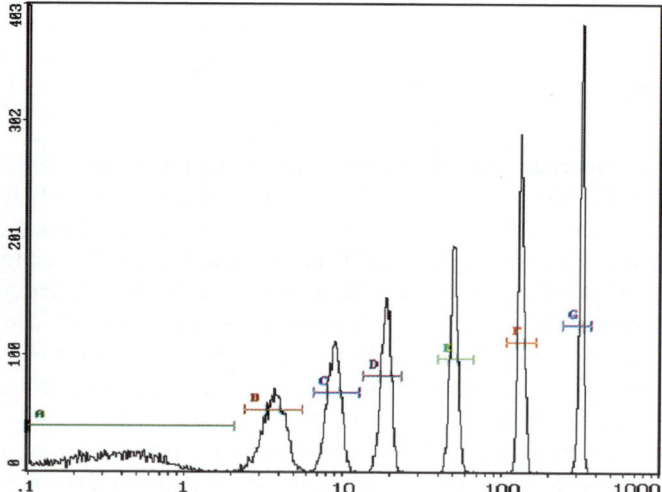

Fig. 9 Flow cytometric histogram of Flow-Cal 575 Fluorospheres. The fluorescence intensity of light-scatter-gated beads (all same size) is displayed as a histogram in which the y-axis represents the number of fluorescent events (count in arbitrary units), and the x-axis shows fluorescence intensity over four decades, 0.1 to 1,000. The negative (blank beads) bead population is positioned in the left quarter of the histogram

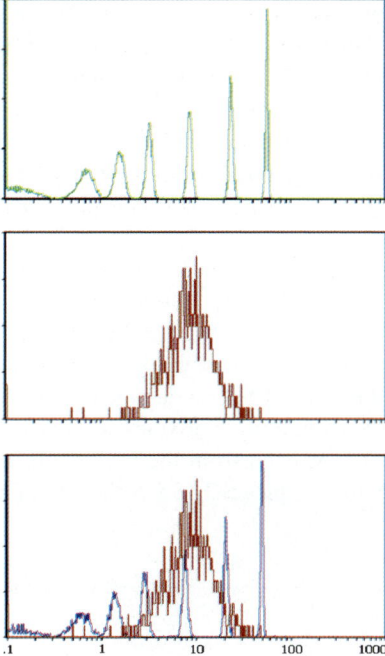

Fig. 10 Top, flow cytometric histogram of Flow-Cal 575 beads as in Fig. 9, but run at settings for CD38 assay. Middle, flow cytometric histogram of whole blood from an AIDS patient stained with CD3-FITC/CD38-PE/CD8-PC5, *x*-axis represents CD-38-PE fluorescence intensity. Bottom, overlay of Flow-Cal 575 and CD38-PE histograms

AIDS patient sample stained with CD3-FITC/CD38-PE/CD8-PC5 (PC5 ≡ PE–Cy5, phycoerythrin–cyanin 5.1) and analyzed in the PE channel.

Other work inspired by Fulwyler's patented ideas includes a multi-analyte detection system using two distinguishable types of particles,[68–71] four types of beads[72] ranging in size from 1 to 10 µm, and a multianalyte assay[73] with extended dynamic range. The advantages of using flow cytometry in bead-based assays can be enumerated as follows: (1) separation-free, (2) multiparameter, multianalyte, (3) extended working range, (4) individual determination of nonspecific binding, (5) high information content, (6) built-in quality assurance, and (7) easily automated.

A note of warning has been sounded about an ambiguity introduced into the use of flow cytometry in bead assays by a saturation phenomenon known as the "hook" effect,[74] which manifests itself in the following way. Measurement accuracy by flow cytometry is directly related to intensity per particle of the signal from the label. Thus, high sensitivity in an assay requires the use of as few binding particles as possible to get a reasonable amount of bound analyte (and then label) per particle. However, to measure high concentrations of analyte, a large number of particles is needed. With each particle having a fixed number of binding sites for analyte, the sum of all particles will represent a fixed binding capacity that defines the upper limit for the highest measurable amount of analyte. When the amount of label is constant, the amount of label per analyte will decrease as the amount of analyte exceeds the maximum capacity. Therefore, when analyte concentrations exceed saturation, the amount of label per particle will decrease as an increasing amount of label will bind the excess free analyte in solution. This is the "hook" effect. Note that this can be avoided by using a larger number of beads and/or using beads with higher binding capacity per particle. The latter solution is preferred since more beads will give lower sensitivity.

Silica-fluorescent dye beads have not been utilized as much as the PS-based ones. As pointed out recently in our efforts[75,76] these are easily prepared in abundance, and have optical and chemical features that lend them suitable for flow cytometry applications. The limiting feature—concentration quenching—of any fluorophore-bead system is illustrated in Fig. 11 for rhodamine 6G in silica. Here, successively higher concentrations of R6G in a layer of silica-R6G around the same core silica showed quenching of R6G emission from the increase to a maximum at S-4 and then decrease in intensity for samples S-1 to S-6 shown in Fig. 11.

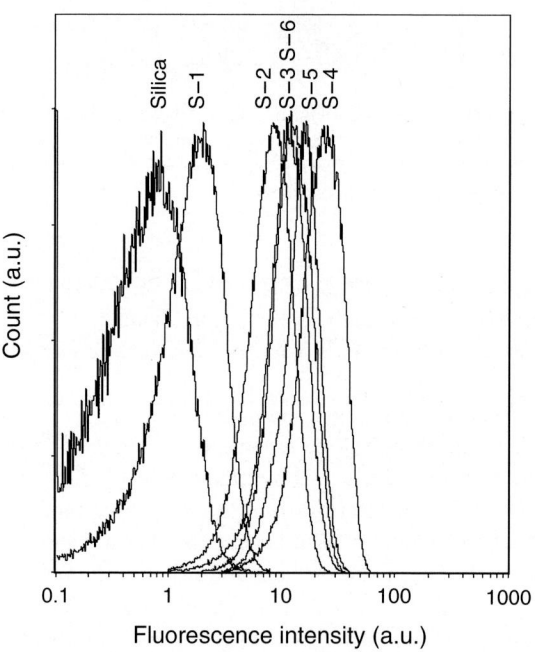

Fig. 11 Composite mean channel fluorescence intensity versus count histogram for singlets of samples S-1 to S-6, containing increasing amounts of rhodamine 6G dye in a thin silica–dye layer around silica-core particles and blank silica-core particles set as negative background in the first quarter of the histogram. Reprinted from M. Bele, O. Siiman, and E. Matijevic, Preparation and Flow Cytometry of Uniform Silica-Fluorescent Dye Microspheres, J. Colloid Interface Sci., Vol. 254, pp. 274–282, Copyright © 2002 with permission from Elsevier

Suspension Arrays

The flow cytometric assays in which reagents are kept in aqueous solution or suspension have an advantage over assays performed with array systems in which reagents are typically dried down before measurements of absorption or fluorescence emission intensity are made. The proteinaceous labels such as phycoerythrin (PE) and its tandem conjugates do not tolerate the dry state. The bright red color of PE in solution is lost when dried on a glass slide. Bilin groups in PE may become rearranged structurally so that concentration quenching due to close bilin–bilin approach occurs (9.89 nm for randomly spaced 34 bilin groups in PE), and dried protein does not show the same bright fluorescence intensity as it does in its native state. Alternatively, PE in the dried state may prefer to, upon excitation from the ground state to the singlet excited electronic state, cross over to the triplet state, which does not exhibit fluorescence emission, but does show phosphorescence.

The steadfast position of many practitioners of microarray and microchip systems brings to mind the allegorical treatise, "Flatland," of the Victorian era by Edwin A. Abbott, in which the narrator preaches the attributes of three-dimensional space after a trip to Spaceland to his countrymen who have only lived in two-dimensional space, *sic* Flatland, and suffers their wrath. "Evolution of the flat-array paradigm" to include bead suspension array technology[77–79] has been proposed and presented. One notable advantage of the bead suspension array is the much greater surface area available for assays that utilize the surface area of microspheres, compared to assays that only use the macroscopic surface on the inside of a typical well. Another significant advantage is the accelerated kinetics of reactions between species on the surface of beads and complementary species in solution. In a fixed well, there is one static entity, whereas in bead assays both partners are free to move. We have pointed this out previously[19] for the reaction between maleimidated beads and thiolated antibody.

The most recent addition to this arena is Bio-Rad's extensive Bio-Plex™ suspension array system,[80] which uses xMultianalyte Profile (xMAP™) technology of Luminex Corp. to multiplex up to 100 different assays from a single sample. A flow cell is utilized for the simultaneous detection and measurement of bioreactions on the surfaces of fluorescent beads. A syringe pump delivers the sample to an array reader flow cell where the beads are aligned in single file and excited with two laser beams, a red (635 nm) classification laser to excite the dyes embedded in each bead for identification and a green (532 nm) reporter laser to excite the fluorescent-labeled reporter molecule specifically bound to the captured analyte on the bead surface. Particle-based immunoassays for flow cytometry have been shown to match the performance of present state of the art in conventional single-parameter assays. The additional multiparameter features of multiparticle immunoassays together with the emphasis on future automated systems should make these assays one of the candidates of choice for the next generation of immunoassay systems.

Low Density Carrier

An alternative to embedding fluorophores inside solid particles is to attach multiple molecular emitters that will not quench the emission intensity of each other when they lie in proximity such as PE, a protein containing multiple bilin groups, onto a large molecular weight polymer. Such is the case with (streptavidin or antibody) – aminodextran –PE, PC5, or ECD ($\equiv$PE – TEXAS RED™) conjugates that we have described.[23,81–84] Large aminodextran polymer molecules of up to about 8 MDa, and 118–153 nm diameter in aqueous media behave as compressed random coils. The importance of using aminodextran with as narrow a size distribution as possible to obtain a homogeneous preparation of conjugates was stressed.[84] Similar narrow size distributions for colloidal particle probes have been important in various assays in which light scattering/emission were monitored. One feature of the molecular conjugates is that they do not exhibit light scattering shifts of targeted cell (lymphocytes) position in FS versus SS histograms. The conjugates are low density (like the aqueous medium) and low refractive index materials of small size relative to the targeted WBCs. They can thus be used in typical biological cell assays just as low molecular weight fluorescent antibodies to detect cells with low numbers[81,85] of antigenic sites per cell, such as lymphocytes expressing CD101 or CD160 on their surface. CD101 (BB27 clone, IgG1 isotype) and CD160 (BY55 clone, IgM isotype) monoclonal antibody conjugates were used. Examples of flow cytometric histograms showing the dim-to-bright staining of these cells with aminodextran-crosslinked antibody and PE are displayed in Figs. 12 and 13, respectively. Two or more markers of the same color but different intensities[82,86] can be combined. In one example, four color markers and three enhanced intensities were used to target seven populations of WBCs simultaneously in the same blood sample.

New Cell/Bead Assay

The flow cytometric investigation of cells in whole blood with soluble, fluorescent antibodies and native, unlabeled antibodies in competitive binding has in one instance led to the conception of a new generalized immunoassay that can be done with analyte bound to the surface of formed bodies and unknown amount of analyte in solution. The formed bodies could be biological cells or colloidal particles such as PS latex beads. It was first demonstrated for CD16 antigen on neutrophils and soluble CD16 in whole blood.[21,87,88]

Two antibodies were investigated. One is the 3G8 antibody (IgG1 isotype) that binds to both isoforms of CD16, FcγRIIIA, the integral, transmembrane form expressed on NK cells, monocytes, and macrophages and FcγRIIIB, the glycosylphosphatidylinositol (GPI)-anchored form expressed only on neutrophils. The other antibody is 1D3 (IgM isotype) that binds only to CD16b, i.e., FcγRIIIB on neutrophils. Initially, an intuitive solution for the concentration of soluble antigen in a whole blood sample was derived from competitive binding curves that overshot

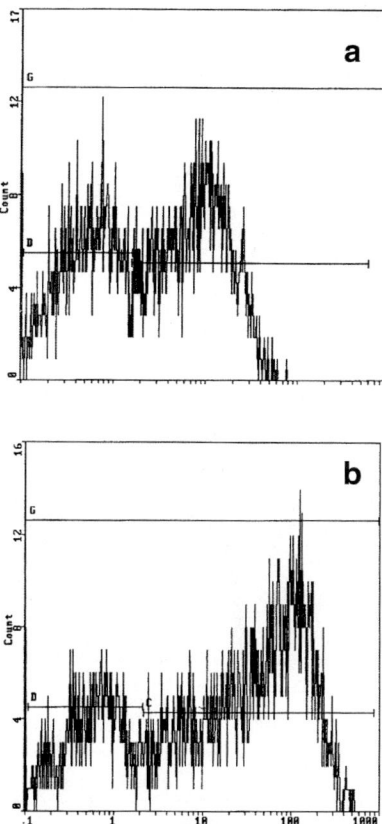

Fig. 12 Flow cytometric histograms showing the staining of scatter-gated lymphocytes with BB27 antibody. In each histogram, the *y*-axis represents the number of PE fluorescent events (count in arbitrary units), and the *x*-axis shows PE fluorescence intensity over four decades, 0.1–1,000. About 100 µL of whole blood from a normal donor was stained with 10 µL containing BB27 antibody as (**a**) BB27-PE and (**b**) BB27-5X-aminodextran-PE. The negative background fluorescence is positioned in the first decade of the histogram on the left

their fluorescent intensities at low concentrations of unlabeled antibody, when compared to the intensity for a control without unlabeled antibody. This is shown in Fig. 14 for the 3G8/3G8-FITC competitive binding trial for which the control had a mean channel 3G8-FITC fluorescence intensity of 31.97, but mean channel intensities of samples containing the lowest concentrations of unlabeled 3G8 were higher at 32–37. More accurate quantitation was achieved by evaluating the intersection of two binding curves of concentration versus mean channel 3G8-FITC fluorescence intensity: (1) 3G8-FITC alone and (2) total antibody, 3G8+3G8-FITC, in 3G8/3G8-FITC competitive binding as shown in Fig. 14. The intersection represents the same 3G8-FITC fluorescence intensity at a total antibody concentration of 3.8×10^{-8} M, or (with a constant 3.1×10^{-8} M concentration of 3G8-FITC in the competitive binding run) an unlabeled 3G8 concentration of 0.7×10^{-8} M.

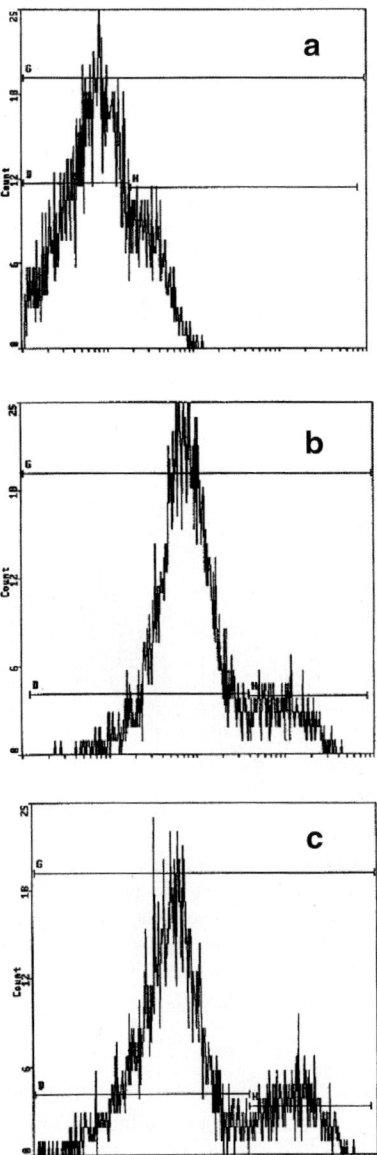

Fig. 13 Flow cytometric histograms showing staining of scatter-gated lymphocytes with BY55 antibody. Histogram axes defined as in Fig. 12. About 100 µL of whole blood from a normal donor was stained with 10 µL containing 0.25 µg of BY55 as (**a**) BY55-PE, (**b**) BY55-5X-aminodextran-PE, 1X/fraction 20, and (**c**) BY55-5X-aminodextran-PE, 2X/fraction 19

Assuming an antigen valence of one and a 1:2 binding stoichiometry between 3G8 (IgG1) antibody and soluble CD16 antigen in the region of antigen excess, we obtain about $7\,\text{nM} \times 2 \times 150\,\mu\text{L}/100\,\mu\text{L} = 21\,\text{nM}$ as the concentration of soluble 3G8 antigen for this blood donor.

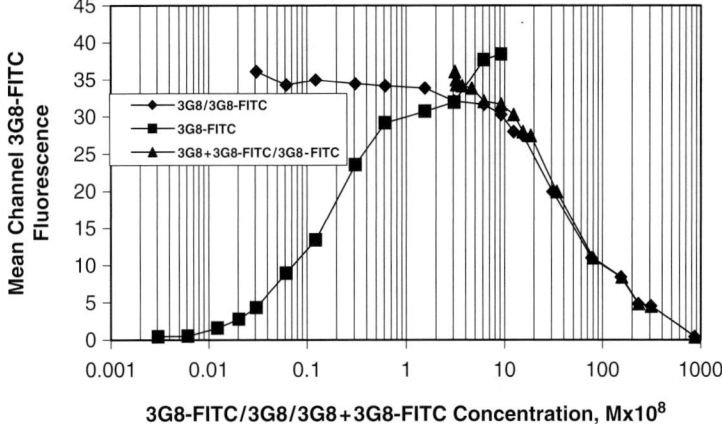

Fig. 14 Linear-log ($y-x$) graphical dependence of mean channel 3G8-FITC fluorescence intensity on molar concentration of 3G8 antibody in competitive binding run, of 3G8-FITC in binding run, and of 3G8 + 3G8-FITC in competitive binding run as determined by flow cytometry of a sample of whole blood from a normal donor. Reprinted from O. Siiman, A. Burshteyn, O. Concepcion, and M. Forman, Competitive Antibody Binding to Soluble CD16b Antigen and CD16b Antigen on Neutrophils in Whole Blood by Flow Cytometry, Cytometry, Vol. 44, pp. 30–37, Copyright © 2001 with permission from John Wiley and Sons

The analytical solution, which required the intersection of two curves, was derived graphically as described above, but never in terms of formulae or equations, which now follow herein. One form of the distribution formula, ratio of bound fraction to unbound fraction (in this case fluorescence-labeled antibody or the combination of labeled + unlabeled antibody) is a version of the Langmuir adsorption isotherm:

$$\theta_2/(1-\theta_2) = Kc_{\text{solution}},$$

which can be rewritten in the form

$$\theta_2 = Kc_{\text{solution}}/(1+Kc_{\text{solution}}),$$

where θ_2 is the fraction of antigenic sites on the targeted cell surface occupied by antibody, K is the equilibrium constant, and c_{solution} is the concentration of antibody in solution, i.e., not bound to the targeted cell. The above expression can therefore be written for labeled antibody, 3G8-FITC, binding alone to targeted cells in whole blood (we will call this the unprimed system) as well as for the combined unlabeled antibody, 3G8, + constant amount of labeled antibody, 3G8-FITC, competitive binding to targeted cells in the same whole blood sample (we will call this the primed system). At the intersection of the two curves, (1) labeled antibody, 3G8-FITC, alone binding to cells, and (2) unlabeled + labeled antibody binding to cells, when the mean channel fluorescence intensities as measured in the flow cytometer are equal, the fractions of bound labeled antibody are also equal, i.e.,

$$\theta_2^{3G8\text{-FITC}} = \theta_2'^{3G8\text{-FITC}}.$$

The equality holds if unlabeled 3G8 antibody does not bind to cell surface antigen. This is highly probable when 3G8 antibody is added first to the whole blood sample in precisely the stoichiometric amount needed to react fully with soluble antigen. Since the latter is much more mobile, i.e., more kinetic energy of translation in particular, than large cells such as the targeted granulocytes (10–15 μm diameter), therefore, the soluble antigen will have more effective collisions per unit time with the added antibody. This effect was also observed previously in our reactions of free antibody versus antibody on beads with targeted cells in whole blood, and in magnetic depletion of targeted neutrophils in whole blood with 1D3 antibody–bead conjugates.

The equality of fractional surface coverage also implies that surface concentrations are equal, i.e.,

$$c_{\text{surf}}^{3G8\text{-FITC}} = c_{\text{surf}}'^{3G8\text{-FITC}} = x.$$

Then, substituting from the Langmuir equation above, we have

$$Kc^{3G8\text{-FITC}} / \left(1 + Kc^{3G8\text{-FITC}}\right) = Kc'^{3G8+3G8\text{-FITC}} / \left(1 + Kc'^{3G8+3G8\text{-FITC}}\right).$$

If equilibrium constant, K, values are equal for the binding of 3G8 and 3G8-FITC, then

$$c^{3G8\text{-FITC}} = c'^{3G8+3G8\text{-FITC}}$$

is a solution to the latter equation.

Separating the two parts on the RHS and rearranging gives

$$c'^{3G8} = c^{3G8\text{-FITC}} - c'^{3G8\text{-FITC}}.$$

Since we do not have values for these concentrations of solution species, but only total concentrations, we can use the equalities

$$c^{3G8\text{-FITC}} = c_{\text{total}} - x \text{ and } c'^{3G8\text{-FITC}} = c'_{\text{total}} - x,$$

followed by substitution gives

$$c'^{3G8} = (c_{\text{total}} - x) - (c'_{\text{total}} - x) = c_{\text{total}} - c'_{\text{total}}$$

or $c'^{3G8} = c^{3G8\text{-FITC}}$ (at intersection of two curves) $- c'$ (constant concentration of 3G8-FITC in competitive binding).

This assay falls into one of the four variations of an immunoassay previously defined,[89] the antibody competition, and one of the three classes of immunoassay previously defined, the antibody capture, but without the presence of pure or partially pure antigen and without an extraneous solid support such as microtiter plates for any component of the reaction, as recommended for quantitating antigen in competition assays.

Although the examples with soluble 1D3 and 3G8, shed from the surface of neutrophilic cells, were taken from the blood of normal donors, probably with nothing more than a seasonal attack of the common cold or flu, there are other cases of shed CD16 that are more serious.[90–92] Among other examples of shed antigen (GPI-linked) associated with serious conditions include CD14 from the surface of myeloid cells[93] (in sepsis) and CD100[94–98] from the spinal fluid of spinal cord injury patients. Additional GPI-linked antigens, which have been identified[99] among the 200+ antigens on the surface of WBCs, are CD24, CD48, CD52, CD55, CD58, CD59, CD66b, CD66c, CD73, CD87, CD90, gp42, LY-6, RT 6, and SCA-2. Most also occur as soluble forms in normal serum. Other GPI-linked molecules are the normal prion[100] protein (CD230), PrP^c, a modified form of which may cause prion diseases or spongiform encephalopathies such as "mad cow disease" and Creutzfeld-Jacob-Disease, and the glycoproteins on the surface of insect-transmitted protozoan parasites[101] linked to African sleeping sickness.

Luminescence Emission: Light Scatter Nanoparticle–Bead Probes

Background

The subject of quantum dots (q-dots™) or semiconductor nanoparticles (~1–10 nm diameter range) has received a lot of recent attention.[102–107] The particle size-dependent colors of luminescence emission observed for these small particles cannot be explained by classical theories. Quantum mechanics must be applied. The simplest form, the one-dimensional "particle-in-a-box," involves application of the Schrödinger equation to a single particle (electron) restricted between 0 and L in a single dimension, and was originally considered an artificial system.[108] A practical example of its use only developed much later for quantum-sized particles. Applying the slightly modified "particle in a sphere" solution,[105] $E_{1s} = h^2/8m_{\text{eff}}a^2$, for the energy of an electron of effective mass, m_{eff}, in the first excited quantum level and in a "spherical box" of radius, a, and with $\lambda = hc/E_{1s}$ for conversion to wavelength, we obtain, for a 4 nm diameter sphere and $m_{\text{eff}} = 0.12 m_{\text{electron}}$, a wavelength (energy) of 1,584 nm (6,300 cm^{-1}) for the first excited state shifted from its position in the bulk material ($E_g \sim 1.7$ eV or 13,700 cm^{-1} for CdSe) giving a confinement energy of 20,000 cm^{-1} (500 nm). Typically, "green" (~530 nm) emitting CdSe q-dots™ have a diameter of 3–4 nm and "red" emitting q-dots™ have a diameter of 5.5–8 nm. Although the "particle-in-a-box" theory does not distinguish different materials within the same quantum level or reproduce emission wavelengths accurately for various sizes of particles, the general trend is correct, i.e., longer wavelength of emission for larger particles. The effective mass model gives closer agreement between experiment and theory at larger sizes of 8–10 nm diameter, but poorer agreement at small sizes.[109] Closer agreement with experimental results can be obtained with more advanced theories[110] that take into account the surface potential for the smaller nanoparticles, which have very high fractions (>50%) of surface atoms.

Early frontier work on semiconductor nanoparticles was forged by Arnim Henglein and his co-workers at the Hahn-Meitner Institute in Berlin. Various sizes/emission colors of CdS particles were produced[111] by using different starting pH, 8.1–9.8, with the precursors to give blue, turquoise, and green CdS particles of 4–6 nm diameter or by fractionating samples of colloidal CdS by exclusion chromatography[112] to give fractions containing particles of mean diameter, 1.86, 3.68, 3.30, 4.26, 4.36, 4.84, and 7.68 nm. Also, CdS particles of 2.2–5.0 nm diameter were fractionated by electrophoresis in acrylamide gels to yield particles of various colors.[113]

Individual semiconductor nanoparticles have caveats in their potential use as luminescent labels for flow cytometric applications. Like single molecules of fluorescent dye, the nanoparticles "blink." The fluctuations in emission intensity follow random, Poisson statistics. To detect any net average luminescence intensity from q-dots™ with on–off periods of the order of seconds, some statistical luminescent intensity averaging is needed for an ensemble of nanoparticles of about 100 or more, which are not all in the off state simultaneously. Further, defect sites on the surface of the nanoparticles need to be suitably capped so that they do not act as electron traps to lower the emission quantum yields, and emission band intensities. Optimizations in the necessarily high temperature, nonaqueous soap-mediated syntheses of capped CdSe nanoparticles to achieve maximum emission intensity were recently addressed.[114] Multiple applications[115–121] of semiconductor nanoparticles in imaging of biological specimens have been demonstrated. At least seven start-up companies [Quantum Dot Corp. (QDC), Hayward, CA; CrystalPlex Corp., Pittsburgh, PA; Evident Technologies, Inc., Troy, NY; BioCrystal Ltd., Westerville, OH; Oxonica, Oxford, UK; Nanoco Technologies Ltd., Manchester, UK; Nanosys, Inc., Palo Alto, CA] have sprung up to provide a variety of semiconductor nanoparticle products. The most advanced in this arena[122,123] is QDC, which is offering several "colors" of CdSe nanoparticles as silica-coated individual dots conjugated with streptavidin, and whose intellectual property portfolio has been most developed.[124]

Bead Systems

Methods have also been devised to satisfy the statistical ensemble (≥100 q-dots™/bead) criteria. The most successful so far[125,126] have been the organic solvent-swollen micron size PS beads into which semiconductor nanoparticles were allowed to diffuse at various concentrations, followed by shrinking in aqueous media and chemical capping of the bead pores. Applications of a different quantum dot-encoded bead system, Qbead™ (8 μm diameter, superparamagnetic polymer beads that are surface-dyed with quantum dots), developed by QDC, have been recently described in a SNP (single nucleotide polymorphism) assay.[127]

The close packing of semiconductor nanoparticles inside microspheres does not show any quenching of luminescent emission intensity unlike the behavior of organic fluorophores such as rhodamine 6G, as shown in our recent flow cytometric work[76] with silica-R6G microparticles (Fig. 11). In addition, a high luminescent intensity and narrowing of emission bandwidths of green and red CdSe nanoparticles inside 1 μm PS latex beads (estimated $\sim 50,000$ per bead) compared to respective, free nanoparticle emission bandwidths was observed. The estimated 50,000 CdSe nanoparticles inside 1 μm diameter bead yields a mean distance apart of nanoparticles inside the bead of 22 nm. The latter is substantially more than an estimate of 6.89 nm for 34 bilin groups inside phycobiliprotein, assuming a random average spacing for the bilins. Individually labelled semiconductor nanoparticles have the potential to reach appropriately large surface densities on biological cells such as WBCs, typically of 8–12 μm diameter, which normally show from 10^4 to 10^6 receptor sites per cell. Saturation at 2×10^5 nanoparticles per cell can bring the nanoparticles suitably close to each other to give a mean square distance apart of 32–47 nm.

The multiplicity (10–12) of different, nonoverlapping emission "colors" of q-dots™ inside beads as well as multiple intensity codes provides an enormous potential for multiplexing applications.

Needs for Cellular Assays

An important consideration[128] for successful use of q-dot™ probes in preparing reagents for flow cytometry is their protection from the deleterious effects (loss of luminescent intensity) of chemicals such as EDAC (1-ethyl-3-(3-dimethylaminopropyl)carbodiimide hydrochloride) or sulfo-SMCC (sulfosuccinimidyl 4-(N-maleimidomethyl)cyclohexane-1-carboxylate) that are used in crosslinking polymer or protein to semiconductor nanoparticles as well as chemical agents (formic acid, saponin, digitonin, etc.) used in lysing and quenching RBCs in whole blood analyses. QDC appears to have solved these problems with their silica coating of q-dots™. An ISAC poster[129] was presented on flow cytometry of 4 μm diameter biotinylated beads labeled with one of three different colors (red, orange, and green with 640, 585, and 520 nm emission) of CdSe q-dots™ coated with silica and streptavidin. The red dots on beads showed intensity almost in the third decade, the orange dots had intensity in the second decade, and the green dots were weakest in the first decade. Only the red and orange dots were used to label cells in whole blood. Indirectly labeled CD3+ cells with red dots showed good separation from background; but similarly labeled CD4+ cells with orange dots gave poor separation. Because of very narrow emission bandwidths of the red and orange dots, no compensation for overlapping emission bands was needed. An equivalent PE marker, compared under the same noncompensation instrument settings, showed a large unresolved distribution in dual color histograms.

Energy Transfer Possibilities

FRET (fluorescence resonance energy transfer) between luminescent semiconductor nanoparticles and other fluorescent molecules such as amino acid residues in proteins, etc. that are attached to the particles has also been observed. Examples include

1. Transfer from CdS nanoparticle donors to PE acceptors: When parallel flow cytometric results with 351–365 nm UV excitation for CD4 antibody–5X-aminodextran–CdS nanoparticle conjugate/SAM-PE were obtained versus the control CD4 antibody/SAM-PE as previously[128] done with 488 nm excitation; then, about 1.5–1.7-fold increase in PE emission intensity was observed.
2. Transfer from tryptophan residue donors on protein to nanoparticle acceptors: The mixed metal, CD4 antibody–5X-aminodextran–(Cd:Hg = 1 : 1)S conjugate prepared[130] by a procedure similar to one already reported for the same –CdS conjugate[128] showed an emission spectrum, with 283.2 nm excitation into the CD4 antibody absorption band (Fig. 15), showing (Fig. 16) an intense emission band centered at 339 nm from tryptophan residues of the CD4 antibody and a medium intensity emission band at 660 nm from the mixed (Cd,Hg)S semiconductor nanoparticles. The predominant excitation peak (Fig. 16, top) was at 281.6 nm when emission was monitored at 656 nm.
3. Transfer from trytophan residue donors of BSA conjugate of CdTe nanoparticles to CdTe nanoparticle acceptors.[131]
4. Transfer from nanoparticle donors to tetramethylrhodamine (TMR) acceptors: This was described[132] for the binding between biotinylated bovine serum albumin (BSA) conjugate with ZnS(CdSe) nanoparticles and TMR-labeled streptavidin by observing enhanced TMR fluorescence.
5. Transfer from BSA conjugate of green-emitting CdTe nanoparticle donors to anti-BSA conjugate of red-emitting CdTe nanoparticle acceptors.[133]

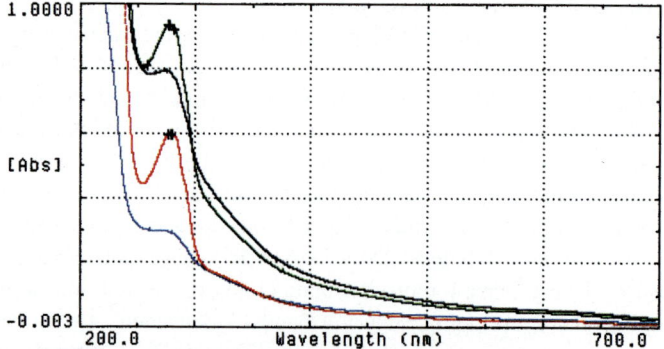

Fig. 15 Absorption spectra of four pooled fractions of CD4 antibody–5X-aminodextran–(CdHg)S nanoparticle conjugate, purified on G-25 Sephadex column (2.5 × 48 cm). Top to bottom: pooled fractions 41–45, 36–40, 46–50, and 31–35, respectively

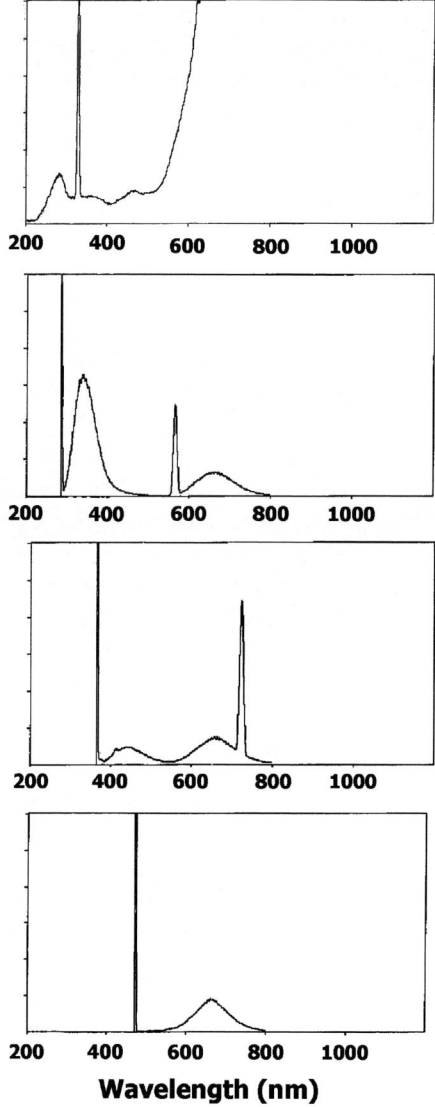

Fig. 16 Excitation-emission spectra of pooled fractions 31–35 of CD4 antibody–5X-aminodextran–(CdHg)S nanoparticle conjugate. Top to bottom: excitation spectrum, 656.0 nm emission; emission spectra, 283.2, 361.6, and 467.2 nm excitation, respectively

Raman Scattering: Light Scatter Nanoparticle Probes

Background

Raman scattering, a discrete quantum effect, observed as normal Raman scattering with an excitation wavelength far from an allowed electronic absorption band

and as resonance Raman scattering with an excitation wavelength into an allowed absorption band, has been combined with the surface plasmon resonance of metal nanoparticles, explained by classical Mie scattering theory, to give SERS (surface enhanced Raman scattering) or SERRS (surface enhanced resonance Raman scattering). Although SERS was first recognized by Jeanmaire and Van Duyne[134] in 1977 on a roughened silver electrode surface, the first observation of SERS from a suspension of gold nanoparticles was reported from Alan Creighton's laboratory[135] at the University of Kent, Canterbury, UK, in 1979. However, some years ago[136] we pointed out that a speculation[137] on such a phenomena was made by Michael Faraday about 122 years earlier in his last Bakerian Lecture on "Experimental Relations of Gold (and other metals) to Light." Faraday recounted that "At one time I hoped that I had altered one coloured ray into another by means of gold," but admitted that "I have not confirmed that result as yet" and "could not find any marked difference between the color or character of the ray reflected and the impinging ray, except in quantity." The "gold" that he was examining was colloidal gold, or finely divided gold as Faraday called it. Upon my bringing this to the attention of Milton Kerker back in the mid-1980s, his interests in the history of science prompted a postscript in a historical note[138] as well as a retrospective note.[139] By the mid-1850s, Faraday was in failing health and his work with gold colloids being his last major work, the speculation was never pursued further. The same reference to Faraday's speculation was more recently mentioned[140] in a biography of Faraday by Thomas, written when he was director of the Royal Institution of Great Britain, and occupant of the chair of Chemistry created for Michael Faraday. Thomas interpreted the speculation as a suggestion for the Raman effect before its 1928 discovery in India by Sir C.V. Raman.

The Raman-activated flow cytometer/cell sorter (RAFC/RACS) has yet to be constructed. However, the hypothetical construction of one was alluded to by Milton Kerker in the final section labeled "A Speculation" of his Paul Mullaney Memorial Lecture[141] of 1982. The first demonstrated use of surface-enhanced Raman scattering in a biological assay was only recently made[142] for gold nanoparticles coated with oligonucleotides and Raman-active dyes. Six different SERRS probes could be used simultaneously to target samples containing DNA/RNA in a stationary measurement on a solid phase. The event was suitably recorded in the September 2, 2002, issue of C&EN with the quote, "After 25 years and more than 4,000 papers, the phenomenon of surface Raman enhancement is finally becoming a useful method for chemical and biological studies," from Professor Shuming Nie, of the Wallace H. Coulter Department of Biomedical Engineering at Emory University and Georgia Institute of Technology. Also, only recently has a firm structural basis for SERS been presented for single particles[143] as well as fractal structures[144] with convincing experimental evidence, although the latter structures were 15 years earlier[145,146] surmised to be responsible for the high observed SERS intensities. Both SERS[143] and enhanced Mie/Rayleigh[38] scattering results for silver and gold nanoparticles have indicated that maximum scattering intensity is obtained with particles of about 100–200 nm in diameter.

In the stationary sampling mode, Raman signals were integrated over a time period of 10–100 s, as in various direct measurements[147–149] on biological specimens, such as eosinophils, cell nuclei, etc., by the Dutch group at the University of Twente. A summary of this and other work appeared in a chapter[150] of a book. A major part of the review summarized more recent work on the application of SERS to microscopic examination of biomedical samples. A recent entire issue, volume 33, July 2002, of the journal of Raman Spectroscopy was devoted to Medical Applications of Raman Spectroscopy, with a preface by G.J. Puppels and an article[151] describing use of a SERS-active label to detect DNA sequences of genes associated with carcinogenesis. Moreover, Intel and the Fred Hutchinson Cancer Research Center have announced[152] a collaborative research effort to build a new instrument, the Intel Raman Bioanalyzer System™, which uses Raman spectroscopy to "help researchers understand the molecular differences between healthy cells and diseased cells for better diagnostic methods."

Observation of Raman scattering intensity in these measurements does not guarantee the ability to obtain enough signal intensity from similar SERS probes in a conventional flow cytometric setting in which the residence time of the probe and cellular and/or particle target in the light beam is at most 10 μs. The approximately 10^6-fold difference in sampling times will require a similar 10^6-fold amplification of Raman signal. This has been shown to be feasible – SERS intensity of R6G on gold or silver nanoparticles of 100 nm diameter was greater than the fluorescence intensity of R6G in a micro-Raman experiment.[143] For some applications that do not require high throughput of labeled cells, it may be advantageous to incorporate a slow-flow and/or a stopped-flow system so that longer integration times for accumulating SERS signals can be used. The Coulter® Vi-CELL™ Analyzer already uses a stopped-flow system for video imaging of cells stained with trypan blue dye to determine cell viability.

More recently, dye-embedded core-shell nanoparticles that are highly SERS-active and may have promise as spectroscopic tags in multiplexed analyses were reported.[153] These core-shell particles contain a single particle, metallic gold core for optical enhancement, a reporter molecule (dye) for spectroscopic signature, and an outer silica shell for protection and conjugation of protein. Also, Natan and collaborators[154] have presented a marker system based on glass-coated, analyte-tagged nanoparticles (GANs) of gold or silver.

In our enhanced, Mie/Rayleigh, elastic light scattering work[36–39] with fractally arrayed structures of silver and gold nanoparticles, formed in situ on the surface of aminodextran-coated PS beads, we also predicted[36,38] enormous enhancements of inelastically scattered light or SERS from molecules located on the surface of these silver/gold particles. High-resolution SEM photos of these silver/gold structures on aminodextran-coated PS beads are shown in Fig. 17. In many of our SEM photos, small silver and gold structures on PS beads tend to have liquid-like properties by assuming the shape of liquid droplets, and chains of droplets, rather than the nanocrystalline structures with sharp edges and vertices that were previously[146] observed in fractal aggregates. Moreover, Raman spectral measurements of single beads, coated with silver- and gold-nanoparticles, to which rhodamine 6G (at concentrations of $1 \times 10^{-5} – 1 \times 10^{-4}$ M) had been adsorbed and the excess washed away, were made with micro-Raman instrumentation (compact Raman system originally marketed by

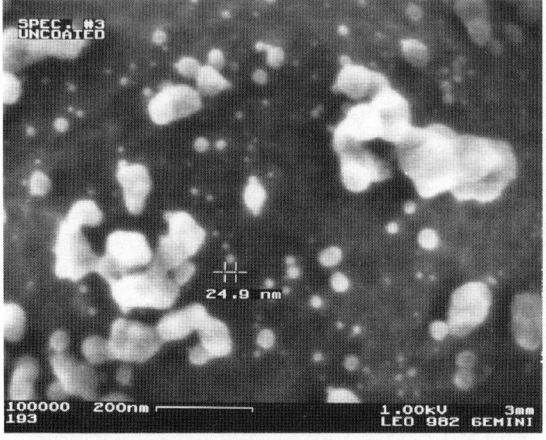
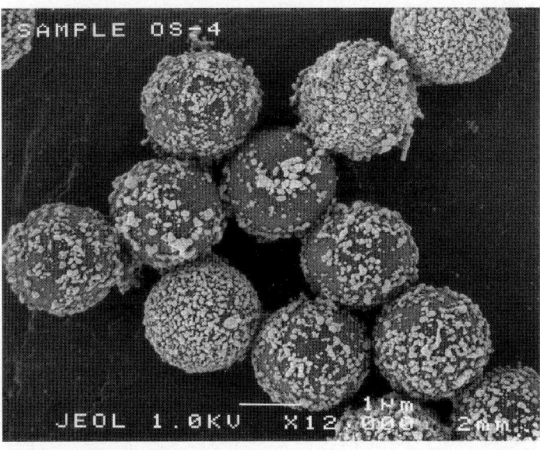

Fig. 17 SEM micrographs of silver-coated microsphere (top) and gold-coated microspheres (bottom)

Detection Limit, equivalent of Advantage 200A system now offered by DeltaNu, Laramie, WY). A 3 mW 633 nm He/Ne laser was used for excitation, the spot size on the sample was 10–20 μm in diameter, and a one second integration time was used to accumulate spectra. Intense SERS spectra of R6G on silver nanoparticle/PS beads and on gold nanoparticle/PS beads were observed as shown in Figs. 18 and 19, respectively, for representative runs. SERS intensities for R6G on silver nanoparticle/PS beads (Fig. 18, top three spectra) increased markedly with increasing R6G concentrations from 1×10^{-5} to 1×10^{-4} M, starting at 1,500 counts/s and reaching and surpassing the saturation point (65,000 counts/s) of the detector with some beads. On the other hand, SERS intensities for R6G on gold nanoparticle/PS beads changed very little (1,500–2,500 counts/s) for R6G concentrations between 1×10^{-5} and 1×10^{-4} M. The SERS band intensity differences for silver versus gold samples may be explained by two factors. First, a much more

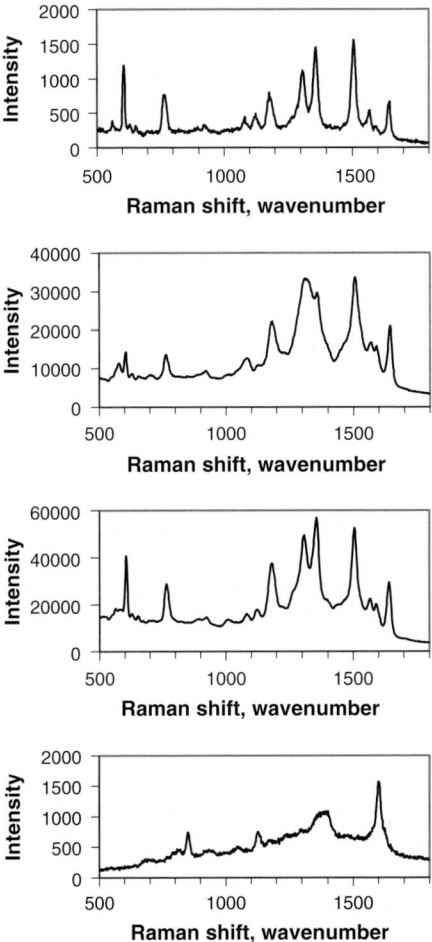

Fig. 18 Raman spectra of silver nanoparticle-5X-aminodextran-PS bead with rhodamine 6G (top 3 spectra, top to bottom, 1% w/v solids bead suspension, equilibrated with 1×10^{-5} M, 5×10^{-5} M, and 1×10^{-4} M R6G, respectively, followed by wash and 10× dilution with deionized water, was applied to microscope slide) and without R6G (bottom) (1% w/v solids bead suspension was applied to slide)

uniform coating of silver nanoparticles as compared to the often spotty coating of gold nanoparticles around the 2 μm diameter PS beads was obtained as seen in our published[38] and patented[36,37] work. Second, a red-shifted wavelength of incident radiation is required to excite surface plasmon resonance bands of gold nanoparticles in the same size range as the silver nanoparticles for maximum SERS intensity, since the extinction maximum[36] lies between 800 and 900 nm for gold/PS beads (G-5s and G-15s trials[38]) and at ~600 nm for silver/PS beads (S7 trial[38]). The positions and relative intensities of SERS bands of R6G observed in this work are similar to those previously reported[155] for R6G on colloidal silver particles of 30 nm

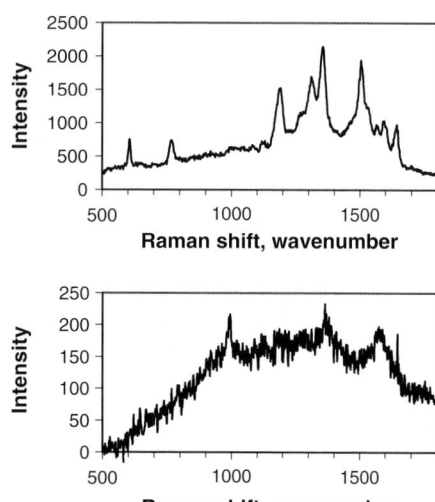

Fig. 19 Raman spectra of gold nanoparticle-5X-aminodextran-PS bead with rhodamine 6G (top), (1% w/v solids bead suspension, equilibrated with 5×10^{-5} M R6G, followed by wash, but no dilution, was applied to microscope slide) and without R6G (bottom) (1% w/v solids bead suspension was applied to slide)

diameter in suspension with red, 647 nm Kr^+ laser excitation, far from the strong absorption band maximum of R6G at 530 nm.

Furthermore, the dense, saturation packing of R6G or other dye molecules on colloidal gold/silver surface does not limit Raman intensities in the way that fluorescence intensities are lowered by excited state dipole–dipole interactions. In comparison with fluorescent dyes and quantum dots, the enhanced Raman probes have an intrinsic amplification mechanism, with improved encoding and multiplexing capabilities. If multiple (4 or even more) sizes of gold/silver nanoparticles in the 50–300 nm diameter range with a narrow size distribution can be prepared, and then loaded with 20 different dye molecules with unique SERS fingerprints, this would provide 80 probes for each metal nanoparticle.

Acknowledgments The author is thankful to Shuming Nie (Emory University/Georgia Institute of Technology) for his hospitality on a visit to Emory University to perform micro-Raman measurements, and for providing a copy of an article prior to publication and other clarifications, to Douglas Stuart (Emory University) for help with the Raman measurements, to Jorge Quintana (Beckman Coulter, Inc.) for providing powerpoint slides of his presentation on Flow-Cal beads, and to John Maples (Beckman Coulter, Inc.) for his photographs of a Wright's stain of a blood sample that had been incubated with CD4 antibody–PS latex beads.

References

1. Howard M. Shapiro, *Practical Flow Cytometry*, 4th edn., Wiley-Liss, New York, NY, 2003.
2. *Diagnostic Applications of Latex Technology: Theory and Practice*, sponsored by Bangs Laboratories, Inc., and Emerald Diagnostics, Inc., May 19–21, 1999, Indianapolis, IN.
3. The Latex Course™ 2002—Designing Microsphere-Based Tests and Assays, Bangs Laboratories, Inc., June 10–12, 2002, Indianapolis, IN.

4. L.J. Kricka and P. Fortina, *Clin. Chem.* **47** (2001) 1479; **48** (2002) 662; **48** (2002) 1620.
5. M.J. Fulwyler, J.D. Perrings, and L.S. Cram, *Rev. Sci. Instrum.* **44** (1973) 204.
6. M.J. Fulwyler, British Patent No. 1,561,042 (1976).
7. P.K. Horan and L.L. Wheeless, Jr., *Science* **198** (1977) 149.
8. Thomas M. McHugh and Mack J. Fulwyler, in *Clinical Flow Cytometry, Principles and Application*, K.D. Bauer, R.E. Duque, and T.V. Shankey, eds., Williams & Wilkins, Baltimore, MD, 1993, Chap. 32, p. 535.
9. K.H. Kortright, W.H. Coulter, C. Rodriguez, T. Russell, and R. Paul, U.S. Patent No. 5,223,398 (Jun 29, 1993).
10. T. Russell, K.H. Kortright, W.H. Coulter, C.M. Rodriguez, R. Paul, C.M. Hajek, and J.C. Hudson, U.S. Patent No. 5,231,005 (Jul 27, 1993).
11. T. Russell, C.M. Hajek, C.M. Rodriguez, and W.H. Coulter, U.S. Patent No. 5,260,192 (Nov 9, 1993).
12. K.H. Kortright, W.H. Coulter, C. Rodriguez, T. Russell, and R. Paul, U.S. Patent No. 5,464,752 (Nov 7, 1995).
13. J.C. Hudson, R.F. Brunhouse, C. Garrison, C.M. Rodriguez, R. Zwerner, and T.R. Russell, *Cytometry* **22** (1995) 150.
14. O. Siiman, A. Burshteyn, and R.K. Gupta, U.S. Patent No. 5,169,754 (Dec 8, 1992).
15. O. Siiman, A. Burshteyn, and R.K. Gupta, U.S. Patent No. 5,466,609 (Nov 14, 1995).
16. O. Siiman, A. Burshteyn, and R.K. Gupta, U.S. Patent No. 5,639,620 (Jun 17, 1997).
17. O. Siiman, A. Burshteyn, and R.K. Gupta, U.S. Patent No. 5,707,877 (Jan 13, 1998).
18. O. Siiman, A. Burshteyn, and R.K. Gupta, U.S. Patent No. 5,776,706 (Jul 7, 1998).
19. O. Siiman, A. Burshteyn, and M.E. Insausti, *J. Colloid Interface Sci.* **234** (2001) 44.
20. O. Siiman and A. Burshteyn, *Cytometry* **40** (2000) 316.
21. O. Siiman, A. Burshteyn, O. Concepcion, and M. Forman, U.S. Patent No. 5,814,468 (Sep 29, 1998).
22. O. Siiman and A. Burshteyn, U.S. Patent No. 5,062,991 (Nov 5, 1991).
23. O. Siiman, J. Wilkinson, A. Burshteyn, P. Roth, and S. Ledis, *Bioconjugate Chem.* **10** (1999) 1090.
24. C. Liu, B. Zou, A.J. Rondinone, and Z.J. Zhang, *J. Phys. Chem.* **104** (2000) 1141.
25. Z.J. Zhang, Z.L. Wang, B.C. Chakoumakos, and J.S. Yin, *J. Am. Chem. Soc.* **120** (1998) 1800.
26. J.A. Maples, R.H. Raynor, O. Siiman, M.J. Stiglitz, S.F. Healy, Jr., U.S. Patent No. 5,763,204 (Jun 9, 1998) and U.S. Patent No. 5,342,754 (Aug 30, 1994).
27. R.J. Schmittling, O. Siiman, N. Kenyon, and W. Bolton, *Ann. NY Acad. Sci.* **677** (1993) 447.
28. N.S. Kenyon, R.J. Schmittling, O. Siiman, A. Burshteyn, and W.E. Bolton, *Cytometry* **16** (1994) 175.
29. W.H. Coulter, R.K. Zwerner, R.J. Schmittling, and T.R. Russell, U.S. Patent No. 5,576,185 (Nov 19, 1996).
30. O. Siiman, A. Burshteyn, J.A. Maples, and J.K. Whitesell, *Bioconjugate Chem.* **11** (2000) 549.
31. J.K. Whitesell and H.K. Chang, *Science* **261** (1993) 73.
32. R.-M. Bohmer and N.J.C. King, *Cytometry* **5** (1984) 543.
33. R. Festin, B. Bjorklund, and T.H. Totterman, *J. Immunol. Methods* **101** (1987) 23.
34. Thomas H. Totterman and Roger Festin, in *Colloidal Gold: Principles, Methods and Applications, Vol. 2*, M. A. Hayat, ed., Academic Press, San Diego, 1989, Chap. 22, p. 431.
35. C. Neagu, K.O. van der Werf, C.A.J. Putman, Y.M. Kraan, B.G. de Grooth, N.F. van Hulst, and J. Greve, *J. Struct. Biol.* **112** (1994) 32.
36. O. Siiman, A. Burshteyn, and M. Cayer, U.S. Patent No. 5,552,086 (Sep 3, 1996).
37. O. Siiman, K. Gordon, C.M. Rodriguez, A. Burshteyn, J.A. Maples, and J.K. Whitesell, U.S. Patent No. 5,945,293 (Aug 31, 1999).
38. O. Siiman and A. Burshteyn, *J. Phys. Chem. B* **42** (2000) 9795.
39. O. Siiman, K. Gordon, A. Burshteyn, J.A. Maples, and J.K. Whitesell, *Cytometry* **41** (2000) 298.
40. Max Born and Emil Wulf, *Principles of Optics: Electromagnetic Theory of Propagation, Interference and Diffraction of Light*, 7th edn., Cambridge University Press, Cambridge, 1999.

41. Julius Adams Stratton, *Electromagnetic Theory*, McGraw-Hill, New York, 1941.
42. Craig F. Bohren and Donald F. Huffman, *Absorption and Scattering of Light by Small Particles*, Wiley-Interscience, New York, 1983.
43. S.R. Nicewarner-Pena, R.G. Freeman, B.D. Reiss, L. He, D.J. Pena, I.D. Walton, R. Cromer, C.D. Keating, and M.J. Natan, *Science* **294** (2001) 137.
44. I.D. Walton, S.M. Norton, A. Balasingham, L. He, D.F. Oviso, Jr., D. Gupta, P.A. Raju, M.J. Natan, and R.G. Freeman, *Anal. Chem.* **74** (2002) 2240.
45. J. Yguerabide and E.E. Yguerabide, *Anal. Biochem.* **262** (1998) 157.
46. J. Yguerabide, E.E. Yguerabide, D.E. Kohne, and J.T. Jackson, U.S. Patent No. 6,214,560 B1 (Apr 10, 2001).
47. J.C. Hulteen, D.A. Treichel, M.T. Smith, M.L. Duval, T.R. Jensen, and R.P. Van Duyne, *J. Phys. Chem. B* **103** (1999) 3854.
48. T.R. Jensen, M.D. Malinsky, C.L. Haynes, and R.P. Van Duyne, *J. Phys. Chem. B* **104** (2000) 10549.
49. C.L. Haynes and R.P. Van Duyne, *J. Phys. Chem. B* **105** (2001) 5599.
50. C.L. Haynes, A.D. McFarland, L. Zhao, R.P. Van Duyne, G.C. Schatz, L. Gunnarsson, J. Prikulis, B. Kasemo, and M. Kall, *J. Phys. Chem. B* **107** (2003) 7337.
51. J.T. Krug, II, G.D. Wang, S.R. Emory, and S. Nie, *J. Am. Chem. Soc.* **121** (1999) 9208.
52. R.J. Fulton, U.S. Patent No. 5,736,330 (Apr 7, 1998).
53. V.S. Chandler, R.J. Fulton, and M.B. Chandler, U.S. Patent No. 5,981,180 (Nov 9, 1999).
54. M.B. Chandler and D.J. Chandler, U.S. Patent No. 6,268,222 B1 (Jul 31, 2001).
55. D.J. Chandler, B.A. Lambert, J.J. Reber, and S.L. Phipps, U.S. Patent No. 6,514,295 B1 (Feb 4, 2003).
56. D.J. Chandler, U.S. Patent No. 6,528,165 B2 (Mar 4, 2003).
57. R. Fulton, R. McDade, P. Smith, L. Kienker, and J. Kettman, Jr., *Clin. Chem.* **43** (1997) 1749.
58. C. Camilla, J.P. Defoort, M. Delaage, R. Auer, J. Quintana, T. Lary, R. Hamelik, S. Prato, B. Casano, M. Martin, and V. Fert, *Cytometry Suppl.* **8** (1998) 132.
59. R. Carson and D. Vignali, *J. Immunol.* Methods **227** (1999) 41.
60. K.L. Kellar, R.R. Kalwar, K.A. Dubois, D. Crouse, W.D. Chafin, and B.-E. Kane, *Cytometry* **45** (2001) 27.
61. M.C. Earley, R.F. Vogt, Jr., H.M. Shapiro, F.F. Mandy, K.L. Kellar, R. Bellisario, K.A. Pass, G.E. Marti, C.C. Stewart, and W.H. Hannon, *Cytometry* **50** (2002) 239.
62. I.G. Loscertales, A. Barrero, I. Guerrero, R. Cortijo, M. Marquez, and A.M. Ganan-Calvo, *Science* **295** (2002) 1695.
63. I. Tsagkatakis, S. Peper, and E. Bakker, *Anal. Chem.* **73** (2001) 315.
64. I. Tsagkatakis, S. Peper, R. Retter, M. Bell, and E. Bakker, *Anal. Chem.* **73** (2001) 6083.
65. M. Telting-Diaz and E. Bakker, *Anal. Chem.* **74** (2002) 5251.
66. R. Retter, S. Peper, M. Bell, I. Tsagkatakis, and E. Bakker, *Anal. Chem.* **74** (2002) 5420.
67. J. Quintana, presented at CD38 Quantitation Conference, Francis F. Mandy, org., National Laboratory for HIV Immunology, Health Canada, Ottawa, Canada, November 1997.
68. T. Lindmo, O. Bormer, J. Ugelstad, and K. Nustad, *J. Immunol. Methods* **126** (1990) 183.
69. J. Frengen, R. Schmid, B. Kierulf, K. Nustad, E. Paus, A. Berge, and T. Lindmo, *Clin. Chem.* **39** (1993) 2174.
70. J. Frengen, T. Lindmo, E. Paus, R. Schmid, and K. Nustad, *J. Immunol. Methods* **178** (1995) 141.
71. J. Frengen, K. Nustad, R. Schmid, and T. Lindmo, *J. Immunol. Methods* **178** (1995) 131.
72. J. Frengen, T. Lindmo, R. Schmid, and J. Ugelstad, 70th Colloid and Surface Science Symposium, American Chemical Society, Clarkson University, June 16–19, 1996.
73. M.L. Bell, U.S. Patent No. 6,551,788 (April 22, 2003).
74. L.E.M. Miles, D.A. Lipschitz, C.P. Bieber, and J.D. Cook, *Anal. Biochem.* **61** (1974) 209.
75. M. Bele, O. Siiman, and E. Matijevic, *Cytometry Suppl.* **11** (2002) 128.
76. M. Bele, O. Siiman, and E. Matijevic, *J. Colloid Interface Sci.* **254** (2002) 274.
77. J.P. Nolan and F.F. Mandy, *Cell. Mol. Biol.* **47** (2001) 1241.

78. A. Goodey, J.J. Lavigne, S.M. Savoy, M.D. Rodriguez, T. Curey, A. Tsao, G., Simmons, J. Wright, S.-J. Yoo, Y. Sohn, E.V. Anslyn, J.B. Shear, D.P. Neikirk, and J.T. McDevitt, *J. Am. Chem. Soc.* **123** (2001) 2559.
79. J.P. Nolan and L.A. Sklar, *Trends in Biotechnology* **20** (2002) 9.
80. E. Willis, S. Allauzen, and S. Vlasenko, *BioRadiations* **111** (2003) 30, and other reports and articles in same issue on the Bio-Plex system, as distributed by Bio-Rad Laboratories, Inc., Hercules, CA 94547.
81. O. Siiman, C. Smith, P. Roth, A. Burshteyn, and R. Raynor, U. S. Patent No. 5,891,741 (Apr 6, 1999).
82. O. Siiman, A. Burshteyn, J. Wilkinson, and R. Mylvaganam, U. S. Patent No. 5,994,089 (Nov 30, 1999).
83. O. Siiman, A. Burshteyn, R. Mylvaganam, R. Raynor, P. Roth, C. Smith, and J. Wilkinson, U. S. Patent No. 6,387,622 (May 14, 2002).
84. S. Ledis, C. Healy, and O. Siiman, U.S. Patent application filed (Aug 1, 2003).
85. C. Smith, J. Wilkinson, P. Roth, and O. Siiman, *Cytometry Suppl.* **9** (1998) 56.
86. R. Mylvaganam, J. Wilkinson, C. Healy, W. Bolton, and O. Siiman, *Cytometry Suppl.* **9** (1998) 117.
87. O. Siiman, A. Burshteyn, O. Concepcion, and M. Forman, *Cytometry* **44** (2001) 30.
88. O. Siiman, U.S. Patent Application Pub. No. US 2002/0142289 A1 published (Oct 3, 2002).
89. Ed Harlow and David Lane, *Antibodies – A Laboratory Manual*, Cold Spring Harbor Laboratory, 1988, Chap. 14.
90. T.W.J. Huizinga, M. de Haas, M. Kleijer, J.H. Nuijens, D. Roos, and A.E.G. Kr. von dem Borne, *J. Clin. Invest.* **86** (1990) 416.
91. T.W.J. Huizinga, R.W.A.M. Kuijpers, M. Kleijer, T.W.J. Schulpen, H.T.M. Cuypers, D. Roos, and A.E.G. Kr. von dem Borne, *Blood* **76** (1990) 1927.
92. H.B. Fleit, C.D. Kobasiuk, C. Daly, R. Furie, P.C. Levy, and R.O. Webster, *Blood* **79** (1992) 2721.
93. P. Antal-Szalmas, I. Szollosi, G. Lakos, E. Kiss, I. Csipo, A. Sumegi, S. Sipka, J.A.G. van Strijp, K.P.M. van Kessel, and G. Szegedi, *Cytometry* **45** (2001) 115.
94. S. Delaire, A. Elhabazi, A. Bensussan, and L. Boumsell, *Cell. Mol. Life Sci.* **54** (1998) 1265.
95. A. Elhabazi, S. Delaire, A. Bensussan, L. Boumsell, and G. Bismuth, *J. Immunol.* **166** (2001) 4341.
96. S. Delaire, C. Billard, R. Tordjman, A. Chedotal, A. Elhabazi, A. Bensussan, and L. Boumsell, *J. Immunol.* **166** (2001) 4348.
97. X. Wang, A. Kumanogoh, C. Watanabe, W. Shi, K. Yoshida, and H. Kikutani, *Blood* **97** (2001) 3498.
98. A. Elhabazi, A. Marie-Cardine, I. Chabbert-de Ponnat, A. Bensussan, and L. Boumsell, *Crit. Rev. Immunol.* **23** (2003) 65.
99. A.N. Barclay, M.H. Brown, S.K.A. Law, A.J. McKnight, M.G. Tomlinson, and P.A. van der Merwe, *The leukocyte antigen facts book*, 2nd edn., Academic Press, San Diego, 1997, Sect. II, pp. 132–593.
100. N. Stahl, D.R. Borchelt, K. Hsaio, and S.B. Prusiner, *Cell* **51** (1987) 229.
101. M.A.J. Ferguson, *J. Cell Sci.* **112** (1999) 2799.
102. A.P. Alivisatos, *Scientific American* **285** (2001) 66.
103. H. Mattoussi, *J. Am. Chem. Soc.* **122** (2000) 12142.
104. S. Empedocles and M. Bawendi, *Acc. Chem. Res.* **32** (1999) 389.
105. M. Nirmal and L. Brus, *Acc. Chem. Res.* **32** (1999) 407.
106. L.E. Brus and J.K. Trautman, *Phil. Trans. R. Soc. Lond.* **353A** (1995) 313.
107. X. Peng, M.C. Schlamp, A.V. Kadavanich, and A.P. Alivisatos, *J. Am. Chem. Soc.* **119** (1997) 7019.
108. J.W. Linnett, *Wave Mechanics and Valency*, Wiley, New York, NY, 1960, Chap. II.
109. Christopher B. Murray, *Synthesis and Characterization of II–VI Quantum Dots and Their Assembly into 3D Quantum Dot Superlattices*, Ph. D. thesis, Massachusetts Institute of Technology (Sep 1995).

110. P.E. Lippens and M. Lannoo, *Phys. Rev. B* **39** (1989) 10935.
111. L. Spanhel, M. Haase, H. Weller, and A. Henglein, *J. Am. Chem. Soc.* **109** (1987) 5649.
112. Ch.-H. Fischer, J. Lilie, H. Weller, L. Katsikas, and A. Henglein, *Ber. Bunsenges. Phys. Chem.* **93** (1989) 61.
113. A. Henglein, 70th Colloid and Surface Science Symposium, American Chemical Society, Clarkson University, June 16–19, 1996.
114. L. Qu and X. Peng, *J. Am. Chem. Soc.* **124** (2002) 2049.
115. M. Bruchez, M. Moronne, P. Gin, S. Weiss, and A.P. Alivisatos, *Science* **281** (1998) 2013.
116. W.C.W. Chan and S. Nie, *Science* **281** (1998) 2016.
117. M.E. Ackerman, W.C.W. Chan, P. Laakkonen, S.N. Bhatia, and E. Ruoslahti, *Proc. Natl. Acad. Sci. USA* **99** (2002) 12617.
118. X. Wu, H. Liu, J. Liu, K.N. Haley, J.A. Treadway, J.P. Larson, N. Ge, F. Peale, and M.P. Bruchez, *Nature Biotechnol.* **21** (2003) 41.
119. J.K. Jaiswal, H. Mattoussi, J.M. Mauro, and S.M. Simon, *Nature Biotechnol.* **21** (2003) 47.
120. D.R. Larson, W.R. Zipfel, R.M. Williams, S.W. Clark, M.P. Bruchez, F.W. Wise, and W.W. Webb, *Science* **300** (2003) 1434.
121. B. Dubertret, P. Skourides, D.J. Norris, V. Noireaux, A.H. Brivanlou, and A. Libchaber, *Science* **298** (2002) 1759.
122. A. Watson, X. Wu, and M. Bruchez, *BioTechniques* **34** (2003) 296.
123. T.M. Jovin, *Nature Biotechnol.* **21** (2003) 32.
124. See, for example, their website at www.qdots.com and their newsletter article, K. Barovsky, *Intellectual Property: Illuminating the Path to Licensing Compliance*, Quantum Dot eVision, May 19, 2003.
125. M. Han, X. Gao, J.Z. Su, and S. Nie, *Nature Biotechnol.* **19** (2001) 631.
126. W.C.W. Chan, D.J. Maxwell, X. Gao, R.E. Bailey, M. Han, and S. Nie, *Curr. Opinion Biotech.* **13** (2002) 40.
127. H. Xu, M.Y. Sha, E.Y. Wong, J. Uphoff, Y. Xu, J.A. Treadway, A. Truong, E. O'Brien, S. Asquith, M. Stubbins, N.K. Spurr, E.H. Lai, and W. Mahoney, *Nucleic Acids Res.* **31** (2003) e43.
128. I. Sondi, O. Siiman, S. Koester, and E. Matijevic, *Langmuir* **16** (2000) 3107.
129. W. Hyun, R.H. Daniels, C.Z. Hotz, and M. Bruchez, *Cytometry Suppl.* **10** (2000) 182.
130. O. Siiman, unpublished results.
131. N.N. Mamedova, N.A. Kotov, A.L. Rogach, and J. Studer, *NanoLetters* **1** (2001) 281.
132. D.M. Willard, L.L. Carillo, J. Jung, and A. van Orden, *NanoLetters* **1** (2001) 469.
133. S. Wang, N. Mamedova, N.A. Kotov, W. Chen, and J. Studer, *NanoLetters* **2** (2002) 817.
134. D.L. Jeanmaire and R.P. Van Duyne, *J. Electroanal. Chem.* **84** (1977) 1.
135. J.A. Creighton, C.G. Blatchford, and M.G. Albrecht, *J. Chem. Soc., Faraday Trans. 2* **75** (1979) 790.
136. O. Siiman and W.P. Hsu, *J. Chem. Soc., Faraday Trans. 1* **82** (1986) 851.
137. M. Faraday, *Philos. Trans. Roy. Soc. London* **147** (1857/58) 145.
138. M. Kerker, *J. Colloid Interface Sci.* **112** (1986) 302.
139. M. Kerker, *Proc. Roy. Inst. London* **61** (1989) 229.
140. John M. Thomas, *Michael Faraday and the Royal Institute (The Genius of Man and Place)*, Adam Hilger (IOP Publishing Ltd.), Bristol, England, 1991, p. 81.
141. M. Kerker, *Cytometry* **4** (1983) 1.
142. Y.-W. C. Cao, R. Jin, and C.A. Mirkin, *Science* **297** (2002) 1536.
143. S. Nie and S.R. Emory, *Science* **275** (1997) 1102.
144. Z. Wang, S. Pan, T.D. Krauss, H. Du, and L.J. Rothberg, *Proc. Natl. Acad. Sci. USA* **100** (2003) 8638.
145. H. Feilchenfeld and O. Siiman, *J. Phys. Chem.* **90** (1986) 4590.
146. O. Siiman and H. Feilchenfeld, *J. Phys. Chem.* **92** (1988) 453.
147. G.J. Puppels, F.F.M. de Mul, C. Otto, J. Greve, M. Robert-Nicoud, D.J. Arndt-Jovin, and T.M. Jovin, *Nature* **347** (1990) 301.
148. G.J. Puppels, H.S.P. Garritsen, G.M.J. Segers-Nolten, F.F.M. de Mul, and J. Greve, *Biophys. J.* **60** (1991) 1046.

149. G.J. Puppels, H.S.P. Garritsen, J.A. Kummer, and J. Greve, *Cytometry* **14** (1993) 251.
150. Michel Manfait and Igor Nabiev, Applications in Medicine, in *Raman Microscopy Developments and Applications*, G. Turrell and J. Corset, eds., Academic Press Ltd., London, UK, 1996, Chap. 9, p. 379.
151. T. Vo-Dinh, L.R. Allain, and D.L. Stokes, *J. Raman Spectrosc.* **33** (2002) 511.
152. *Intel and Fred Hutchinson Cancer Research Center to Explore the Use of Nanotechnology Tools for Early Disease Detection*, News Release, Stanford, CA, Oct 23, 2003, as posted on website www.fhcrc.org.
153. W.E. Doering and S. Nie, *Anal.Chem.*, web release date: 03 Oct 2003; DOI: 10.1021/ac034672u.
154. S.P. Mulvaney, M.D. Musick, C.D. Keating, and M.J. Natan, *Langmuir* **19** (2003) 4784.
155. P. Hildebrandt and M. Stockburger, *J. Phys. Chem.* **88** (1984) 5935.

Medical and Technological Application of Monodispersed Colloidal Silica Particles

Herbert Giesche

Introduction

Monodispersed submicrometer-sized silica particles have been studied for a variety of application. Most of these publications are based on the same synthesis technique, the so-called Stöber-process,[1] which is modified to some degree in each case in order to "adjust" the particles for the different applications. Stöber silica particles have some very exceptional properties. They can be produced with outstanding uniformity; less than 5% standard deviation of the particle size distribution is standard and values as low as 1% can be achieved (in the particle size range between 100 and 1,000 nm) when the synthesis process is controlled more carefully. This property is only been rivaled by the control over particle size in latex systems. Another major advantage is the ease with which somewhat larger quantities of the material can be produced. This fact makes the commercial aspect of moving from the laboratory results to production interesting for companies. And last but not least, the chemistry of silica has been extensively studied, and a vast pool of information is available, in addition to numerous silane precursor compounds, which are commercially available in many cases. Those silane precursors allow grafting specific functional groups on the surface of the silica particles and, thus, make it possible to modify the chemical nature of the surface, as well as the interactions of the particles with their environment. The high degree of chemical and physical control allows one to fine-tune and tailor these particles for particular applications.

Preparation

The standard Stöber process[1] employs tetraethoxysilane (TEOS) in a mixture of water, ammonia, and ethanol. The use of other alkoxysilanes as well as replacing ethanol with other solvents are possible modifications of the process. The specific conditions will have a pronounced effect on the particle size, uniformity, reaction rate, particle density, etc. For example, longer and/or branched alcohol groups in the alkoxide result in slower reaction (hydrolysis and condensation) rates, which then lead to larger particles. Similar effects are noted, when the solvent (ethanol) is replaced with "larger" alcohols, when the ammonia concentration is raised, or when the temperature is lowered. More detailed descriptions of these effects are given in several review articles.[2–10] The particle size can be changed by more than an order of magnitude within the range of the reaction conditions. Another elegant way of controlling the final particle size was described in several other papers,[4, 10–13] in which small seed particles are produced by the "regular" Stöber process and then further grown by a controlled growth reaction. The latter can be achieved with either as a step by step addition, or by a slow continuous addition of the alkoxide to the seed suspension. Exceptional monodispersity and larger particle sizes has been achieved through the seeded growth process. In addition, the volume fraction of the particles in the solution or in other words the yield of the process can be increased to about 10 vol%. Silica compares favorably to all other monodispersed powders because of the higher concentration that can be achieved making the production process more economical. Several companies have actually adopted this process and are producing those silica particles commercial in larger quantities. An important additional improvement was the development of a continuous process as described by Giesche[4] and Kaiser et al.[14]

The Stöber process was also studied by several research groups with respect to the formation and reaction mechanisms.[3, 5–10] Further details can be found in those publications. For example the overall particle growth reaction was described as a function of TEOS, water, and ammonia concentrations, as well as reaction temperatures by the following equation:[3]

$$d[\text{SiO}_2 \text{ particle}]/dt = 2.36\,\text{s}^{-1} \left(\text{mol dm}^{-3}\right)^{-2.15} \times \exp\left\{\frac{-3256\,\text{K}}{T}\right\} \times [\text{H}_2\text{O}]^{1.18} \times [\text{NH}_3]^{0.97} \times [\text{TEOS}]$$

In addition, two models have been proposed and evaluated, which would allow to explain the formation of monodispersed particles: a LaMer type nucleation and growth mechanism or an aggregation growth mechanism[15] as initially proposed by Bogush and Zukoski.[9] Each mechanism can be proven right or wrong depending on the experimental conditions. The reaction may actually start as an aggregation process (assuming a sufficiently fast initial reaction rate), and then change to the LaMer-type monomer addition mechanism as the silane concentration decreases and the reaction slows down. Transmission electron micrographs show a grainy substructure within the particles under some reaction conditions, whereas the particles appear smooth under others. Likewise, the measured porosity within the particles varies, which is an indirect hint for different microstructures. Van Helden and Vrij[16] and Emmett et al.[17] studied the aggregation behavior of monodispersed silica particles in organic solvents and as part of those experiments the preformed light scattering tests and the SAXS analysis provided a clear indication that the original (primary) silica particles were actually porous aggregates, a result that was further verified by density or gas adsorption measurements.[3]

Exceedingly monodispersed nanometer-sized silica particles can also be produced by a so-called microemulsion technique,[18] which employs essentially the same chemicals as the Stöber process. However, the presence of a surfactant and the oil-phase provide means of creating small (nanometer sized) water droplets (microemulsion), within which the particles are formed. These droplets can be seen as size-limiting reaction vessels, producing very small but rather uniform nanoparticles. Since this technique is not limited to silica, it has been applied to the synthesis of many different nanopowders.

Surface and Bulk Modifications

A review article by Yoshinaga[19] and the Ph.D. thesis by Kinkel[20] provide excellent overviews on the various possibilities of surface modifications. Most of the time these reactions are performed on already formed silica particles and the functional groups are grafted onto the existing silica surface (surface modification), but it is also possible to apply these reactions during the growth phase of the Stöber silica particles and, thus, incorporate the functional groups into the particle. Typically, these functional groups are relative "simple" groups, which are then further used as an anchoring group for other molecules (e.g., polymers or bio molecules). A few examples are shown in Fig. 1. Another study by Markowitz et al.[21] showed how the surface acidity and basicity is affected by various functional groups attached to the silica particles. Acidity and basicity are important factors for the interaction between the particles as well as interaction of other probe molecules with those surfaces. Consequently, it then affects the general suspension behavior or the adsorption characteristics in chromatography, medical, or diagnostic applications.

When sufficiently large alkyl groups, as well as sufficient quantities of an alkyl silane (in combination with TEOS), are incorporated into the growing particles, a controlled porosity in the final particle can be created after burnout of the organic groups at 823 K.[14,22] Pore diameter (3–16 nm) and pore volume (up to $0.6 \text{ cm}^3 \text{ g}^{-1}$) could be adjusted depending on the type and amount of alkyl silane.

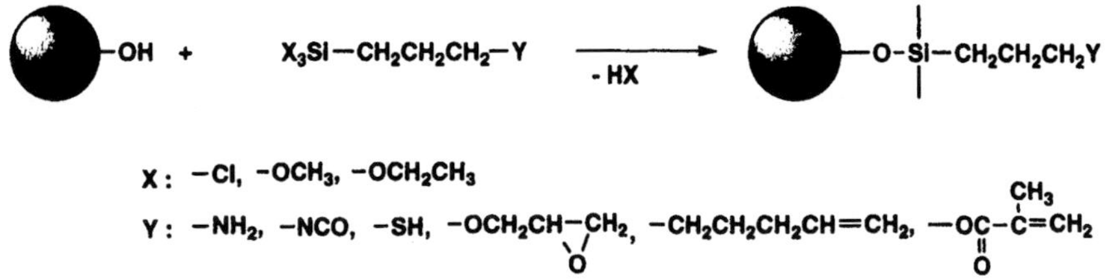

Fig. 1 Functional groups attached to hydroxyl surfaces.[19] Reprinted from Kohji Yoshinaga; Surface modifications of inorganic particles; in: Fine Particles: Synthesis, Characterization, and Mechanism of Growth; Tadao Sugimoto; Marcel Dekker, New York (2000) 626–646 (Chap. 12.1), Copyright © 2000 with permission from Marcel Dekker.

Technological Applications

Particle Packing and Pore Structure

Numerous research groups studied the formation of particle packing structures. For example, Emmett et al.[17] described the formation of ordered and glassy regions formed during the sedimentation of silica particles. These structures are very interesting since they provide means to test the theories and models used for the pore size characterization.[23–26] The size of the "internal" pores as well as the openings (throats) can be predicted from the particle size and the results indicated a reasonablly good agreement between theory and experiment. Yet, the specific close packed structures predicted two sizes of internal pores, the so called octahedral- and tetrahedral-pores. Standard runs with the mercury porosimetry and nitrogen adsorption measurements did not reveal those pores as two different peaks in the measurements. However, when the hysteresis region in mercury porosimetry curves was analyzed in more detail (scanning within the hysteresis), a "fine structure" or, to be more precise, a two step extrusion curve was observed,[25] which could be correlated with the octahedral and tetrahedral pores in the structure. Further details are provided in the Handbook of Porous Solids.[27]

Yin et al.[28] reported on the template-assisted self assembly of particle aggregate structures. Such an initial aggregate or building block can be used to assemble even more complex larger structures. Initially these authors used polymer beads, but the same techniques can also be applied to inorganic particles or specific organic/inorganic hybrid structures may be formed.

Powder Sintering

Sintering of model systems has also been of much interest, with numerous studies of sol–gel derived as well as Stöber silica samples.[29–37] Models of an exceedingly uniform particle size and particle arrangement allows for a better theoretical prediction of the densification behavior. Experiments with Stöber silica particles of 0.09, 0.2, 0.43, and 1.0 μm in diameter indicated that the sintering time was proportional to the square of the particle size.[30] Viscous flow, the usually expected sintering mechanism for a "glass-type" material, would predict a direct relation to the particle size. It is not clear as yet, if a second effect is superimposed, e.g. a surface accumulated impurity may account for this discrepancy, or whether the viscous flow is not the appropriate sintering mechanism in this situation.

Studies by Ciftcioglu et al.[38] and Milne et al.[39] also demonstrated another important fact about sintering. Monodispersed powders will form ordered areas of close packed particles, which are characterized by uniform and small pore sizes. This condition will favor densification of the material at a lower temperature and in shorter times. It will also lead to smaller and uniform grain sizes within those areas. However, it is nearly impossible to create absolutely defect-free particle packing structures and sintering along the boundary lines of these ordered areas will actually lead to pore-coarsening and large remaining defects along those lines. Much higher temperatures (or longer times) are then needed to eliminate these macro-defects and most of the potential benefits of the small and uniform initial particle size are lost.

3-D Photonic-Bandgap Structures

Several research groups have demonstrated that ordered silica particles, also referred to as colloidal crystals, can be used to create 3-D photonic-bandgap structures.[40–59] Especially, silica is being used as place holder or template for this purpose. The particles are packed in an ordered arrangement and then the internal voids are filled with a different, high refractive index material, such as silicon metal or titania. Figure 2 shows such a sample after the silica spheres have been removed by an etching process, which leaves behind an ordered arrangement of uniform sized holes. The uniform size and arrangement of the spherical voids is needed to filter/"stop" light of a specific wavelength. Intentional defects can then be used to allow light to pass through under specific conditions. The key for a successful production of those structures is not only the preparation of ordered silica spheres, which many research groups have done, but to be able to control the position and orientation of those crystal,[40,43,44] as well as making sure that the structures either contain no defects or that they contain defects in specific positions. Ozin et al. have perfected the controlled formation and orientation of those colloidal crystals by using a V-shaped grove. The specific angle of the side walls forces the silica particles to form a per-

Fig. 2 SEM micrographs showing the remaining titania skeleton after infiltration of the silica colloidal crystals and removing the silica by an etching process.[43] Reprinted from Andreas Stein; *Sphere templating methods for periodic porous solids*; Microporous Mesoporous Mater., **44–45** (2001) 227–239, Copyright © 2001 with permission from Elsevier Science

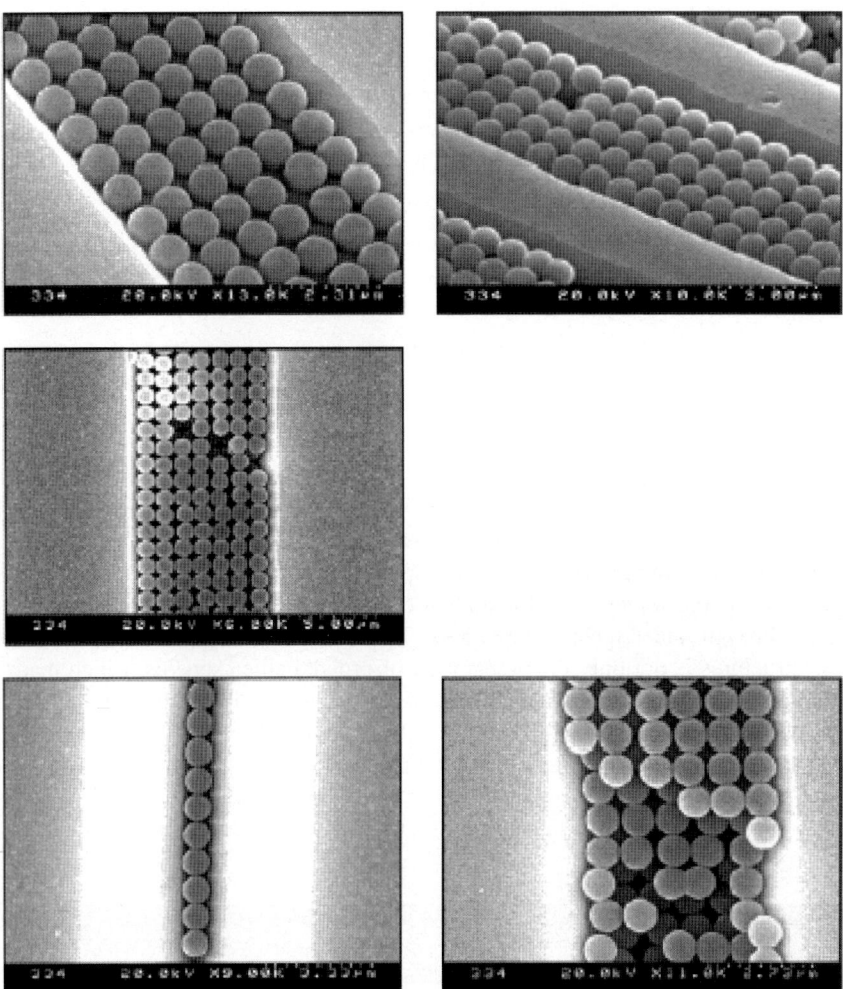

Fig. 3 SEM micrographs showing the formation of oriented silica colloidal crystals formed in a V-shaped channel etched in a silicon wafer. Bottom left shows a 1-D chain of silica spheres and bottom right shows three of the six layered (100) oriented silica colloidal crystal.[42] Reprinted from San Ming Yang and Geoffrey A. Ozin; *Opal chips: vectorial growth of colloidal crystal patterns inside silicon wafers*; Chem. Commun., **2000**(24) (2000) 2507–2508, Copyright © 2000 with permission from The Royal Society of Chemistry

fectly (100)-oriented crystal as illustrated in Fig. 3. Another important parameter is the average size and uniformity of the silica particles. Diameters of ∼1 μm are needed to push the operating wavelength into an acceptable range (near infrared wavelength of 1.5 μm), and the polydispersity has to be <1% in order to form the highly ordered structures as well as to achieve nearly 100% blocking of a specific wavelength.

Sensors Utilizing Ordered Structures

Ordered particle packings are also being used in other sensor type applications. Asher et al.[59,60] described a system, where colloidal arrays of polymer particles were embedded in a gel-matrix. Changes in the environment of the particles caused the structures to shrink or expand, which could then be measured through the changes in the optical properties, as exemplified in Fig. 4. The same general idea can be applied to silica particles. This way, very specific sensors could be created when combined with appropriate surface modifications of the silica particles. The latter could be attached to the end of optical fibers or produced as a multisensor array on a chip.

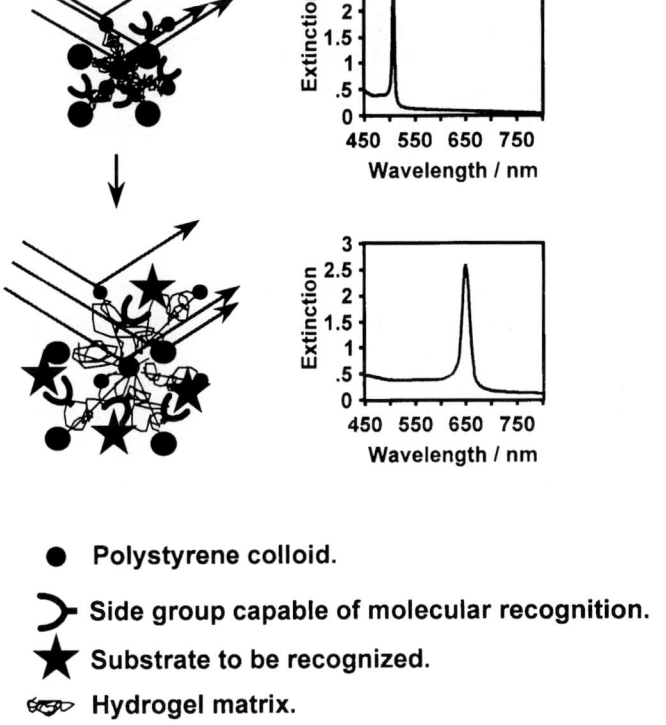

- **Polystyrene colloid.**
- **Side group capable of molecular recognition.**
- **Substrate to be recognized.**
- **Hydrogel matrix.**

Fig. 4 General motif for the intelligent polymerized crystalline colloidal array sensor. The colloidal crystal array Bragg diffraction is a sensitive monitor of the hydrogel volume change induced by the interaction or binding of the molecular recognition agent to a substrate.[60] Reprinted from Sanford A. Asher, John Holtz, Jesse Weissman, and Guisheng Pan; *Mesoscopically periodic photonic-crystal materials for linear and nonlinear optics and chemical sensing*; MRS Bulletin, **23**(10) (1998) 44–50, Copyright © 1998 with permission from Materials Research Society

Rheology of Hard Sphere Suspensions

Stöber-silica particles are of tremendous importance in studies of concentrated colloidal dispersions. For example, light scattering and fluid state theories were tested with those materials.[61–83] Brady and co-workers[62–64] and Maranzano et al.[65–68] provided a number of interesting review reports in the area of frequency-dependent rheology of hard sphere colloidal dispersions. The great degree of control over the properties of Stöber silica particles make these systems an ideal situation for testing of particle–particle interactions in different media. Surface properties can be adjusted by grafting proper functional groups or coating layers. These modifications can lead to particles with primarily electrostatic, steric, or no stabilization at all. For example, So et al.[70,71] studied the rheological behavior of such particles after they were surface modified with methacrylate or aminosilane functional groups. The authors observed shear thinning behavior of the charge stabilized suspensions at low ionic concentrations in water and the contrasting hard sphere behavior under either high ionic strength or when the appropriate refractive index matching solvent (THF) was used. De Kruif et al.,[73–75] van Helden et al.,[16] and Emmet[17] evaluated the viscosity of colloidal "hard sphere" silica suspensions as a function of shear rate and volume fraction in a similar system, i.e., when silica particles were surface modified with octadecyl groups. The particles, suspended in cyclohexane, did not show any long range electrostatic repulsion, but were stabilized due to the steric effects of the octadecyl chains on the particle surface. The viscoelastic properties of concentrated dispersions above a volume concentration of 50% indicated interesting shear-thickening/shear-thinning phenomena and, at even higher concentrations, the formation of various ordered structures. Further details of those model studies are described in the corresponding publications.

Wagner et al.[83] coated a nanometer-sized magnetic $CoFe_2O_4$-core with silica and studied the formation of ordered colloidal silica crystals under the influence of an external magnetic field.

Chromatography

The nonporous Stöber-silica spheres have been widely used as support material in fast HPLC (High Performance Liquid Chromatography). In particular, the separation and detection of biopolymers could be demonstrated by means of Reversed Phase, Hydrophobic Interaction, Ion Exchange, or Affinity Chromatography. The group of Unger[84–103] initially developed these silica powders and established their superior properties for a variety of HPLC-applications since the mid 1980s. Thereafter, other research groups[104–123] joined in and the materials are now commercialized by several companies. The main advantage of smaller and smaller packing materials in chromatography is the increased efficiency of the column and a decreased analysis time. For example, the number of theoretical plates (an indicator for the efficiency of separating two substances) is inversely proportional to the particle size. Shorter analysis time is another important factor in the separation of biomolecules, since the latter molecules have a tendency to unfold and denaturize during prolonged interaction with the adsorbent surface. The short columns also indicated a surprisingly high peak capacity and a fast column regeneration. Thus, a 36×8 mm-column packed with nonporous 1.5 μm n-alkyl silica resolved proteins as well as peptides with peak capacities in the order of 50–200 at a 10 min gradient time in the gradient elution mode using acidic low ionic strength hydro-organic eluents.[84] Unger et al. also demonstrated that the enzymatic activity of proteins could be retained to a higher degree compared with the traditional separation processes when using the nonporous silica columns.[84]

One example of a very fast separation process of proteins is given in Fig. 5 and another example demonstrating the high peak capacity, in which 60 components were resolved within 3 min., is displayed in Fig. 6.

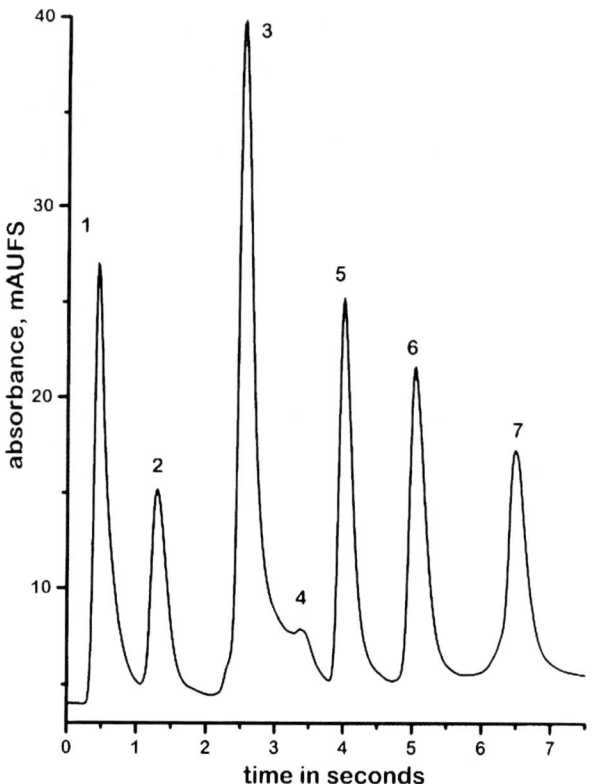

Fig. 5 Separation of a mixture of six proteins on a short Micra NPS-RP column. Conditions: column 15×4.6 mm, flow rate 4 ml min^{-1}, pressure 405 bar, gradient from 30–100% of acetonitrile in water with 0.1% TFA in 5 s; analytes: 1, ribonuclease A; 2, cytochrome c; 3, lysozime; 4, unknown; 5, bovine serum albumin; 6, catalase; 7, albumin egg (ovalbumin).[101] Reprinted from T. Issaeva, A. Kourganov, and K. Unger; *Super-high-speed liquid chromatography of proteins and peptides on non-porous Micra NPS-RP packings*; J. Chromatogr. A, **846** (1999) 13–23, Copyright © 1999 with permission from Elsevier Science

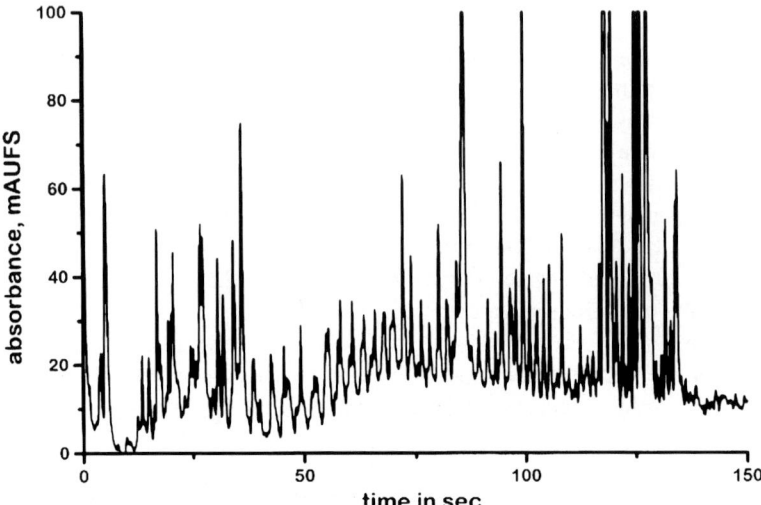

Fig. 6 Separation of an old sample of tryptic digest of hemoglobin A on a Micra NPS-RP column. Conditions as in Fig. 6, besides the temperature 30 °C and the flow rate 1 ml min^{-1}. Conditions: column 33 × 4.6 mm, flow rate 2.5 mm, pressure 495 bar (room temperature) and 250 bar (70 °C), gradient of acetonitrile in water with 0.1% TFA according to program (% acetonitrile, time in seconds): 1%, 0; 20%, 20; 30%, 90; 40%, 100; 40%, 120; 60%, 130.[101] Reprinted from T. Issaeva, A. Kourganov, and K. Unger; *Super-high-speed liquid chromatography of proteins and peptides on non-porous Micra NPS-RP packings*; J. Chromatogr. A, **846** (1999) 13–23, Copyright © 1999 with permission from Elsevier Science

The main disadvantage of smaller particles is the increased back-pressure during the operation of HPLC systems. The pressure can be calculated according to Darcy's law and is inversely proportional to the square of the particle size. For example, a 33 × 4.5 mm column packed with 1.5 μm nonporous silica particles needed a pressure of approximately 500 bar (the limit for most commercial pumps in HPLC units) at a flow rate of 2 ml min^{-1} with acetonitrile and water. A general relation between particle size, pressure drop, plate number, and analysis time is provided in Fig. 7. The assumed specific conditions for viscosity, analyte diffusivity, retention factor, and other parameters are given in the legend. Fast analysis times combined with a limited flow rate also necessitates the need for fast detector systems, small volume detection cells (<1 μl), as well as small volume injection loops, yet, all these challenges have been successfully resolved.

In its original state the surface of Stöber silica is covered with hydroxyl groups, ~8 μmol OH m^{-2}, but the chemical nature of the surface can be dramatically changed through a variety of processes.[60–62] A specific application might require that the surface be modified with different functional groups in order to achieve the required interaction or separation. Silanization is a convenient method to introduce functionalities,[124] such as amino-, mercapto-, or alkyl-groups, which can then be further used to anchor other (bio)-molecules on the particle surface. In addition, polymers can be grafted onto the surface by a direct esterification process,[125] or

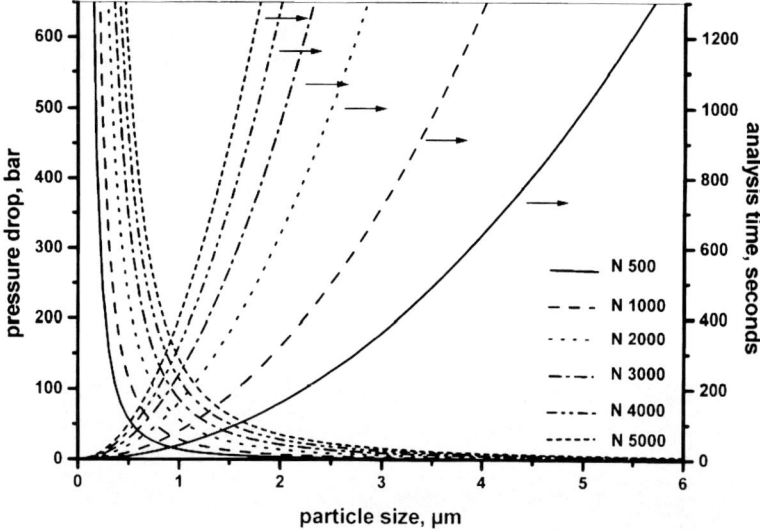

Fig. 7 Dependence of the analysis time and pressure drop of the column on particle size at a given plate number for nonporous packings. $D_m = 5 \times 10^{-7} \text{cm}^2 \text{s}^{-1}$, $\eta = 1 \text{cP}$, $\varepsilon = 0.4$, $\psi = 945$, $k' = 3$ and the required N is shown in the figure.[101] Reprinted from T. Issaeva, A. Kourganov, and K. Unger; *Super-high-speed liquid chromatography of proteins and peptides on non-porous Micra NPS-RP packings*; J. Chromatogr. A, **846** (1999) 13–23, Copyright © 1999 with permission from Elsevier Science

through urethane linkages.[126] Details of some of these methods are described by Yoshinaga,[20] Kinkel,[21] or Bradley et al.[127]

Most recently micrometer sized Stöber silica particles have also been also prepared with controlled intraparticle porosity. The latter powders can be used in many "standard" HPLC applications, taking advantage of the much faster separation times, possible with micrometer-sized particles. These particles can have either a uniform internal pore structure or they might have a porous shell surrounding a dense core. While particles with a porous shell had been already prepared by DuPont in the late 1960s,[111] those particles were relative large with approximately 30 μm in diameter. More recently the same technique was again employed by Kirkland et al., who produced particles of ~3–6 μm in diameter. A modified Stöber silica precipitation/coagulation process was used in the preparation of the porous shell. The size of the Stöber silica particles essentially determined the size of the pores, which could be varied between 9 and 80 nm in diameter. The "Poroshell" particles combined the advantage of faster separation processes when using smaller, micrometer-sized, packing materials, with the larger interaction surface area available through the porous structure. A sufficiently fast diffusion exchange between the thin porous layer and the main solvent flow around the particles enables very fast and efficient separation processes and at the same time reduces the problem of peak broadening, caused by longer and variable residence time of the molecules within particles, which are porous throughout their structure. Figures 8 and 9 show

Fig. 8 Schematic of a Poroshell superficially porous particle.[111] Reprinted from J. J. Kirkland, F. A. Truszkowski, C. H. Dilks Jr., and G. S. Engel; *Superficially porous silica microspheres for fast high-performance liquid chromatography of macromolecules*; J. Chromatogr. A, **890** (2000) 3–13, Copyright © 2000 with permission from Elsevier Science

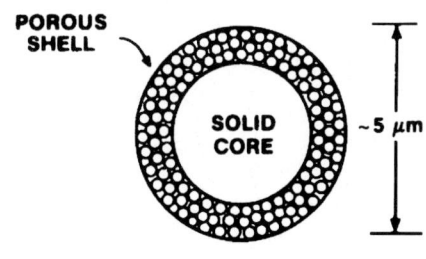

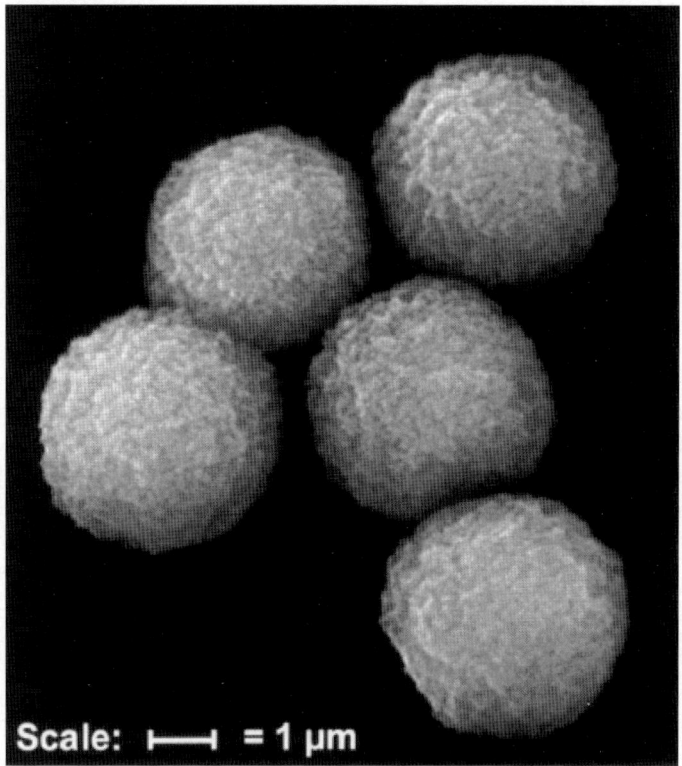

Fig. 9 Scanning electron micrograph of Poroshell particles.[111] Reprinted from J. J. Kirkland, F. A. Truszkowski, C. H. Dilks Jr., and G. S. Engel; *Superficially porous silica microspheres for fast high-performance liquid chromatography of macromolecules*; J. Chromatogr. A, **890** (2000) 3–13, Copyright © 2000 with permission from Elsevier Science

a schematic representation and a SEM micrograph of such materials, while Fig. 10 provides an example of the separation of Triton X-114 oligomers, achieved with the latter porous shell particles.

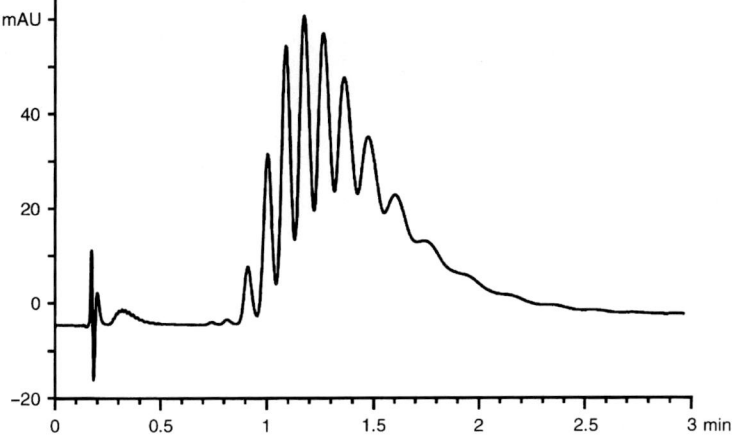

Fig. 10 Separation of Triton X-114 oligomers. Column: 75×2.1 mm, $5\,\mu$m Poroshell 300 SB-C_{18}, 0.25-μm porous shell (surface area: $3.0\,m^2\,g^{-1}$); mobile phase: A, methanol–water (45:55); B, methanol; gradient: delay 1.0 min, then 0–5% B in 2.0 min; flow rate: $0.5\,ml\,min^{-1}$; temperature: 24 °C; UV detector: 225 nm.[111] Reprinted from J. J. Kirkland, F. A. Truszkowski, C. H. Dilks Jr., and G. S. Engel; *Superficially porous silica microspheres for fast high-performance liquid chromatography of macromolecules*; J. Chromatogr. A, **890** (2000) 3–13, Copyright © 2000 with permission from Elsevier Science

Aiming for a similar product, Büchel and co-workers[19, 128–131] added *n*-octadecyl trimethoxysilane (or similar silanes) or *n*-alkylamines (acting as a nonionic template) to TEOS during the "standard" Stöber synthesis process. The alkyl groups essentially acted as space holders during the precipitation of the silica particles, which could be removed later on by a calcination/burnout step. The observed pore diameters were between 2 and 4 nm and the specific pore volumes could be adjusted between 0.1 and $0.8\,ml\,g^{-1}$. Ma et al.[132] and Vacassy et al.[133] have further modified this process, using various pore forming additive, such as surfactants different alkoxides, or the addition/adsorption of glycerol. Ma achieved pore sizes of 3–10 nm and pore volumes of up to $1.74\,ml\,g^{-1}$ (Fig. 11). However, the overall particle size was somewhat larger, i.e. 3–5 μm, and the particles were not quite as uniform, as usually observed with the Stöber process. Nevertheless, the materials showed superior separation characteristics in chromatography tests performed by these research groups.

Pigments

It is of great interest to find new chromophores and otherwise improved pigments for a range of new printing and diagnostic processes. The preparation of "well-defined" submicron pigment particles is of major concern to their use for optical and rheological reasons and it was quite obvious that the above described silica particles

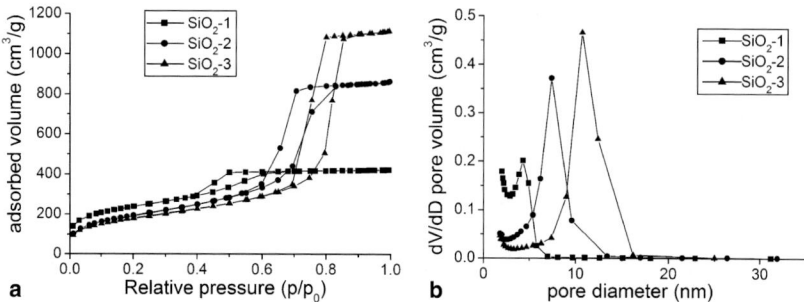

Fig. 11 N$_2$ adsorption–desorption isotherms (**a**) and pore size distribution curves from the adsorption branch (**b**) of mesoporous silica spheres.[132] Reprinted from Yurong Ma, Limin Qi, Jiming Ma, Yongqing Wu, Ou Liu, and Humin Cheng; *Large-pore mesoporous silica spheres: Synthesis and application in HPLC*; Colloids Surf. A: Physicochem. Eng. Asp., **229** (2003) 1–8, Copyright © 2003 with permission from Elsevier Science

were seen as ideal materials to produce pigments in a predictable way. One can either incorporate different dyes or adsorb them on silica particles of controlled size. Several publications demonstrated how Stöber silica particles can be used for this purpose.[104, 134–145] For example, a variety of so-called "reactive" dyes, containing a sulfonic acid group, could be attached to the surface or incorporated into the growing silica particles, using an amino silane anchor group.[136–139]

The resulting pigment particles had greater color intensities due to the controlled particle morphology. If used in the correct mixture with TEOS the dyes could even be incorporated in a silica-dye layer around the original silica seed or core particles during the growth process. The latter incorporation process allowed for an increased color intensity and the control of particle size further enhanced the color-intensity and purity of the pigment. It turned out that in some cases the silica pigments even had a higher color intensity than the original (pure) dye-powder, which was due to the lack of its particle size control. Applying different surface coatings to the silica pigments allowed to modify the electrical (charge) properties independently of the color of the particles, a process, which further enhanced the applicability of these new pigments. Using the same procedures it was possible to couple fluorescent dyes, for example through an aminopropylsilane–thioisocyanate mechanism.[140] The latter particles were primarily designed for diagnostic applications, where the silica particles were used as carrier.[141] In another approach, several organic dyes could be attached to Stöber silica particles by using aluminum impurities as an intermediate coupling agent.[142]

In a different study, it was shown that titania-coated silica could be used as a substitute for pure titania in paper whiteners.[143–145] The particle size and coating thickness could be adjusted based on calculations using the Mie theory for concentric spheres. Again the control and uniformity of the particle size and of the coating layer was essential in achieving the same "hiding power" as with comparable pure TiO$_2$ (rutile) pigments. The process also allowed to lower cost of the whitener pigment by substituting the titania component with less expensive silica.

The uniform size silica spheres have also found some surprising applications, such as dispersing or lubrication aids. Added to plate-like pigment powders, e.g., pearl luster pigments, the silica spheres act as lubricating roller bearings between the pigment particles, thus making the pigments easier to disperse and apply as layered coatings on surfaces.

Medical Applications/Controlled Drug-Release Function

The ability to prepare Stöber silica of different properties makes it possible to use such particles as "storage" containers for other molecules. For example, drugs can be stored inside the particle and specific functional groups on the outer surface of the particle enable one to target specific areas in the body where the drugs should be applied. The controlled pore structure provides a means to slowly release the drug into the designed environment. Controlling the pore structure at different levels throughout the particle now enables one to manipulate the release rate. For example, an outermost layer of smaller pores may act as a hindrance layer and provide the ability to continuously and constantly release the drug over a prolonged period of time. Special surface coatings can even be designed to act like a valve, turning the release on or off depending on which chemicals are present in the environment surrounding the particle or stimulated by other external factors.

The incorporation of magnetic cores, e.g., iron oxides, can have additional advantages. For one, it enables one to easily collect or remove the particles and secondly the magnetic core can also affect the local environment of the particle. For example, it was demonstrated that exposing the particles to an alternating magnetic field, will cause a slight heat-up, which then can be used to control other properties of the system. For instance, the release of a drug will be accelerated at higher temperatures, or certain cells might affected by temperature changes. An example of the latter is the greater heat sensitivity of cancer cells as compared to "normal" cells, which has been used in preliminary studies on cancer treatment. Gao et al.[146] describe similar bio-catalyst carriers, when they encapsulate magnetic cores in silica and then attached β-lactamase on the surface. Their study demonstrated the possibility to utilize enzyme activity during a process when the enzymes are attached to (magnetic)-carrier particles. In addition they were able to remove the enzyme afterwards by a magnetic separation process. Another example of an encapsulated magnetic particle was described by Cao et al.[147] These authors used particles, which were labeled with rhenium-188, in a "targeted" radiotherapy application.

The preparation of porous hollow silica nanoparticles for a drug delivery applications was described by Chen et al.[148] The authors used the time release profile of cefradine as a test example. Caruso et al.[149] prepared hollow silica spheres and silica–polymer layered composite spheres. A controlled pore structure remained after decomposition of the organic material and possible applications in medical and pharmaceutical applications are described in their publication. Another drug release example was described by Charnay et al.,[150] who incorporated ibuprofen, a well studied and frequently used anti-inflammatory drug, in meso porous silica through successive loading processes. Their process lead to an improved initial drug loading and a complete release of the drug was observed under various simulated biological fluid conditions. This opens the possibility of a specifically targeted drug release when the carrier beads are deposited in a given area or organ. Moreover, certain glass compositions can be made bio-resorbable. In that case, the carrier beads are designed in such a way that they are initially too large to be removed by the blood stream or they are otherwise fixed to the area where they are applied, but after some time they will gradually dissolve, be washed out, and no further residues will remain in the body.

Xu et al.[151] coated Stöber silica particles with polyethylene glycol (PEG) in order to improve the biocompatibility of those particles and verified this improvement using a protein adsorption test. The article then further describes possible applications of encapsulation reagents for diagnosis, analysis or other measurements inside active biological systems.

Figure 12 shows an idealized drug release particle, which may contain a magnetic core (●) in order to be collected after its useful cycle or for thermal activation purposes. In addition, the particle could contain dyes or radioactive tracers (○). The porous interior structure primarily functions as a storage reservoir for the drug and specific surface coatings or surface groups (⌐ and ⌐), which can be light, pH, or otherwise sensitive, might facilitate the loading and release of the drug under controlled conditions. Moreover, the outer particle surface can have pores of smaller size, to restrict the release rate. The entire particle must also have a bio-compatible surface layer in order not to be rejected by the biological system. Last but not least,

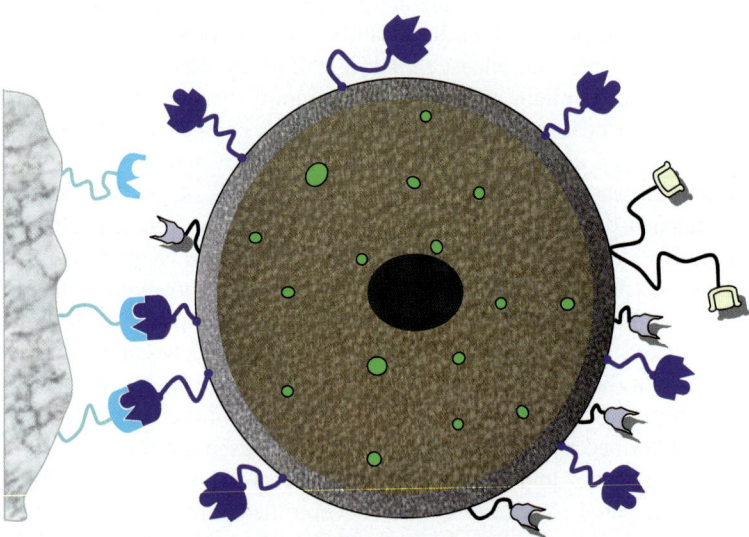

Fig. 12 Schematic drawing of a drug release particle with various key–lock, recognition, release, or other sensor functions

"key–lock" recognition groups ( ) provide a means to have the particles attached to specific cells or organs, which will allow for a localized application of the drug at higher concentrations and without affecting the rest of the body. Obviously, all these functions may be possible with other materials as well. However, the ease of control over physical and chemical properties of Stöber silica particles makes them a particular suitable system.

References

1. Werner Stöber, Arthur Fink, and Ernst Bohn; *Controlled growth of monodisperse silica spheres in the micron size range*; J. Colloid Interface Sci., **26** (1968) 62–69
2. Herbert Giesche; *Hydrolysis of Silicon Alkoxides in Homogeneous Solutions*; in: Fine Particles: Synthesis, Characterization, and Mechanism of Growth(Tadao Sugimoto, ed); Marcel Dekker, New York (2000) 126–146 (Chap. 2.1)
3. Herbert Giesche; *Synthesis of monodispersed silica powders. I. particle properties and reaction kinetics*; J. Eur. Ceram. Soc., **14** (1994) 189–204
4. Herbert Giesche; *Synthesis of monodispersed silica powders. II. Controlled growth reaction and continuous production process*; J. Eur. Ceram. Soc., **14** (1994) 205–214
5. T. Matsoukas and E. Gulari; *Dynamics of growth of silica particles from ammonia-catalyzed hydrolysis of tetra-ethyl-orthosilicate*; J. Colloid Interface Sci., **124** (1988) 252–261
6. T. Matsoukas and E. Gulari; *Monomer-addition growth with a slow initiation step: A growth model for silica particles from alkoxides*; J. Colloid Interface Sci., **132** (1989) 13–21
7. G.H. Bogush, G.L. Dickstein, P. Lee, K.C. Zukoski, and C.F. Zukoski IV; *Studies of the hydrolysis and polymerization of silicon alkoxides in basic alcohol solutions*; in: Materials Research Society Symposium Proceedings, Vol. 121, Better Ceramics Through Chemistry III (Jeffrey Brinker C., Clark D.E. & Ulrich D.R., eds); Materials Research Society, Pittsburgh, PA (1988) 57–65
8. G.H. Bogush and C.F. Zukoski IV; *Studies of the kinetics of the precipitation of uniform silica particles through the hydrolysis and condensation of silicon alkoxides*; J. Colloid Interface Sci., **142** (1991) 1–18
9. G.H. Bogush and C.F. Zukoski IV; *Uniform silica particle precipitation: An aggregative growth model*; J. Colloid Interface Sci., **142** (1991) 19–34
10. A. van Blaaderen, J. van Geest, and A. Vrij; *Monodisperse colloidal silica spheres from tetraalkoxysilanes: Particle formation and growth mechanism*; J. Colloid Interface Sci., **154** (1992) 481–501
11. G.H. Bogush, C.J. Brinker, P.D. Majors, and D.M. Smith; *Evolution of surface area during the controlled growth of silica spheres*; in: Materials Research Society Symposium Proceedings, Vol. 180: Better Ceramics Through Chemistry IV(Zelinski B.J.J., Jeffrey Brinker C., Clark D.E. & Ulrich D.R, eds); Materials Research Society, Pittsburgh, PA (1990) 491–494
12. S. Coenen and C.G. De Kruif; *Synthesis and growth of colloidal silica* particles; J. Colloid Interface Sci., **124** (1988) 104–110
13. A.P. Philipse; *Quantitative aspects of the growth of (charged) silica spheres*; Colloid Polym. Sci., **266** (1988) 1174–1180
14. C. Kaiser, M. Hanson, H. Giesche, J. Kinkel, and K.K. Unger; *Nonporous Silica Microsheres in the Micron and Submicron Range: Manufacture, Characterization and Application*; in: Fine Particle Science and Technology From Micro to Nanoparticles, NATO AST Series (E. Pelizetti, ed); Klüwer Academic Publishers, Dordrecht, NL (1996) 71–84
15. Kangtaek Lee, Arun N. Sathyagal, Alon V. McCormick; *A closer look at an aggregation model of the Stöber process*; Colloids Surf. A: Physicochem. Eng. Asp., **144** (1998) 115–125
16. A.K. van Helden and A. Vrij; *Contrast variation in light scatting: Silica spheres dispersed in apolar solvent mixtures*; J. Colloid Interface Sci., **76** (1980) 418–433.

17. S. Emmett, S.D. Lubetkin, and B. Vincent; *The growth of ordered sediments of monodispersed hydrophobic silica particles*; Colloids Surf., **42** (1989) 139–153
18. K. Osseo-Asare and F.J. Arriagada; *Synthesis of nanosize particles in reverse microemulsions*; in: Ceramic Transactions, Vol. 12, Ceramic Powder Science III (Messing G.L., Hirano S.-i. & Hausner H., eds); The American Ceramic Society, Westerville, OH (1990) 3–16
19. Kohji Yoshinaga; *Surface modifications of inorganic particles*; in: Fine Particles: Synthesis, Characterization, and Mechanism of Growth (Tadao Sugimoto, ed); Marcel Dekker, New York (2000) 626–646 (Chap. 12.1)
20. Joachim N. Kinkel; *Darstellung and Charakterisierung von Siliziumdioxidträgermaterialien zur Trennung von Biopolymeren durch Hochdruckflüssigchromatographie*, Dissertation, Universität Mainz, Germany (1984)
21. Michael A. Markovitz, Paul E. Schoen, Paul Kust, and Bruce P. Gaber; *Surface acidity and basicity of functionalized silica particles*; Colloids Surf. A: Physicochem. Eng. Asp., **150** (1999) 85–94
22. Gunter Büchel, Michael Grün, and Klaus K. Unger; *Tailored syntheses of nanostructured silicas: Control of particle morphology, particle size and pore size*; Supramol. Sci., **5** (1998) 2530–2539
23. H. Giesche, K.K. Unger, U. Müller, and U. Esser; *Hysteresis in nitrogen sorption and mercury porosimetry on mesoporous model adsorbents made of aggregated monodisperse silica spheres*; Colloids Surf., **37** (1989) 93–113
24. S. Bukowiecki, B. Straube, and K.K. Unger; *Pore structure analysis of close-packed silica spheres by means of nitrogen sorption and mercury porosimetry*; in: Principles and Applications of Pore Structural Characterization (Haynes J.M. & Rossi-Doria P., eds); Arrowsmith, Bristol (1985) 43–55
25. H. Giesche; *Interpretation of hysteresis 'fine-structure' in mercury-porosimetry measurements*; in: Materials Research Society Symposium Proceedings, Volume 371, Advances in Porous Materials (Komarneni S., Smith D.M. & Beck J.S., eds); Materials Research Society, Pittsburgh, PA (1995) 505–510
26. D.M. Smith, T.E. Holt, D.P. Gallegos, and D.L. Stermer; *Pore structure analysis via NMR, mercury porosimetry, and dynamic methods*; in: Advances in Ceramics, Vol. 21, Ceramic Powder Science(Messing G.L., (Joe) Mazdiyasni K.S., McCauley J.W. & Haber R.A., eds); The American Ceramic Society, Columbus, OH (1987) 779–791
27. Herbert Giesche, *Mercury Porosimetry*; in: Handbook of Porous Materials (F. Schüth and K.S.W. Sing, eds); Wiley-VCH, Weinheim (2002) 309–351 (Chap. 2.7)
28. Y. Yin, Y. Lu, B. Gates, and Y. Xia; *Template-assisted self-assembly: A practical route to complex aggregates of monodispersed colloids with well-defined sizes, shapes, and structures*; J. Am. Chem. Soc., **123**(36) (2001) 8718–8729
29. Hiroshi Suzuki, Shigeyuki Takagi, Hideki Morimitsu, and Shin-ichi Hirano; *Microstructure control of porous silica glass with monodispersed spherical silica particles*; J. Ceram. Soc. Jpn., **100** (1992) 284–287
30. H. Giesche and K.K. Unger; *Sintering of monodispersed silica*; in: Ceramic Powder Processing Science, Proceedings of the Second International Conference, Berchtesgaden, 1988 (H. Hausner, G.L. Messing, & S. Hirano, eds.); Deutsche Keramische Gesellschaft, Köln (1989) 755–764
31. M.D. Sacks, T.S. Yeh, and S.D. Vora; *Effect of green microstructure on sintering of model powder compacts*; Ceramic Powder Processing Science, Proceedings of the Second International Conference, Berchtsgarden, FRG, 12–14 Oct 1988 (H. Hausner, G.L. Messing, and S. Hirano, eds.); Deutsche Keramische Gesellschaft, Köln (1989) 693–704
32. Michael D. Sacks and Tseung-Yuen Tseng; *Preparation of SiO_2 glass from model powder compacts. I. Formation and characterization of powders, suspensions, and green compacts*; J. Am. Ceram. Soc., **67** (8) (1984) 526–532
33. Michael D. Sacks and Tseung-Yuen Tseng; *Preparation of SiO_2 glass from model powder compacts. II. Sintering*; J. Am. Ceram. Soc., **67** (8) (1984) 532–537

34. Michael D. Sacks and Shailesh D. Vora; *Preparation of SiO$_2$ glass from model powder compacts. III Enhanced densification by sol infiltration*; J. Am. Ceram. Soc., **71** (4) (1988) 245–249
35. Michael D. Sacks, Gary W. Scheiffele, Nazim Bozkurt, and Ramesh Raghunathan; *Fabrication of ceramics and composites by viscous and transient viscous sintering of composite particles*; in: Ceramic Transactions, Ceramic Powder Science IV (Shin-ichi Hirano, Gary L. Messing, & Hans Hausner, eds.); The American Ceramic Society, Westerville, OH (1991) 437–455
36. T. Shimohira, A. Makishima, K. Kotani, and M. Wakakuwa; *Sintering of monodispersed amorphous silica particles*; in: Proceedings of International Symposium of Factors in Densification and Sintering of Oxide and Non-Oxide Ceramics (S. Somiya & S. Saito, eds.); Tokyo Institute of Technology, Tokyo (1978) 119–127
37. D.W. Johnson Jr., E.M. Rabinovich, J.B. MacChesney, and E.M. Vogel; *Preparation of high-silica glasses from colloidal gels. II. Sintering*; J. Am. Ceram. Soc., **66** (10) (1983) 688–693
38. Muhsin Ciftcioglu, Douglas M. Smith, and Steven B. Ross; *Sintering studies on ordered monodisperse silica compacts: Effect of consolidation*; Powder Technol., **69** (2) (1992) 185–193
39. Steven J. Milne, Mohammed Patel, and Eric Dickinson; *Experimental studies of particle packing and sintering behaviour of monosize and bimodal spherical silica powders*; J. Eur. Ceram. Soc., **11** (1) (1993) 1–7
40. Michael Freemantle; *Opal chips: Photonic jewels*; Chem. Eng. News, **79** (4), (2001) 55–58
41. Alvaro Blanco, Emmanuel Chomski, Serguei Gratchak, Marta Ibisate, Sajeev John, Stephen W. Leonard, Cefe Lopez, Francisco Meseguer, Hernan Miguez, Jessica P. Mondia, Geoffrey A. Ozin, Ovidiu Toader, and Henry M. van Driel; *Large-scale synthesis of a silicon photonic crystal with a complete three-dimensional bandgap near 1.5 micrometres*; Nature, **405** (2000) 437–440
42. San Ming Yang and Geoffrey A. Ozin; *Opal chips: Vectorial growth of colloidal crystal patterns inside silicon wafers*; Chem. Commun., **2000**(24) (2000) 2507–2508
43. Andreas Stein; *Sphere templating methods for periodic porous solids*; Microporous Mesoporous Mater., **44–45** (2001) 227–239
44. Preston B. Landon and R. Glosser; *Self-assembly of spherical colloidal silica along the [100] direction of the FCC lattice and geometric control of crystallite formation*; J. Colloid Interface Sci., **276** (2004) 92–96
45. F. Meseguer, A. Blanco, H. Miguez, F. Garcia-Santamaria, M. Ibisate, and C. Lopez; *Synthesis of inverse opals*; Colloids Surf., **202** (2002) 281–290
46. V.N. Astratov, Yu. A. Vlasov, O.Z. Karimov, A.A. Kaplyanskii, Yu. G. Musikhin, N.A. Bert, V.N. Bogomolov, and A.V. Prokofiev; *Photonic band gaps in 3D ordered FCC silica matrices*; Phys. Lett. A, **222**() (1996) 349–353
47. V.N. Bogomolov, A.V. Prokofiev, S.M. Samoilovich, E.P. Petrov, A.M. Kapitonov, and S.V. Gaponenko; *Photonic band gap effect in a solid state cluster lattice*; J. Luminescence, **72–74** (1997) 391–392
48. P.V. Braun and P. Wiltzius; *Macroporous materials – Electrochemically grown photonic crystals*; Curr. Opin. Colloid Interface Sci., **7**(1–2) (2002) 116–123
49. Jes Broeng, Stig E. Barkou, Anders Bjarklev, Jonathan C. Knight, Tim A. Birks, and Philip St. J. Russell; *Highly increased photonic band gaps in silica/air structures*; Opt. Commn., **156**(4–6), (1998) 240–244
50. S.V. Gaponenko, V.N. Bogomolov, E.P. Petrov, A.M. Kapitonov, A.A. Eychmueller, A.L. Rogach, I.I. Kalosha, F. Gindele, and U. Woggon; *Spontaneous emission of organic molecules and semiconductor nanocrystals in a photonic crystal*; J. Luminescence, **87–89** (2000) 152–156
51. Michail I. Samoilovich, Svetlana M. Samoilovich, Andrey V. Guryanov, Michail Yu. Tsvetkov; *Artificial opal structures for 3D-optoelectronics*; Microelectron. Eng., **69**(2–4) (2003) 237–247
52. V.M. Shelekhina, O.A. Prokhorov, P.A. Vityaz, A.P. Stupak, S.V. Gaponenko, and N.V. Gaponenko; *Towards 3D photonic crystals*; Synth. Met., **124**(1) (2001) 137–139

53. S. Tsunekawa, Yu. A. Barnakov, V.V. Poborchii, S.M. Samoilovich, A. Kasuya, and Y. Nishina; *Characterization of precious opals: AFM and SEM observations, photonic band gap, and incorporation of CdS nano-particles*; Microporous Mater., **8**(5–6) (1997) 275–282
54. Yu. A. Vlasov, K. Luterova, I. Pelant, B. Hönerlage, and V.N. Astratov; *Optical gain and lasing in a semiconductor embedded in a three-dimensional photonic crystal*; J. Crystl. Growth, **184–185** (1998) 650–653
55. H.M. Yates, M.E. Pemble, H. Míguez, A. Blanco, C. López, F. Meseguer, and L. Vázquez; *Atmospheric pressure MOCVD growth of crystalline InP in opals*; J. Crystl. Growth, **193** (1–2) (1998) 9–15
56. Anvar A. Zakhidov, Ray H. Baughman, Ilyas I. Khayrullin, Igor A. Udod, Mikhail Kozlov, Nayer Eradat, Valy Z. Vardeny, Mihail Sigalas, and Rana Biswas; *Three-dimensionally periodic conductive nanostructures: Network versus cermet topologies for metallic PBG*; Synth. Met., **116**(1–3) (2001) 419–426
57. Anvar A. Zakhidov, Ilyas I. Khayrullin, Ray H. Baughman, Zafar Iqbal, Katsumi Yoshino, Yoshiaki Kawagishi, and Satoshi Tatsuhara; *CVD synthesis of carbon-based metallic photonic crystals*; NanoStruct. Mater., **12** (1999) 1089–1095
58. Alfons van Blaaderen; *From the de Broglie to visible wavelengths: Manipulating electrons and photons with colloids*; MRS Bull., **23**(10) (1998) 39–43
59. Chad E. Reese, Carol D. Guerrero, Jesse M. Weissman, Kangtaek Lee, and Sanford A. Asher; *Synthesis of highly charged, monodisperse polystyrene colloidal particles for the fabrication of photonic crystals*; J. Colloid Interface Sci., **232** (2000) 76–80
60. Sanford A. Asher, John Holtz, Jesse Weissman, and Guisheng Pan; *Mesoscopically periodic photonic-crystal materials for linear and nonlinear optics and chemical sensing*; MRS Bull., **23**(10), (1998) 44–50
61. J.W. Bender, N.J. Wagner; *Reversible shear thickening in monodisperse and bidisperse colloidal dispersions*; J. Rheol., **40** (1996) 899–916
62. J.F. Brady, J.F. Morris; *Microstructure of strongly sheared suspensions and its impact on rheology and diffusion*; J. Fluid Mech., **348** (1997) 103–139
63. J.F. Brady; *The rheological behavior of concentrated colloidal dispersion*; J. Chem. Phys., **99** (1993) 567–581
64. John F. Brady; *Computer simulation of viscous suspensions*; Chemical Eng. Sci., **56**(9) (2001) 2921–2926
65. B.J. Maranzano, N.J. Wagner; *The effects of interparticle interactions and particle size on reversible shear thickening: Hard sphere colloidal dispersions*; J. Rheol., **45** (2001) 1205–1222
66. B.J. Maranzano, N.J. Wagner, G. Fritz, and O. Glatter; *Surface charge of 3-(trimethoxysilyl)propyl methacrylate (TPM) coated Stöber silica colloids by zeta-phase analysis light scattering and small angle neutron scattering*; Langmuir, **16** (2000) 10556–10558
67. B.J. Maranzano and N.J. Wagner; *The effects of particle size on reversible shear thickening of concentrated colloidal dispersions*; J. Chem. Phys., **114** (2001) 10514–10527
68. G. Fritz, B.J. Maranzano, N.J. Wagner, and N. Willenbacher; *High frequency rheology of hard sphere colloidal dispersions measured with a torsional resonator*; J. Non-Newtonian Fluid Mech., **102** (2002) 149–156
69. Cécile Gehin, Jacques Persello, Daniel Charraut, and Bernard Cabane; *Electrorheological properties and microstructure of silica suspensions*; J. Colloid Interface Sci., **273** (2004) 658–667
70. Jae-Hyun So, Seung-Man Yang, Jae Chun Hyun; *Microstructure evolution and rheological responses of hard sphere suspensions*; Chem. Eng. Sci., **56** (2001) 2967–2977
71. Jae-Hyun So, Seung-Man Yang, Chongyoup Kim, and Jae Chun Hyun; *Microstructure and rheological behaviour of electrosterically stabilized silica particle suspensions*; Colloids Surf. A: Physicochem. Eng. Asp., **190** (2001) 89–98
72. Hans M. Wyss, Elena Tervoort, Lorenz P. Meier, Martin Müller, and Ludwig J. Gauckler; *Relation between microstructure and mechanical behavior of concentrated silica gels*; J. Colloid Interface Sci., **273** (2004) 455–462

73. C.G. de Kruif, E.M.F. van Iersel, and A. Vrij; *Hard sphere colloidal dispersions: Viscosity as a function of shear rate and volume fraction*; J. Chem. Phys., **83** (1985) 4717–4725
74. J.C.v. der Werff and C.G. de Kruif; *Hard-sphere colloidal dispersions: The scaling of rheological properties with particle size, volume fraction, and shear rate*; J. Rheol., **33** (1989) 421–454
75. J.C. van der Werff, C.G. de Kruif, C. Blom, and J. Mellema; *Linear viscoelastic behavior of dense hard-sphere dispersions*; Phys. Rev. A, **39** () (1989) 795–807
76. Bruce J. Ackerson; *Shear induced order and shear processing of model hard sphere suspensions*; J. Rheol., **34**() (1990) 553–590
77. D. Andrew R. Jones, Bruce Leary, and David V. Boger; *The rheology of a concentrated colloidal suspension of hard spheres*; J. Colloid Interface Sci., **147**(2) (1991) 479–495
78. David Andrew Ross Jones; *Depletion flocculation of sterically-stabilized particles*; PhD thesis, Bristol (1988)
79. L. Marshall and C.F. Zukoski IV; *Flow of dispersion near close packing*; Material Research Society Symposium, Vol. 155: Processing Science of Advanced Ceramics (I.A. Aksay, G.L. McVay, and D.R. Ulrich, eds.); Materials Research Society, Pittsburgh, PA (1989) 65–72
80. Louise Marshall and Charles F. Zukoski IV; *Experimental studies on the rheology of hard-sphere suspensions near the glass transition*; J. Phys. Chem., **94**() (1990) 1164–1171
81. P.N. Pusey and W. van Megen; *Phase behaviour of concentrated suspensions of nearly hard colloidal spheres*; Nature, **320** (1986) 340–342
82. William B. Russel; *Controlling the rheology of colloidal dispersions through the interparticle potential*; Ceramic Transactions, Vol. 12, Ceramic Powder Science III (Garry L. Messing, Shin-ichi Hirano, and Hans Hausner, eds.); The American Ceramic Society, Westerville, Ohio (1990) 361–373
83. Joachim Wagner, Tina Autenrieth, and Rolf Hempelmann; *Core shell particles consisting of cobalt ferrite and silica as model ferrofluids [$CoFe_2O_4$-SiO_2 core shell particles]*; J. Magn. Magn. Mater., **252** (2002) 4–6
84. K.K. Unger, G. Jilge, Janzen R., Giesche H., and Kinkel J.N.; *Non-porous microparticulate supports in high-performance liquid chromatography of biopolymers - concepts, realization and prospects*; Chromatographia, **22** (1986) 379–80.
85. K.K. Unger, O. Jilge, J.N. Kinkel, and M.T.W. Hearn; *Evaluation of advanced silica packings for the separation of biopolymers by high-performance liquid chromatography. II. Performance of non-porous monodisperse 1.5-μm Silica beads in the separation of proteins by reversed-phase gradient elution high-performance liquid*; J. Chromatogr. A, **359** (1986) 61–72
86. G. Jilge, R. Janzen, H. Giesche, K.K. Unger, J.N. Kinkel, and M.T.W. Hearn; *Evaluation of advanced silica packings for the separation of biopolymers by high-performance liquid chromatography. III. Retention and selectivity of proteins and peptides in gradient elution on non-porous monodisperse 1.5-μm reversed-phase silicas*; J. Chromatogr. A, **397** (1987) 71–80
87. R. Janzen, K.K. Unger, H. Giesche, J.N. Kinkel, and M.T.W. Hearn; *Evaluation of advanced silica packings for the separation of biopolymers by high-performance liquid chromatography. IV. Mobile phase and surface-mediated effects on recovery of native proteins in gradient elution on non-porous, monodisperse 1.5-μm reversed-phase silicas*; J. Chromatogr. A, **397** (1987) 81–89
88. R. Janzen, K.K. Unger, H. Giesche, J.N. Kinkel, and M.T.W. Hearn; *Evaluation of advanced silica packings for the separation of biopolymers by high-perforamnce liquid chromatography. V. Performance of non-porous monodisperse 1.5-μm bonded silicas in the separation of proteins by hydrophobic-interaction chromatography*; J. Chromatogr. A, **397** (1987) 91–97
89. B. Anspach, K.K. Unger, J. Davies, and M.T.W. Hearn; *Affinity chromatography with triazine dyes immobilized onto activated non-porous monodisperse silicas*; J. Chromatogr. A, **457** (1988) 195–204
90. G. Jilge, K.K. Unger, U. Esser, H.-J. Schäfer, G. Rathgeber, and W. Müller; *Evaluation of advanced silica packings for the separation of biopolymers by high-performance liquid*

chromatography. VI. Design, chromatographic performance and application of non-porous silica-based anion exchangers; J. Chromatogr. A, **476** (1989) 37–48

91. H. Giesche, K.K. Unger, U. Esser, B. Eray, U. Trüdinger, and J.N. Kinkel; *Packing technology, column bed structure and chromatographic performance of 1–2-μm non-porous silicas in high-performance liquid chromatography*; J. Chromatogr. A, **465**() (1989) 39–57
92. F.B. Anspach, A. Johnston, H.-J. Wirth, K.K. Unger, and M.T.W. Hearn; *High-performance liquid chromatography of amino acids, peptides and proteins. XCII. Thermodynamic and kinetic investigations on rigid and soft affinity gels with varying particle and pore sizes*; J. Chromatogr. A, **476** (1989) 205–225
93. M. Hanson, K.K. Unger, and G. Schomburg; *Non-porous polybutadiene-coated silicas as stationary phases in reversed-phase chromatography*; J. Chromatogr. A, **517** (1990) 269–284
94. G. Stegeman, R. Oostervink, J.C. Kraak, H. Poppe, and K.K. Unger; *Hydrodynamic chromatography of macromolecules on small spherical non-porous silica particles*; J. Chromatogr. A, **506** (1990) 547–561
95. F.B. Anspach, A. Johnston, H.-J. Wirth, K.K. Unger, and M.T.W. Hearn; *High-performance liquid chromatography of amino acids, peptides and proteins. XCV. Thermodynamic and kinetic investigations on rigid and soft affinity gels with varying particle and pore sizes: Comparison of thermodynamic parameters and the adsorption behaviour of proteins evaluated from bath and frontal analysis experiments*; J. Chromatogr. A, **499** (1990) 103–124
96. H.J. Wirth, K.K. Unger, and M.T.W. Hearn; *High-performance liquid chromatography of amino acids, peptides and proteins. CIX. Investigations on the relation between the ligand density of Cibacron Blue immobilized porous and non-porous sorbents and protein-binding capacities and association constants*; J. Chromatogr. A, **550** (1991) 383–395
97. Michael Hanson and Klaus K. Unger, Colin T. Mant, and Robert S. Hodges; *Polymer-coated reversed-phase packings with controlled hydrophobic properties. I. Effect on the selectivity of protein separations*; J. Chromatogr. A, **599**(1–2), (1992) 65–75
98. Michael Hanson, Klaus K. Unger, Renaud Denoyel, and Jean Rouquerol; *Interactions of lysozyme with hydrophilic and hydrophobic polymethacrylate stationary phases in reversed phase chromatography (RPC)*; J. Biochem. Biophys. Methods, **29**(3–4) (1994) 283–294
99. Béatrice de Collongue-Poyet, Claire Vidal-Madjar, Bernard Sebille, and Klaus K. Unger; *Study of conformational effects of recombinant interferon γ-adsorbed on a non-porous reversed-phase silica support*; J. Chromatogr. B: Biomed. Sci. Appl., **664**(1), (1995) 155–161
100. Michael Hanson, Klaus K. Unger, Colin T. Mant, and Robert S. Hodges; *Optimization strategies in ultrafast reversed-phase chromatography of proteins*; TrAC Trends Anal. Chem., **15**(2), (1996) 102–110
101. T. Issaeva, A. Kourganov, K. Unger; *Super-high-speed liquid chromatography of proteins and peptides on non-porous Micra NPS-RP packings*; J. Chromatogr. A, **846** (1999) 13–23
102. K. Wagner, K. Racaityte, K.K. Unger, T. Miliotis, L.E. Edholm, R. Bischoff, G. Marko-Varga; *Protein mapping by two-dimensional high performance liquid chromatography*; J. Chromatogr. A, **893** (2000) 293–305
103. B.A. Grimes, S. Ludtke, K.K. Unger, A.I. Liapis; *Novel general expressions that describe the behavior of the height equivalent of a theoretical plate in chromatographic systems involving electrically-driven and pressure-driven flows*; J. Chromatogr. A, **979** (2002) 447–466
104. A. van Blaaderen and A. Vrij; *Synthesis and characterization of colloid dispersions of fluorescent, monodispersed silica spheres*; Langmuir, **8** (1992) 2921–2931
105. J.D. Wells, L.K. Koopal, and A. de Keizer; *Monodisperse, nonporous, spherical silica particles*; Colloids Surf. A: Physicochem. Eng. Asp., **166** (2000) 171–176
106. Howard A. Ketelson, Robert Pelton, and Michael A. Brook; *Surface and colloidal properties of hydrosilane-modified Stöber silica*; Colloids Surf. A: Physicochem. Eng. Asp., **132** (1998) 229–239
107. Hoon Choi and I-Wei Chen; *Surface-modified silica colloid for diagnostic imaging*; J. Colloid Interface Sci., **258** (2003) 435–437

108. Yoshio Kobayashi, Kiyoto Misawa, Masaki Kobayashi, Motohiro Takeda, Mikio Konno, Masanobu Satake, Yoshiyuki Kawazoe, Noriaki Ohuchi, and Atsuo Kasuya; *Silica-coating of fluorescent polystyrene microspheres by a seeded polymerization technique and their photobleaching property*; Colloids Surf. A: Physicochem. Eng. Asp., **242** (2004) 47–52
109. Yinhan Gong, Yanqiao Xiang, Bingfang Yue, Guoping Xue, Jerald S. Bradshaw, Hian Kee Lee, and Milton L. Lee; *Application of diaza-18-crown-6-capped b-cyclodextrin bonded silica particles as chiral stationary phases for ultrahigh pressure capillary liquid chromatography*; J. Chromatogr. A, **1002** (2003) 63–70
110. Mitsuhiro Nakamura, Kouseki Hirade, Tadashi Sugiyama, and Yoshihiro Katagiri; *High-performance liquid chromatographic assay of zonisamide in human plasma using a non-porous silica column*; J. Chromatogr. B, **755** (2001) 337–341
111. J.J. Kirkland, F.A. Truszkowski, C.H. Dilks Jr., and G.S. Engel; *Superficially porous silica microspheres for fast high-performance liquid chromatography of macromolecules*; J. Chromatogr. A, **890** (2000) 3–13
112. Duš an Berek, Son Hoai Nguyen, and Gérard Hild; *Molecular characterization of block copolymers by means of liquid chromatography: I. Potential and limitations of full adsorption–desorption procedure in separation of block copolymers*; Eur. Polym. J., **36**(6) (2000) 1101–1111
113. Klaus Rissler; *Separation of polyesters by gradient reversed-phase high-performance liquid chromatography on a 1.5 μm non-porous column*; J. Chromatogr. A, **871** (2000) 243–258
114. Anja P. Kohne and T. Welsch; *Coupling of a microbore column with a column packed with non-porous particles for fast comprehensive two-dimensional high-performance liquid chromatography*; J. Chromatogr. A, **845** (1999) 463–469
115. Wen-Chien Lee; *Protein separation using non-porous sorbents*; J. Chromatogr. B, **699** (1997) 29–45
116. Fabrice Mangani, Genevieve Luck, Christophe Fraudeau, and Eric Verette; *On-line column-switching high-performance liquid chromatography analysis of cardiovascular drugs in serum with automated sample clean-up and zone-cutting technique to perform chiral separation*; J. Chromatogr. A, **762** (1997) 235–241
117. E. Venema, J.C. Kraak, H. Poppe, and R. Tijssen; *Packed-column hydrodynamic chromatography using 1-μm non-porous silica particles*; J. Chromatogr. A, **740**(2) (1996) 159–167
118. Wen-Chien Lee and Chien-Yi Chuang; *Performance of pH elution in high-performance affinity chromatography of proteins using non-porous silica*; J. Chromatogr. A, **721**(1) (1996) 31–39
119. Qi-Ming Mao, Ian G. Prince, and Milton T.W. Hearn; *High-performance liquid chromatography of amino acids, peptides and proteins. CXXXIX. Impact of operating parameters in large-scale chromatography of proteins*; J. Chromatogr. A, **691**(1–2) (1995) 273–283
120. Vittorio Brizzi and Danilo Corradini; *Rapid analysis of somatostatin in pharmaceutical preparations by HPLC with a micropellicular reversed-phase column*; J. Pharm. Biomed. Anal., **12**(6) (1994) 821–824
121. Q.M. Mao, R. Stockmann, I.G. Prince, and M.T.W. Hearn; *High-performance liquid chromatography of amino acids, peptides and proteins. CXXVI. Modeling of protein adsorption with non-porous and porous particles in a finite bath*; J. Chromatogr. A, **646**(1) (1993) 67–80
122. Gerrit Stegeman, Johan C. Kraak, and Hans Poppe; *Dispersion in packed-column hydrodynamic chromatography*; J. Chromatogr. A, **634**(2) (1993) 149–159
123. Noriyuki Nimura, Hiroko Itoh, Toshio Kinoshita, Norikazu Nagae, and Mitsugu Nomura; *Fast protein separation by reversed-phase high-performance liquid chromatography on octadecylsilyl-bonded non-porous silica gel: Effect of particle size of column packing on column efficiency*; J. Chromatogr. A, **585**(2) (1991) 207–211
124. Karl C. Vrancken, Luc De Coster, Pascal Van Der Voort, Piet J. Grobert, and Etienne F. Vansant; *The Role of Silanols in the Modification of Silica Gel with Aminosilanes*; Journal of Colloid and Interface Science, **170**(1) (1995) 71–77
125. H. Ben Ouada, H. Hommel, A.P. Legrand, H. Balard, and E. Papirer; *Organization of the layers of polyethylene oxide grafted with different densities on silica*; Journal of Colloid and Interface Science, **122**(2) (1988) 441–449

126. K. Bridger and B. Vincent; *The terminal grafting of poly(ethylene oxide) chains to silica surfaces;* European Polymer Journal, **16**(10) (1980) 1017–1021
127. R.D. Badley, W.T. Ford, F.J. McEnroe, and R.A. Assink; *Surface modification of colloidal silica;* Langmuir, **6** (1990) 792
128. Christian Kaiser; *Poröse and unporöse Kieselgelpartikel in Submikrometerbereich;* Dissertation, Universität Mainz, Germany (1996)
129. Ch. Kaiser, G. Buechel, S. Luedtke, I. Lauer, and K.K. Unger; *Processing of microporous/mesoporous submicron-size silica spheres by means of a template-supported synthesis;* Characterization of Porous Solids IV(B. Mc Enaney, T.J. Mays, J. Rouquerol, F. Rogriguez-Reinoso, K.S.W. Sing, & K.K. Unger, eds;The Royal Society of Chemistry, London (1997) 406–412
130. M. Grün, G. Büchel, D. Kumar, K. Schumacher, B. Bidlingmaier, and K.K. Unger; *Rational design, tailored synthesis and characterisation of ordered mesoporous silicas in the micron and submicron size range;* in Characterization of Porous Solids V (K.K. Unger, G. Kreysa and J.P. Baselt, eds), Stud. Surf. Sci. Catal. 128; Elsevier, Amsterdam (2000) 155
131. K.K. Unger, D. Kumar, M. Grün, G. Büchel, S. Lüdtke, Th. Adam, K. Schumacher, and S. Renker; *Synthesis of spherical porous silicas in the micron and submicron size range – Challenges and opportunities for miniaturized high-resolution chromatographic and electrokinetic separations;* J. Chromatogr. A **892** (2000) 47–55
132. Yurong Ma, Limin Qi, Jiming Ma, Yongqing Wu, Ou Liu, and Humin Cheng; *Large-pore mesoporous silica spheres: Synthesis and application in HPLC;* Colloids Surf. A: Physicochem. Eng. Asp., **229** (2003) 1–8
133. R. Vacassy, R.J. Flatt, H. Hofmann, K.S. Choi, and K.S. Singh; *Synthesis of microporous silica spheres;* J. Colloid Interface Sci., **227**(2) (2000) 302–315
134. Shoji Kaneko, Hiroshi Saitoh, Yoshio Maejima, and Motoshi Nakamura; *Adsorption characteristics of organic dyes in aqueous solutions on mixed-oxide gels silica-containing mixed-oxide gels;* Anal. Lett., **22** (6) (1989) 1631–1641
135. Francoise M. Winnik and Barkev Keoshkerian (Xerox); *Ink jet inks containing colored silica particles;* US Pat. 4,877,451 (1989)
136. Francoise M. Winnik, Barkev Keoshkerian, J. Roderick Fuller, and Peter G. Hostra; *New water-dispersible silica-based pigments: Synthesis and characterization;* Dyes Pigments, **14**(2) (1990) 101–112
137. H. Giesche and E. Matijevic; *Well defined pigments. I. monodispersed silica-acid dyes systems;* Dyes Pigments; **17** (1991) 323–340
138. T. Jesionowski; *Synthesis of organic-inorganic hybrids via adsorption of dye on an aminosilane-functionalised silica surface;* Dyes Pigments, **55**(2–3) (2002) 133–141
139. Teofil Jesionowski, Monika Pokora, Modzimierz Tylus, Aleksandra Dec, and Andrzej Krysztafkiewicz; *Effect of N-2-(aminoethyl)-3-aminopropyltrimethoxysilane surface modification and C.I. Acid Red 18 dye adsorption on the physicochemical properties of silica precipitated in an emulsion route, used as a pigment and a filler in acrylic paints;* Dyes Pigments, **57**(1) (2003) 29–41
140. S. Eiden-Assmann, B. Lindlar, and G. Maret; *Synthesis and characterization of colloidal fluorescent mesoporous silica particles;* J. Colloid Interface Sci., **271**(1) (2004) 120–123
141. Marjan Bele, Olavi Siiman, and Egon Matijevic; *Preparation and flow cytometry of uniform silica-fluorescent dye microspheres;* J. Colloid Interface Sci., **254** (2002) 274–282
142. Wan Peter Hsu, Rongchi Yu, and Egon Matijevic; *Well-defined colloidal pigments. II. Monodispersed inorganic spherical particles containing organic dyes;* Dyes Pigments; **19** (1992) 179–201
143. W.P. Hsu, R. Yu, E. Matijevic; *Paper whiteners. I. Titania coated silica;* J. Colloid Interface Sci., **156** (1993) 56–65
144. Qunyan Li and Peng Dong; *Preparation of nearly monodisperse multiply coated submicrospheres with a high refractive index;* J. Colloid Interface Sci., **261** (2003) 325–329
145. Xiao-an Fu and Syed Qutubuddin; *Preparation and characterization of titania nanocoating on monodisperse silica particles;* Colloids Surf., **186** (2001) 245–250

146. X. Gao, K.M. Yu, K.Y. Tam, and S.C. Tsang; *Colloidal stable silica encapsulated nanomagnetic composite as a novel bio-catalyst carrier*; Chem. Commun., **24** (2003) 2998–2999
147. Jinquan Cao, Yongxian Wang, Junfeng Yu, Jiaoyun Xia, Chunfu Zhang, Duanzhi Yin, and U.O. Urs O. Häfeli; *Preparation and radiolabeling of surface-modified magnetic nanoparticles with rhenium-188 for magnetic targeted radiotherapy*; Journal of Magnetism and Magnetic Materials, **277**(1–2) (2004) 165–174
148. J.F. Chen, H.M. Ding, J.X. Wang, and L. Shao; *Preparation and characterization of porous hollow silica nanoparticles for drug delivery applications*; Biomaterials, **25**(4) (2004) 723–727
149. F. Caruso, R.A. Caruso, and H. Mohwald; *Nanoengineering of inorganic and hybrid hollow spheres by colloidal templating*; Science, **282**(5391) (1998) 1111–1114
150. C. Charnay, S. Begu, C. Tourne-Peteilh, L. Nicole, D.A. Lerner, and J.M. Devoisselle; *Inclusion of ibuprofen in mesoporous templated silica: Drug loading and release property*; Eur. J. Biopharm., **57**(3) (2004) 533–540
151. H. Xu, F. Yan, E.E. Monson, and R. Kopelman; *Roomtemperature preparation and characterization of poly(ethylene glycol)-coated silica nanoparticles for biomedical applications*; J. Biomed. Mater. Res., **66**(4) (2003) 870–879

Pharmaceutical Applications of Finely Dispersed Systems

H. William Bosch

Introduction

Pharmaceutical dosage forms comprise active agents (therapeutics or diagnostics) in a variety of physical forms including rigid (crystalline or amorphous) solids, liquids, homogeneous solutions, micellar solutions, emulsions, lipid complexes, and polymer complexes to name a few. The form in which the active agent is present depends on the physical and chemical characteristics of the agent, such as solubility and melting point as well as the intended route of administration. Fine-particle active agents (defined as having volume mean diameters of approximately 10 μm or less) have been incorporated into finished dosage forms for more than 30 years. The use of active agent nanoparticles in pharmaceutical products has appeared more recently and is a rapidly growing area of interest. This manuscript will discuss the preparation of fine-particle active agents and their uses in pharmaceutical dosage forms for oral, parenteral, and pulmonary administration and will be primarily limited to active agents that are in solid crystalline form.

Methods of Preparing Finely Dispersed Solid Therapeutics and Diagnostics

The techniques that have been reported for preparing finely dispersed drug particles can be broadly categorized as either particle size reduction or controlled precipitation. There are variations on each theme, some of which are described later.

Particle Size Reduction

Comminution of large particles into smaller ones is a common unit operation in a wide variety of industries. It generally produces polydisperse systems with approximately log-Gaussian particle size distributions. One of the key advantages of particle size reduction is the scale at which it can be viably performed, at or above the multiton scale in commercial applications.

Within this category, the processes can be divided into those that utilize attrition media and those that do not. The simplest approach to particle size reduction is that of air-jet milling in which particles are fluidized into a high-velocity airstream.[1] Collisions between drug particles and impaction of particles into the mill lead to size reduction. In general, jet milling can achieve particle distributions with mean diameters of ca. 2–10 μm. The powders are often highly amorphous and possess poor flow properties because of their sizes and the electrostatic cohesive forces that are induced during the jet milling process. Nevertheless, air-jet milling has been routinely used in pharmaceutical dosage form manufacturing for many years, especially for drugs that are poorly water-soluble and intended for oral delivery. It is also applied extensively to powders used in pulmonary delivery products.

Somewhat more recently, homogenization has been described as a method for preparing fine-particle dispersions of pharmaceutical agents.[2–4] Devices such as the Microfluidizer® (MFIC™ Corporation, Newton, MA) can be used to process aqueous dispersions of poorly water-soluble compounds into the submicron size range. This approach offers several advantages over conventional air-jet milling, namely the smaller average particle sizes that can be achieved and the absence of amorphous material and electrostatic charge buildup. However, it does require that the active agent be dispersed in a liquid phase (usually water), and a steric or electrostatic stabilizer must be included to prevent aggregation of the active agent particles.

Milling methods that use attrition media can be divided into low-energy processes (e.g., ball milling) and high-energy processes. Ball milling generally involves charging a container with active agent and high-density beads made of glass, metal, or ceramic materials.[5] The active agent may be milled as a dry powder or dispersed in a liquid medium, the latter usually requiring addition of one or more stabilizing agents. Ball milling has the added advantage that it can be performed over a wide range of scales, from milligrams to multikilogram quantities of active agent per batch.

In contrast, high-energy milling usually involves two-phase systems. There is a wide variety of commercially available media mills, which may be of vertical or horizontal design. Generally, a slurry of active agent in a liquid is recirculated through the chamber of a media mill that comprises the attrition media and a rotating agitator shaft. The shear forces generated by the agitator shaft and media cause fracturing of the active agent particles. Traditional high-energy milling was not used for pharmaceutical manufacturing until relatively recently,[6] because of the high levels of contamination caused by deterioration of the glass, metal, or ceramic media. However, the advent of highly cross-linked polystyrene media has enabled high-purity pharmaceutical formulations to be manufactured in this way.[7] To date, four commercial pharmaceutical products including Rapamune® (Wyeth Inc.), Emend® (Merck & Co., Inc.), TriCor® 148 mg (Abbott Laboratories), and Megace ES® (Par Pharmaceuticals, Inc.) utilize high-energy media milling with cross-linked polystyrene attrition media in the manufacturing process. All four were developed using NanoCrystal® Technology (Elan Pharma International Ltd).

Controlled Precipitation

In contrast to size reduction techniques, controlled precipitation allows much more monodisperse systems to be prepared.[8] Judicious choice of reaction conditions can allow a wide range of particle sizes, morphologies, and crystal forms to be prepared. In general, the active agent must be dissolved and then precipitated by concentration, dilution in a nonsolvent, or by pH shift. Depending on the experimental conditions, crystalline or amorphous particles may result. For pharmaceutical applications, the choices of solvent may limit the applicability of this approach, as some active agents may not be soluble in pharmaceutically acceptable solvents (e.g., ethanol). The use of supercritical fluids, especially supercritical CO_2, has also been shown to be an effective way to produce fine particles of active agents.[9] One of the key benefits of this approach is the ease with which the solvent (or antisolvent) can be removed. Controlled precipitation also tends to be more effective under dilute conditions; hence, it is best suited to production of low-volume specialized products where monodispersity is critical (such as powders for inhalation). Examples of active agents that have been prepared as monodisperse systems by controlled precipitation include loratadine,[10] danazol, cyclosporine,[11] budesonide,[12] and the ethyl ester of diatrizoic acid.[13] Scanning electron micrographs of budesonide particles prepared by controlled precipitation are shown in Fig. 1a–c.

Oral Delivery of Fine-Particle Active Agents

Fine-particle technology has been in use in the pharmaceutical industry primarily to improve the oral bioavailability of poorly water-soluble drugs. Many if not most biologically active compounds have lipophilic properties that enable them to permeate cell membranes and bind to lipophilic receptor sites, but which also render them poorly water-soluble. Low aqueous solubility can lead to significant formulation development and oral drug delivery challenges, because the absolute bioavailability of orally administered drugs often depends on their rate of dissolution. Other factors associated with poor water solubility include variable absorption associated with the presence or absence of food at the time of administration (the "food effect") and delayed onset of therapeutic activity associated with slow active agent dissolution and absorption.

Increased Oral Bioavailability

Oral bioavailability is defined as the fraction of orally administered drug substance that enters systemic circulation. It is generally accepted that for a drug to be absorbed it must be present in molecular form (i.e., dissolved) and must permeate the gastrointestinal membrane before entering the bloodstream. There are a variety of

Fig. 1 (**a**) Amorphous budesonide particles produced by evaporating an ethanolic solution of budesonide and hydroxypropylcellulose at atmospheric pressure, (**b**) Crystalline budesonide particles prepared by addition of water to a sonicated ethanolic solution of budesonide, (**c**) Crystalline budesonide particles prepared by addition of water to a sonicated ethanolic solution of budesonide and hydroxypropylcellulose

chemical and physiological factors that can impede the absorption of orally administered drugs. Although a small fraction of molecules or ionic species may be actively transported across gastrointestinal membranes, in most cases the process is one of passive transcellular diffusion. The latter favors lipophilic molecules that have adequate thermodynamic affinity for the membrane lipid bilayer. Molecules that are highly polar lack the needed affinity for a passive diffusion process, and therefore may be very poorly bioavailable. For example, alendronate sodium, a bisphosphonate ester used to treat osteoporosis, is highly water-soluble, but is reported to have an absolute oral bioavailability of less than 1%.[14]

Another factor influencing drug absorption is the residence time of the drug dose within the gastrointestinal "window of absorption." It is well known that many drug molecules are well absorbed only in a relatively small section of the gastrointestinal tract, usually at the upper end of the small intestine (duodenum) where the intestinal physiology is most favorable for absorption. The residence time of the drug in this window depends on the length of the absorption window and the gastrointestinal rate of motility. Thus, if the total drug dissolution time is greater than the GI residence time, less than 100% of the dose will be absorbed. In these instances, it is desirable to decrease the drug dissolution time (e.g., increase the dissolution rate) which is most commonly done by decreasing the average particle size of the drug substance. It is here that fine particle technology plays its most important role in pharmaceutical development.

The rate of dissolution of a particle is given by the Noyes–Whitney equation (1):

$$-dM/dt = DS(Cs - C)/h, \qquad (1)$$

where D is the diffusivity of the molecule, S is the total amount of surface area, C_s is the concentration of the drug at saturation, C is the concentration in the bulk, and h is the distance between the particle surface and the point in the bulk at which the concentration $= C$ (the "boundary layer thickness").[15] Decreasing the average particle size of a drug powder sample will increase the surface area, which leads to an increase in dissolution rate. However, Pruitt and Ryde[16] have recently shown that the Noyes–Whitney equation does not accurately predict dissolution rates at active agent particle sizes below ca. 1 μm, but that a surface reaction phenomenon may limit dissolution rates in active agent nanoparticles and is responsible for the increases in oral bioavailability observed by reducing active agent particle sizes to the nanometer range.

For many years, pharmaceutical formulators have used size reduction techniques to enhance the dissolution and absorption properties of poorly water-soluble drugs. These techniques may include conventional hammer milling, dry ball milling, and air-jet milling. The latter is frequently used and can generate drug particle distributions with average sizes of ca. 5 μm or less.

An early report by Prescott et al.[17] compares the oral bioavailability of phenacetin particles that are coarse (>250 μm), medium (150–180 μm), or fine (<75 μm) and clearly shows how bioavailability increases with decreasing active agent particle size. Phenacetin (N-(4-ethoyxphenyl)acetamide) is an analgesic and antipyretic with an aqueous solubility of less than 1 mg/mL and the chemical structure shown in Fig. 2.

Liversidge and Cundy reported a dramatic increase in the oral bioavailability of a nanoparticulate danazol composition relative to a micronized composition when administered to fasted beagle dogs.[18] Danazol, sold under the trade name Danocrine®, is a synthetic steroid used to treat endometriosis. It has an aqueous solubility of approximately 10 μg/mL and has the chemical structure shown in Fig. 3.

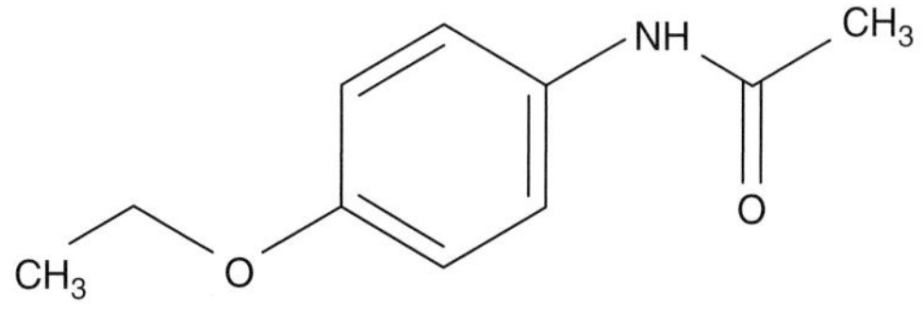

Fig. 2 Chemical structure of phenacetin

Fig. 3 Chemical structure of danazol

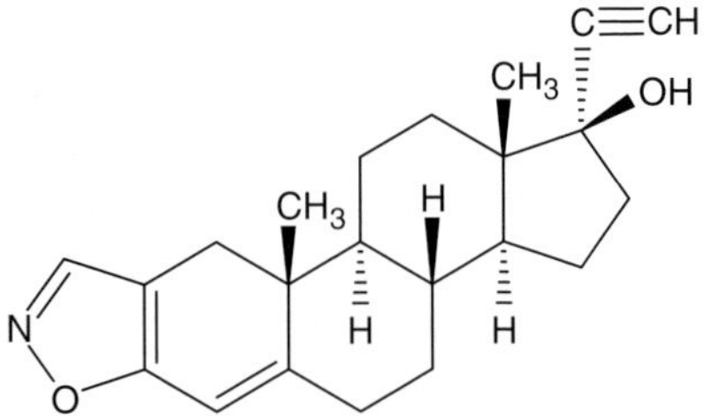

Table 1 Pharmacokinetic parameters following oral and intravenous administration of danazol formulations to fasted male beagle dogs ($n = 5$)

Formulation	C_{max} (μg/mL)	t_{max} (h)	AUC (μg h/mL)	Absolute bioavailability (%)
Cyclodextrin solution (oral)	3.94 ± 0.14	1.2 ± 0.2	20.4 ± 1.9	106.7 ± 12.3
Nanoparticulate dispersion	3.01 ± 0.80	1.5 ± 0.3	16.5 ± 3.2	82.3 ± 10.1
Micronized suspension	0.20 ± 0.06	1.7 ± 0.4	1.0 ± 0.4	5.1 ± 1.9
Cyclodextrin solution, i.v.			19.8 ± 0.6	100

In the Liversidge study, fasted beagle dogs were dosed with each of three different formulations of danazol in a randomized crossover study. The three formulations were (1) an aqueous nanoparticulate dispersion comprising 5% danazol and 1.5% polyvinylpyrrolidone (PVP) with a mean particle size of 169 nm, (2) an aqueous solution comprising 1% danazol in 50% hydroxypropyl-β-cyclodextrin (HP-β-CD), and (3) an aqueous suspension comprising 5% micronized danazol and 1.5% PVP with a mean particle size of 10 μm. The dogs were administered 200 mg of danazol for each formulation, with a 1-week washout period between doses. Bioanalysis was performed on the plasma samples to determine maximum plasma concentrations (C_{max}) and extent of absorption (AUC) for each formulation; these results were then compared to the plasma levels obtained from an intravenously administered dose of danazol. The results indicated that the nanoparticulate danazol formulation was more than 16 times more orally bioavailable than the micronized dosage form (Table 1). Furthermore, there was no statistically significant difference between the oral bioavailability of the nanoparticulate formulation and the HP-β-CD solution. Thus, the results confirm that the oral bioavailability of danazol is dissolution-rate limited, and that conversion of the drug substance into fine particles could substantially enhance oral absorption.

Food Effects

The absorption of insoluble drugs can be substantially influenced by the presence or absence of food in the digestive tract. In the fasted state, gastrointestinal (GI) transit is relatively rapid; thus, the residence time of the drug in the region of absorption

Fig. 4 Chemical structure of megestrol acetate

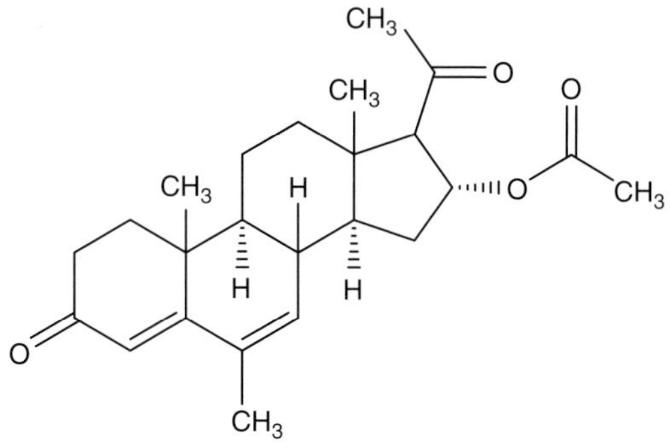

is minimized and oral bioavailability is often minimized. Under fed conditions, a variety of factors that enhance drug uptake are present and active. In general, the GI transit is slowed down, favoring absorption. Additionally, the presence of fatty material in the food and the production of bile salts in response to the fed condition both contribute to solubilization of the drug and lead to more complete absorption. As a result, lipophilic drugs with poor aqueous solubility are often reasonably well absorbed in the fed state, but very poorly absorbed under fasted conditions. Such fed-fasted variability can lead to dosing limitations and require product labeling to indicate that a drug product must be taken with food. In extreme cases, especially those where a drug has a narrow therapeutic index (difference between toxic and therapeutic plasma concentrations), the food effect may preclude development and commercialization of the drug.

One example of a drug that can have highly variable absorption is megestrol acetate ($C_{24}H_{32}O_4$). It has an aqueous solubility of approximately $2\,\mu g/mL$ at $37\,°C$ and has the chemical structure shown in Fig. 4.

Megestrol acetate is the active ingredient in two commercial prescription products, one being an aqueous suspension of micronized active agent (megestrol acetate oral suspension), and the other a nanoparticulate colloidal dispersion of megestrol acetate (Megace® ES). It is an appetite-enhancing agent used to treat anorexia, cachexia, or unexplained significant weight loss in patients who have been diagnosed with acquired immune deficiency syndrome (AIDS). A pharmacokinetic study in human subjects[19] showed that a 625-mg dose of the nanoparticulate formulation of megestrol acetate was bioequivalent to an 800-mg dose of the micronized suspension when both were administered to patients under fed conditions. The bioavailability of the two products was also compared at various doses under fed and fasted conditions; selected data from the study are presented in Table 2.

The data indicate that the oral bioavailability of nanoparticulate megestrol acetate was 54.8% higher in the fed state when dosed at 450 mg and 43.3% higher when dosed at 675 mg in the fed state than the respective values in the fasted state. However, the 800-mg dose of drug in the micronized form was more than 100% more bioavailable in the fed state than in the fasted state. The reduced variability

Table 2 Pharmacokinetic data for three formulations of megestrol acetate

	Dose					
	450 mg[a]		675 mg[a]		800 mg[b]	
	Fast	Fed	Fast	Fed	Fast	Fed
C_{max} (ng/mL)	955	1,079	1,044	1,616	187	1,364
$AUC_{0-\infty}$ (ng h/mL)	9,483	11,800	11,879	17,029	8,942	18,625
T_{max} (h)	1.74	3.16	1.96	2.76	5.89	3.85

[a]Megace ES (nanoparticulate)
[b]Megestrol acetate oral suspension (micronized)

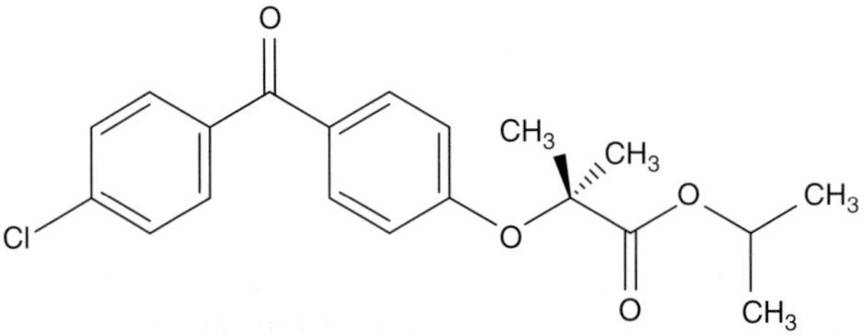

Fig. 5 Chemical structure of fenofibrate

seen in the nanoparticulate form is attributed to more rapid dissolution of the drug nanoparticles and associated improved absorption in the fasted state.

Variable oral bioavailability has also been observed with fenofibrate, a lipid-lowering agent that has the chemical structure shown in Fig. 5.

Fenofibrate is metabolized to the active metabolite fenofibric acid, the pharmacological effect of which is attributed to activation of peroxisome proliferator-activated receptor α (PPARα).[20]

Fenofibrate has been marketed in the United States by Abbott Laboratories, Inc. since the 1990s under the trade name TriCor®. It had been available as capsules containing 67, 134, or 200 mg of micronized fenofibrate, and the bioavailability of the drug was reported to be increased by approximately 35% under fed as compared to fasting conditions.[21] As a result, the prescribing information for Tricor indicated that it should be given with meals. In 2001, Abbott received approval to market Tricor (fenofibrate) tablets of 54 and 160 mg.[22] The 160-mg tablets were found to be bioequivalent to the 200-mg capsules and were also 35% more bioavailable in the fed state, and therefore needed to be taken with food. However, in 2004, Abbott received approval for a new formulation of Tricor, which contained fenofibrate (48 or 145 mg) in NanoCrystal® form (Elan Pharma International Ltd.).[23] These tablets (3 × 48 mg or 145 mg) were bioequivalent to one 200-mg capsule under fed conditions; furthermore, there was sufficiently little difference in absorption between the fed and fasted states that the nanoparticulate Tricor tablets can be taken without regard to meals. Since 2004, two additional fenofibrate products have been introduced in the United States: Antara™ (fenofibrate) capsules, 43, 87, and 130 mg (Reliant Pharmaceuticals, Inc.),[24] and Triglide™ (fenofibrate) tablets, 50 and 160 mg (First Horizon Pharmaceutical® Corporation).[25] The 130-mg Antara capsules are reported to be equivalent under low-fat fed conditions to 200-mg fenofibrate capsules after multiple dose administration, and thus have achieved an increase in bioavailability relative to the reference product. However, in the presence of a high-fat meal, the extent of absorption is approximately 26% greater relative to the fasting state, and therefore the Antara capsules are recommended to be taken with meals. The fenofibrate in Triglide tablets is formulated using IDD®-P MicroParticle technology (SkyePharma), which uses phospholipids to prevent drug particle aggregation and enhance absorption and bioavailability. The 160-mg tablets contain the same quantity of fenofibrate as the original Tricor tablets, yet are reported to have comparable extents of absorption (AUC) in fed and fasted conditions. As a result, they may be administered with or without food.

The improvements in absorption and reductions in variability observed with megestrol acetate and fenofibrate illustrate the significant role fine particle technology can play in these types of products. Incorporation of microparticles or nanoparticles can lead to products that require lower doses of active ingredient and can eliminate the need to take certain products with meals, facilitating patient compliance.

Rapid Absorption

Slow drug dissolution and absorption can also lead to slow onset of therapeutic effect. For drugs that are taken on an "as needed" basis, slow onset of action can be a serious disadvantage. For instance, it is highly desirable that drugs used to treat acute pain be rapidly absorbed so that the onset of pain relief is not delayed. Many drugs that fall into the category of nonsteroidal anti-inflammatory drugs (NSAIDS) have relatively poor water solubility and are thus absorbed rather slowly. For instance, the time needed to reach maximum plasma concentrations after administration of a dose of meloxicam (Mobic®) is approximately 4–5 h in the fasted state and 5–6 h in the fed state.[26] The systematic name of meloxicam is 4-hydroxy-2-methyl-*N*-(5-methyl-2-thiazolyl)-2H-1,2-benzothiazine-3-carboxamide 1,1 dioxide; structure shown in Fig. 6.

A recent study by Pruitt compared the rates of absorption of three different formulations of meloxicam when administered to beagle dogs.[27] The three test

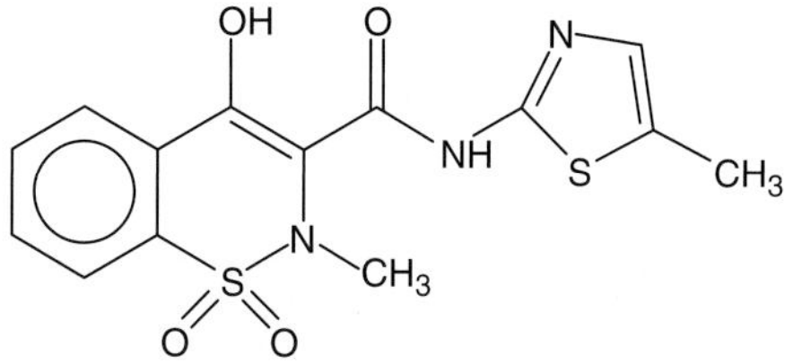

Fig. 6 Chemical structure of meloxicam

articles were Mobic®, a 10% NanoCrystal® colloidal dispersion of meloxicam, and a NanoCrystal solid dosage form of meloxicam. The average particle size of the meloxicam in both nanoparticulate formulations was 111 nm. Figure 7 shows scanning election micrographs of the meloxicam drug substance before milling (a) and in nanoparticulate form (b).

In the study, 7.5 mg of meloxicam was given orally to each of eight dogs with a 7-day washout period between formulations. Plasma samples were taken at 10, 20, and 45 min, and at 1, 1.5, 2, 2.5, 3, 4, 6, 12, 18, 24, and 48 h postdosing.

Analysis of the plasma samples for meloxicam showed that the drug was much more rapidly absorbed after administration in NanoCrystal form than after being given in micronized form. The time to peak plasma levels (t_{max}) was 0.75 and 1.29 h for the liquid and solid NanoCrystal formulations, respectively, but was 3.38 h for the commercial product. Figure 8 shows the meloxicam plasma profiles for the three

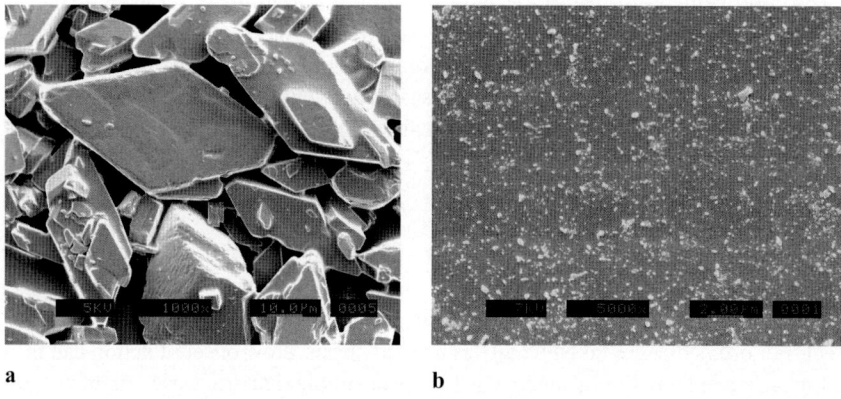

Fig. 7 (a) Meloxicam raw drug substance and (b) in NanoCrystal® form

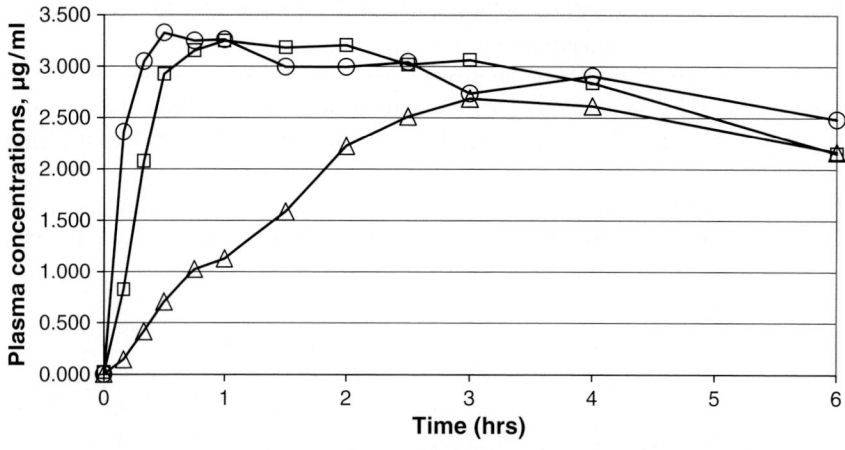

Fig. 8 Pharmacokinetic profiles of three formulations of meloxicam

Table 3 Summary of meloxicam pharmacokinetic data

	NanoCrystal® dispersion	NanoCrystal® solid dosage form	Mobic®
C_{max} (μg/mL)	3.50	3.42	2.77
t_{max} (h)	0.75	1.29	3.38
AUC (μg h/mL)	118.23	106.64	99.87

test articles from $t = 0$ to 6 h, and the summary pharmacokinetic data are presented in Table 3.

The results of this study illustrate how much more rapidly a poorly water-soluble drug can be orally absorbed when formulated in nanoparticulate form instead of conventional micronized form. Furthermore, the closely similar pharmacokinetic profiles of the two nanoparticulate test articles confirm that the benefits of small particles (high surface area and rapid dissolution) can be realized from solid products as well as liquid dispersions. The application of fine particle technology to other poorly water-soluble analgesics such as naproxen and ketoprofen can be expected to produce qualitatively similar results.

Topical Administration of Active Agents to the Gastrointestinal Tract

Certain diseases of the digestive tract may be best treated by delivery of the therapeutic active agent directly to the affected areas instead of indirectly through the systemic circulation. In instances where the active agent is poorly water-soluble, it may be highly desirable to administer the agent in fine particle form (nanoparticles or microparticles). One such commercial product is Entocort® EC, a capsule containing the anti-inflammatory corticosteroid budesonide in micronized form. Entocort® EC (Prometheus Laboratories, Inc.) is indicated for the treatment of mild to moderate Crohn's disease.

The efficacy of a topically administered fine particle active agent likely depends on how well the agent covers the affected physiological area. In any such system, the number of particles per unit dose is a function of particle size and increases proportionally to $1/r^3$ as the particle size is reduced (spherical particles assumed). Thus, a given dose of smaller particles will be able to cover a much greater surface area than the same dose of larger particles. It has further been demonstrated that fine (<4 μm) particles that have cationic surface charges are highly bioadhesive compared with larger particles or to those with neutral or anionic surface charges.[28] Bioadhesive properties can be used to enhance coverage of the intestinal mucosa and prolong the residence time of the active agent in the gastrointestinal tract.

Interaction of a bioadhesive nanoparticulate active agent with the intestinal mucosa is shown in Fig. 9. In this study by Ruddy et al.,[29] a bioadhesive formulation

Fig. 9 Flat-film X-ray image after oral administration of nanoparticulate BaSO$_4$

of nanoparticulate barium sulfate was administered orally to a dog. Examination of the flat-film X-ray image shows a uniform thin coating of the BaSO$_4$ throughout the small intestine. It can be expected that topical delivery of therapeutic active agents to the GI tract for treatment of gastric ulcers, infections, and inflammatory conditions could benefit substantially from the use of fine particle technology.

Parenteral Administration of Fine Particle Active Agents

Intravenous Injection

Therapeutic and diagnostic compounds intended for intravenous administration have traditionally been formulated as homogeneous solutions. This approach is trivial for compounds with good aqueous solubility, but can be problematic for active agents that are poorly water-soluble. Some of the formulation strategies that have been employed with poorly soluble actives include the use of organic solvents, solubilizing agents, and pH extremes. One such example is phenytoin sodium injection USP, an antiepileptic therapeutic comprising an aqueous solution of phenytoin sodium with propylene glycol and alcohol at pH = 10.0–12.3.[30]

The cytotoxic oncology agent Taxotere® (Sanofi-Aventis Inc.; active ingredient docetaxel) is supplied as a concentrate containing 1,040 mg of polysorbate 80 per 40 mg of docetaxel and must be diluted in a 13% ethanol in water solution before administration.[31]

In many cases, the toxicity of the excipients in these systems can limit the amount of drug that can be safely administered. Moreover, these formulations often have to be administered slowly to avoid precipitation of the active agent in vivo. An alternative approach that has recently become available is the use of nanoparticle technology to formulate these problematic molecules. In order for solid particles to be intravenously injected safely, they must first of all be small enough to pass through the narrowest capillaries without causing occlusion (red blood cells, which are deformable, are about 8 μm in diameter). Injected particles that do not dissolve rapidly can also be taken up by the mononuclear phagocytic system (MPS) and transported to organs such as the liver and spleen. The application of nanoparticle technology to intravenous drug product development can enable poorly soluble active agents to be formulated in aqueous-based vehicles without toxic cosolvents, solubilizing agents, or pH extremes. Furthermore, nanoparticles (<200 nm) are much smaller than capillary diameters and offer the possibility of injecting solid, nondeformable particles without risking capillary occlusion. A report by Merisko-Liversidge et al.[32] describes how several different poorly water-soluble oncology agents (camptothecin, etoposide, and paclitaxel) were formulated as nanoparticulate colloidal dispersions and administered to animals that had been implanted with pancreatic, lung, or mammary tumors. In one specific experiment, a nanoparticulate formulation of paclitaxel was compared to a soluble formulation of the drug that contained ethanol and polyethoxylated castor oil (Cremaphor® EL). Both formulations were tested in animals that had been implanted with the MV-522 human lung xenograft murine tumor model. The maximum tolerated dose of the soluble formulation was 30 mg/kg, which resulted in a 22% death rate due to test article toxicity. The nanoparticulate formulation was given at 90 mg/kg with no deaths attributed to the test article, and the efficacy of the nanoparticulate formulation (as measured by reduction in tumor weight) was much greater for the nanoparticulate composition than for the soluble formulation.

There are relatively few examples of human exposure to intravenous administration of solid fine-particle active agents. Several particle-based products for intravenous injection are currently available in the United States; these include Ambisome® (Astellas Pharma US, Inc.), which is a liposomal formulation of amphotericin B; Abelcet® (Enzon Pharma, Inc.), an amphotericin B lipid complex; Doxil® (Tibotec), a liposomal formulation of doxorubicin hydrochloride; and Abraxane® (American Bioscience, Inc.), a paclitaxel product in which the active agent molecules are protein bound. The particles in liposome and lipid complexes are generally not rigid solid particles, and the protein-bound paclitaxel particles are also expected to be deformable.

Several years ago, Donnelly et al.[33] described a clinical study in which human subjects were administered an intravenous formulation of nanoparticulate itraconzole, an antifungal agent. The patients were all hematopoietic stem cell transplant recipients receiving cyclosporine, and as such were at risk for invasive fungal infections and mucositis (fungal infection of the oral cavity). Patients were given 200 mg of intravenous NanoCrystal itraconazole daily for 14 days. The treatment was well tolerated, and it afforded plasma itraconazole levels of $\geq 500\,\mu g/L$ for 3 weeks, which was deemed sufficient for oral mucositis to subside such that oral itraconazole treatment could be continued if necessary.

Intramuscular Administration of Finely Dispersed Systems

Commercial drug products containing fine-particle active agents for intramuscular, intrasynovial, or intralesional administration are relatively common. These include products for the treatment of a variety of inflammatory conditions (methyprednisolone acetate), bacterial infections (inipenem), acromegaly (octreotide acetate), endometriosis (leuprolide acetate), prostate cancer (triptorelin pamoate), and schizophrenia (risperidone). In some cases, these products are intended to provide therapeutic blood levels of the active agent for extended periods of time (the "depot" effect). Fine particles that are injected other than intravenously are not subject to the same particle size constraints as those that are injected directly into the bloodstream.

One advantage of very fine (i.e., nanoparticulate) dispersions of active agents is the high concentration of active agent that can be delivered in a relatively small volume. Intramuscular injections are usually limited to volumes of ≤ 2 mL because of the pain associated with larger volumes of injection. Thus, drugs that are soluble in water even to 5% w/v, in principle, can only be administered at doses of approximately 100 mg or less per injection via the IM route. In contrast, nanoparticulate colloidal dispersions can contain solids contents of $\geq 40\%$ w/v while still having very low viscosities. This allows much higher doses of active agent to be administered (up to ca. 800 mg per 2 mL injection).

In one such example, a study was performed to compare the pharmacokinetics of orally administered micronized naproxen suspension (Naprosyn®) with an intramuscular injection of nanoparticulate naproxen colloidal dispersion (average particle size ca. 300 nm, concentration 400 mg naproxen/mL).[34] In a crossover experiment, the patients ($n = 3$ per dose) were given oral and intramuscular naproxen at doses of 2.5 and 5.0 mg/kg, corresponding to approximately 175 and 350 mg per injection, respectively. A summary of the pharmacokinetic data is presented in Table 4 later, and the pharmacokinetic profiles are shown in Fig. 10. For each dose, the overall bioavailabilities were comparable, but the absorption was somewhat slower for the intramuscular product in both cases.

Table 4 Summary of pharmacokinetic data for oral and IM naproxen

Dose (mg/kg)	AUC (µg h/mL)		C_{max} (µg/mL)		t_{max} (h)	
	Oral	IM	Oral	IM	Oral	IM
2.5	594 ± 21	545 ± 40	34 ± 4	19 ± 8	2 ± 1	11 ± 11
5.0	998 ± 40	957 ± 30	57 ± 3	29 ± 8	2 ± 1	9 ± 5

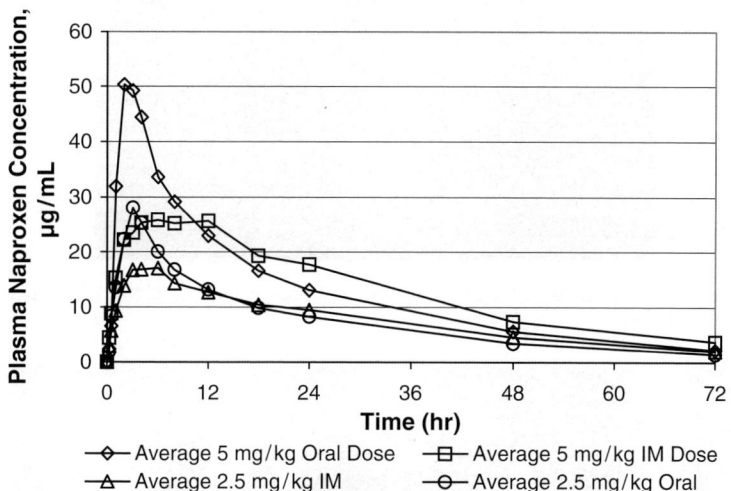

Fig. 10 Pharmacokinetic profiles of oral micronized naproxen and intramuscular nanoparticulate naproxen

Diagnostic Imaging Applications

Fine particle technology has also been applied to the development of diagnostic imaging agents for enhancement of X-ray images of soft tissue and the vasculature.[35] Nanoparticulate compositions of poorly soluble esters of diatrizoic acid (3,5-diactamido-2,4,6-triiodobenzoic acid) have been shown to enhance visualization of the lymph nodes after subcutaneous administration,[36] and can also be used to enhance contrast between the vasculature and surrounding tissue after intravenous injection. Figure 11 is a computer enhanced X-ray image of a rabbit after injection with a nanoparticulate ester of diatrizoic acid (Imcor Pharmaceuticals, Inc), while Fig. 12 shows enhanced images of lymph nodes, major blood vessels, and hepatic blood vessels after administration of nanoparticulate esters of diatrizoic acid.

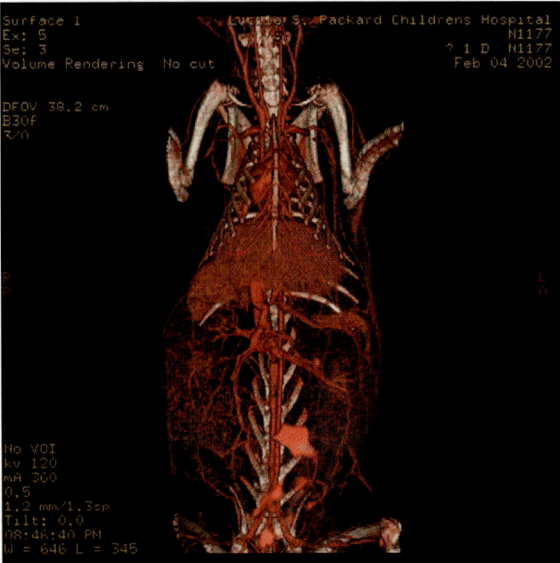

Fig. 11 Computer-enhanced X-ray image after intravenous administration of a nanoparticulate ester of diatrizoic acid

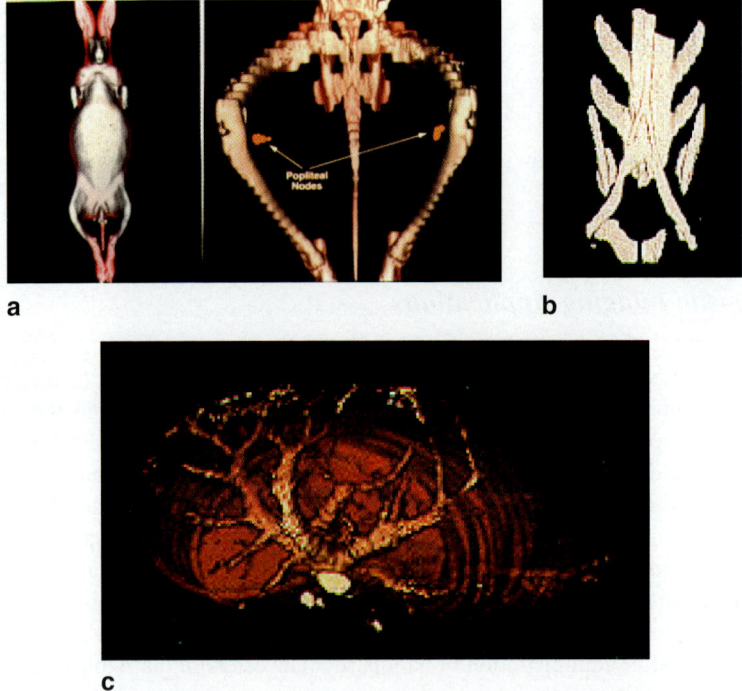

Fig. 12 X-ray image enhancement with a nanoparticulate ester of diatrizoic acid: (**a**) Lymph nodes (subcutaneous injection), (**b**) major arteries (intravenous injection), and (**c**) hepatic arteries (intravenous injection)

Summary

Fine particle technology can be used to formulate poorly water-soluble drugs in aqueous vehicles that have little-to-no toxicity. Intravenous formulations require that the active agent particles be in the nanometer size range (<1 µm and preferably <200 nm). Depending on the size distribution of the particles, long-acting depot formulations can be developed (larger particles) or highly concentrated dispersions may be produced (nanoparticles).

Pulmonary Delivery of Finely Dispersed Therapeutic Agents

Respiratory Aerosols

Delivery of pharmacologically active agents to the lung is important for two reasons. First, it can be used to treat respiratory diseases such as asthma, bronchitis, cystic fibrosis, or emphysema. In these instances, pulmonary delivery is desirable because it delivers the drug directly to the diseased tissue and bypasses the systemic circulation where drug metabolism may occur or unwanted side effects may be produced. Secondly, in cases where systemic delivery is desirable, the lung can serve as a portal to the systemic circulation if the drug substance can be delivered all the way to the alveolar sacs. Once in these structures, dissolved drug, in principle, can diffuse across the alveolar membrane into the systemic circulation. This provides a relatively novel and "user-friendly" alternative to subcutaneous, intramuscular, or intravenous injections of drugs that cannot be administered orally.

Treatment of respiratory diseases such as asthma requires delivery of the active pharmaceutical agent to the conducting airways of the lungs, where inflammation and bronchoconstriction occur. This can be accomplished either by administering the drug systemically (such as via the oral route) or by delivering the drug topically, e.g., directly to the conducting airways of the lung. The latter approach is preferred for anti-inflammatory steroids that are extremely effective in treating the symptoms of asthma, but are associated with adverse side effects when given systemically. For many years, oral prednisone was one of the treatments of choice for severe asthma, but with the availability of more potent glucocorticosteroids such as beclomethasone dipropionate, triamcinolone acetonide, budesonide, flunisolide, and fluticasone, topical delivery of these compounds to the conducting airways of the lung has been preferred and represents a route of administration in which finely dispersed particles are essential for therapeutic efficacy to be achieved.

Inhaled respiratory therapeutic agents are, as a rule, administered as aerosols. The aerosol may be an aqueous solution or suspension of the drug generated by a compressed air or ultrasonic nebulizer. Alternatively, the aerosol may be generated by a pressurized metered-dose inhaler (pMDI), a device that contains a solution or suspension of the drug in a volatile propellant such as chorofluorocarbons (CFCs) or hydrofluoroalkanes (HFAs). A third approach is simply to deliver the drug as a dry powder aerosol.

Regardless of the method used to generate the aerosol, efficient pulmonary deposition of the active agent is critically dependent on the aerodynamic diameter of the inhaled particle. Aerodynamic diameter is the physical property of a particle, which defines how it will behave in an airstream, and depends on the particle geometric size, density, and shape. An in-depth discussion of how particle shape affects aerodynamic diameter is beyond the scope of this review; therefore, the cited examples will assume a spherical particle.

The aerodynamic diameter of a particle is calculated from its equivalent volume diameter and density according to (2), which neglects slip correction associated with shape effects and assumes spherical particles.[37] D_{ae} = aerodynamic diameter, ρ_p = particle density, and D_e = equivalent volume diameter:

$$D_{ae} = (\rho_p)^{0.5} D_e. \qquad (2)$$

Thus, for spherical particles having a density equal to $1\,g/cm^3$, the aerodynamic and equivalent volume diameters are mathematically equal. In practice, all medical aerosols comprise particle size distributions; thus, the value mass median aerodynamic diameter (MMAD) is frequently used to characterize an aerosol.

The fate of an orally inhaled particle is strongly dependent on its aerodynamic diameter. Generally, particles larger than ca. 5 μm will inertially impact the mouth or throat, and be swallowed. Particles in the range of ca. 3–5 μm in diameter will reach the upper or conducting airways of the lung and can deposit on the smooth muscle of these structures. Particles of approximately 1–3 μm may follow the airstream all the way to the alveoli and be deposited, and particles less than about 1 μm may be exhaled. Thus, careful control of the particle size distribution of medical aerosols is essential for effective drug delivery.

Delivery of Dry Powder Aerosols to the Lung

The respiratory steroids mentioned earlier are organic crystalline materials with melting points >200 °C. As such, they can be made into fine particles of ca. 3–5 μm MMAD by conventional air-jet milling. However, powders produced in this way tend to be highly aggregated because of the large surface area and electrostatic cohesive forces that are present. Deaggregation of such powders is critical for effective drug delivery and is accomplished by the dry powder inhaler device (DPI) in which the drug is packaged. One such product is Pulmicort® Turbuhaler,[38] marketed by AstraZeneca. This product contains micronized budesonide and provides 200 μg per metered dose. Another approach that has been used involves dry blending the micronized steroid with an excess of large (ca. 60 μm) lactose particles to which they adhere. The drug–lactose blends are then delivered from the DPI that typically includes a mechanism for detaching the drug particles from the lactose particles.

The micron-sized drug particles are then transported to the conducting airways of the lung, while the lactose particles impact the mouth and throat and are swallowed. One such product using this approach is Flovent® Rotadisk® (used in combination with the Diskhaler® inhalation device), which is marketed by GlaxoSmithKline and delivers 50, 100, or 250 µg of fluticasone propionate, blended with lactose.[39]

A novel approach to dry powder formulation development has been described by Edwards et al.[40] Edwards produces powders of very low densities ($\rho < 0.4\,\text{g/cm}^3$) that have mean geometric diameters exceeding 5 µm, but have aerodynamic diameters suitable for delivery to the deep lung. Thus, inhalation of large porous particles containing testosterone or insulin was shown to produce high systemic bioavailability of the active agents in both cases.

Respirable powders for inhalation have also been prepared by spray-drying aqueous nanoparticulate dispersions of glucocorticosteroids.[41] The dispersions, which comprise drug particles of ca. 100–200 nm in diameter and water-soluble polymers, are spray-dried to produce spherical aggregates of drug nanoparticles. The aggregate diameters can be controlled by adjusting the spray-drying parameters. Figure 13 shows scanning electron micrographs of nanoparticulate budesonide aggregates of approximately 1.4 µm average diameter at $2,500\times$ magnification (a) and micronized budesonide (b) at the same magnification. The spray-dried aggregates appear to be uniformly spherical in shape, while the micronized material contains a broad range of particle sizes and morphologies.

A comparative study of the spray-dried nanoparticulate budesonide versus micronized budesonide was performed: Each powder was blended with lactose and then filled into a Clickhaler® (ML Laboratories) DPI, and the aerodynamic diameters of the delivered powders were characterized on an Andersen eight-stage cascade impactor. The results, which are summarized graphically in the following section, show that the MMAD of the nanoparticulate aggregate aerosol was much smaller than the MMAD of the micronized drug aerosol. A substantial fraction of

a   b

Fig. 13 Scanning election micrographs of (**a**) spray-dried nanoparticulate budesonide and (**b**) micronized budesonide

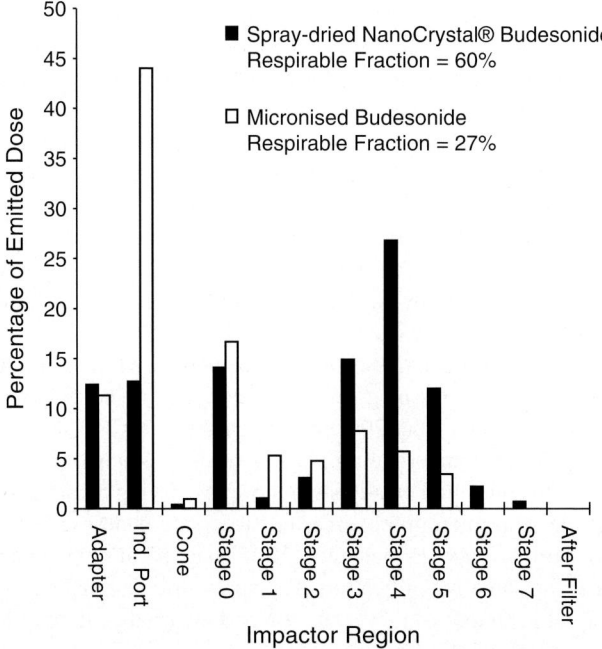

Fig. 14 Cascade impactor data for spray-dried nanoparticulate budesonide and micronized budesonide delivered from a multidose dry powder inhaler

the micronized budesonide drug substance was captured in the induction port of the impactor, indicating poor powder redispersion. The relatively larger amounts of budesonide delivered to impactor stages 2–7 for the spray-dried nanoparticulate budesonide are attributed to more complete deaggregation of the particles. The respirable fraction of drug delivered in nanoparticulate form, defined as the percentage of drug substance deposited on impactor stages 2 and below, was more than twice as great as the respirable fraction delivered in pure micronized form. Thus, the spray-dried nanoparticle approach would appear to offer the potential for improved in-vivo delivery efficiency at lower doses relative to the conventional approach (Fig. 14).

Delivery of Fine Particles with Pressurized Metered-Dose Inhalers

Air-jet milled drug substance can also be incorporated into pressurized metered-dose inhalers (pMDIs), which contain volatile inert propellants. There are numerous examples of such products. These drug/device products offer a high level of convenience, but are generally very inefficient in delivering drug to the lung. Upon actuation, the drug particles exit the device outlet at high velocities because of the propellant pressure, and tend to impact the mouth and throat. Only about 15–30% of the delivered dose is generally thought to reach the conducting airways.

Nebulization of Aqueous Suspensions

Solid particulate drug compounds can also be delivered to the lung via aerosolization of aqueous suspensions. One commercial product that is designed to be used this way is Pulmicort® Respules®, an aqueous suspension of budesonide marketed by AstraZeneca.

The product contains 0.25 or 0.50 mg of budesonide in 2 mL of aqueous polysorbate 80 solution and must be aerosolized with a jet nebulizer for approximately 5 min. Ultrasonic nebulizers cannot be used.

In a system such as Pulmicort® Respules®, the average particle size of the micronized budesonide is approximately the same as the average particle size of the aerosolized liquid droplets (ca. 3–5 µm). Since the suspension is very dilute (approximately 1/2,000 to 1/4,000), delivery efficiency of the budesonide is likely to be poor (only 1 droplet in 2,000 or 4,000 will contain active agent).

Delivery efficiency of aqueous suspensions can be improved by significantly reducing the average particle size of the drug. If the drug particles are in the nanometer size range (and therefore 1–2 orders of magnitude smaller than the nebulized liquid droplets), the aerosolization behavior of the dispersion is much closer to that of a true homogeneous solution because each aerosolized droplet can contain many drug particles. This phenomenon was illustrated by Ostrander et al., when the aerosol properties of a nanoparticulate dispersion of beclomethasone dipropionate (1.25% w/w) were characterized by cascade impaction.[42] In a similar study, a nanoparticulate dispersion of beclomethasone dipropionate (10% w/w) with an average particle size of 94 nm was aerosolized in an Omron MicroAir® ultrasonic nebulizer and collected on the plates of an eight-stage Andersen cascade impactor. Each section of the impactor apparatus was then analyzed for drug substance by high-pressure liquid chromatography. The study demonstrated that over 200 µg of active agent could be delivered in respirable droplets (3–5-µm MMAD) in only 2 s, the approximate time needed for one inhalation. Thus, because the drug particles were much smaller than the aqueous droplets produced by the nebulizer, essentially every aqueous droplet could contain active ingredient, and the delivery efficiency was substantially improved. Most of the drug substance was contained in droplets of aerodynamic diameters ranging from 2.1 to 5.8 µm, which is ideal for topical delivery of drugs such as corticosteroids to the conducting airways of the lung (Fig. 15).

In cases where the drug is intended to be systemically active, it must be delivered to the deep lung. It is generally accepted that aerosolized aqueous droplets of about 1–3 µm in size are ideal for deep lung delivery. Thus, a nebulizer that is capable of generating very small droplets is needed for this approach. Aerosolization of suspensions of micronized drug would not be expected to be very efficient in such systems because the average particle size of the drug substance would likely be larger than the size of the aerosolized droplets. However, drug particles in the nanometer size range do not have this constraint as has been demonstrated by aerosolization of a nanoparticulate formulation of beclomethasone dipropionate with a suitable jet nebulizer.[43] A NanoCrystal colloidal dispersion of the drug (2 mg/mL) with an average particle size of 139 nm was aerosolized with a Circulaire™ Aerosol Drug

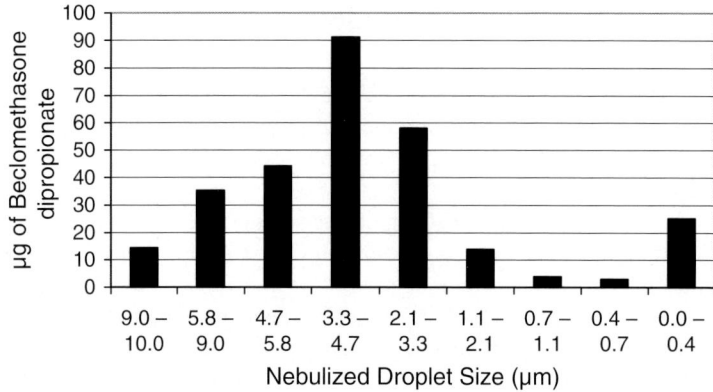

Fig. 15 Cascade impactor data for a 10% w/w NanoCrystal® beclomethasone dipropionate colloidal dispersion nebulized for 2 s from an Omron MicroAir ultrasonic nebulizer (average of two runs)

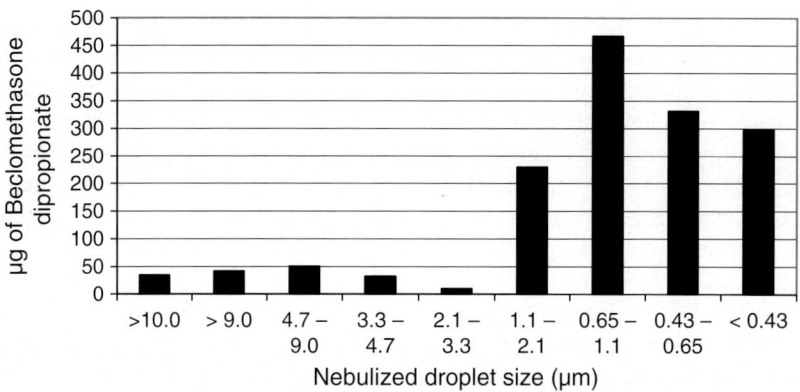

Fig. 16 Cascade impactor data for a 2 mg/mL NanoCrystal® beclomethasone dipropionate colloidal dispersion nebulized by a Circulaire™ Aerosol Drug Delivery System compressed air nebulizer (average of two runs)

Delivery System jet nebulizer (Westmed, Inc.), which is reported to generate an average aqueous droplet size of 0.95 µm. The nebulizer was run using a 4-s on, 4-s off cycle for approximately 4 min. The nebulized droplets were collected on the plates of an eight-stage Andersen cascade impactor and analyzed for beclomethasone content by high-pressure liquid chromatography. Results of the analyses (average of two runs) are presented in Fig. 16. Nearly 90% of the emitted dose (ex-device) was contained in aqueous droplets 2.1 µm in diameter or less, consistent with the reported average droplet size generated by the Circulaire device.

The results of the two studies described earlier illustrate a key advantage to the use of fine particle technology for pulmonary delivery of poorly water-soluble drugs. Nanoparticulate colloidal dispersions of poorly soluble drugs can be aerosolized into droplets suitable for deposition in the conducting airways of the lung (3–5 µm) or for deep lung delivery and systemic activity (1–3 µm). Thus, in vivo deposition of the active agent depends only on the choice of nebulizer and its associated aerosol properties, and is essentially independent of the formulation itself. In contrast, the in-vivo drug deposition achieved with other formulation and delivery approaches such as dry powder inhalers and pressurized metered-dose inhalers is highly formulation-dependent and often much more difficult to control.

Summary

Drugs administered by oral inhalation may be intended for local or systemic delivery. In either case, the aerodynamic particle size of the inhaled drug will determine its in-vivo site of deposition. Drug particles that are greater than about 1 µm but less than about 3 µm will be delivered to the alveoli, where they may enter systemic circulation. Larger particles (3–5 µm) will impact the conducting airways; hence, particles of this size are ideal for treating local respiratory conditions. Particles larger than 5 µm will impact the mouth and throat, where they may be systemically absorbed, metabolized, or otherwise inactivated. Thus, fine particle technology is an essential component of pulmonary drug delivery systems. Nanoparticulate drug formulations offer additional advantages, as spray-dried nanoparticle dispersions are easily deaggregated, while aqueous colloidal dispersions of drug can be selectively targeted to appropriate regions of the lung according to the choice of nebulizer and its aerosol properties.

Conclusions

Fine particle technology has broad applications in medicine for both therapeutics and diagnostics. The most common use of finely dispersed systems occurs in oral products that contain poorly water-soluble active agents. Fine (micro or nano) particles have increased surface area per unit dose relative to coarse active agent particles. As a result, the fine particles have much more favorable dissolution kinetics in-vivo. This can lead to increases in oral bioavailability, reductions in food-effect-associated variability, and more rapid absorption and onset of therapeutic action. Topical delivery of active agent particles can also be enhanced when they are formulated as finely dispersed systems.

Many of the formulation and delivery challenges associated with parenteral administration of poorly water-soluble drugs can be overcome by converting the active agent particles into finely dispersed systems. Products that are safe for intravenous administration can be prepared without the use of toxic solvents, solubilizing agents, or pH extremes, and nanoparticle colloidal dispersions can permit very high doses of active agent to be administered in very small injection volumes.

Lastly, fine particle technology is critical to the efficient delivery of active agents to the lung. In order for orally inhaled particles to reach the lung, it is essential that their aerodynamic particle size be in the appropriate range of ca. 1–5 μm. Depending on the particle size of the inhaled particles, they may deposit on the smooth muscle of the conducting airways or even reach the deep lung, where systemic absorption is possible.

The fraction of new drug candidate molecules that have poor water solubility has been reported to exceed 40%.[44] During the past decade, the importance of fine particle technology to the pharmaceutical industry has become clear as evidenced by the introduction of several new products that incorporate novel microparticle and nanoparticle technologies. With the growing number of poorly soluble drug candidates in discovery and development, there is no question that fine particle technology will play an increasingly important role in the pharmaceutical product development process in years to come.

References

1. R.J. Lantz, Jr., in *Pharmaceutical Dosage Forms: Tablets, Volume 2*, Marcel Dekker, New York (1990) 117.
2. R.H. Müller, C. Jacobs, O. Kayser, *Adv. Drug Deliv. Rev.* **47** (2001) 3.
3. H.W. Bosch, D.M. Marcera, R.L. Mueller, J.R. Swanson, D.S. Mishra, U.S. Patent No. 5,510,118 (1996).
4. J.R. Swanson, H.W. Bosch, K.J. Illig, D.M. Marcera, R.L. Mueller, U.S. Patent No. 5,543,133 (1996).
5. R.J. Lantz, Jr., in *Pharmaceutical Dosage Forms: Tablets, Volume 2*, Marcel Dekker, New York (1990) 123.
6. G.G. Liversidge, K.C. Cundy, J.F. Bishop, D.A. Czekai, U.S. Patent No. 5,145,684 (1992).
7. J.A. Bruno, B.D. Doty, E. Gustow, K.J. Illig, N. Rajagopalan, P. Sarpotdar, U.S. Patent No. 5,518,187 (1996).
8. E. Matijević, *Chem. Mater.* **5** (1993) 412.
9. R.E. Sievers, U. Karst, U. S. Patent No. 6,095,134 (2000).
10. S.D. Škapin, E. Matijević, *J. Colloid Interface Sci.* **272** (2004) 90.
11. L. Joguet, E. Matijević, *J. Colloid Interface Sci.* **250** (2002) 503.
12. F. Ruch, E. Matijević, *J. Colloid Interface Sci.* **229** (2000) 207.
13. H.W. Bosch, S. Škapin, E. Matijević, *Colloids Surf. A Physicochem. Eng. Aspects* **250** (2004) 43.
14. Fosamax® (alendronate sodium tablets), Merck, in *Physicians' Desk Reference*®, 58 Edition, Thomson PDR, Montvale, NJ (2004) 1990.
15. A. Noyes, W.R. Whitney, *J. Am. Chem. Soc.* **19** (1897) 930.
16. J.D. Pruitt, N.P. Ryde, American Association of Pharmaceutical Scientists, Annual Meeting and Exposition, Baltimore, MD (2004).
17. L.F. Prescott, R.F. Steel, W.R. Ferrier, *Clin. Pharmacol. Ther.* **11** (1970) 496.
18. G.G. Liversidge, K.C. Cundy, *Int. J. Pharm* **125** (1995) 91.
19. Megace ES® (megestrol acetate), Par Pharmaceutical, in *PDR*® *Electronic Library*, Thomson Micromedex, Greenwood Village, CO (2006).
20. Tricor® (fenofibrate tablets), Abbott Laboratories, in *Physicians' Desk Reference*®, 58 Edition, Thomson PDR, Montvale, NJ (2004) 522.
21. U. S. Food and Drug Administration, Center for Drug Evaluation and Research, NDA 019304, supplement no. 001, action date 02/09/1998.

22. U. S. Food and Drug Administration, Center for Drug Evaluation and Research, NDA 021203, supplement no. 000, action date 09/04/2001.
23. U. S. Food and Drug Administration, Center for Drug Evaluation and Research, NDA 021656, supplement no. 000, action date 11/05/2004.
24. U. S. Food and Drug Administration, Center for Drug Evaluation and Research, NDA 021695, supplement no. 000, action date 11/30/2004.
25. U. S. Food and Drug Administration, Center for Drug Evaluation and Research, NDA 021350, supplement no. 000, action date 05/07/2005.
26. Mobic® (meloxicam), Boehringer Ingelheim Pharmaceuticals, in *Physicians' Desk Reference*®, *58 Edition*, Thomson PDR, Montvale, NJ (2004) 1017.
27. J.D. Pruitt, American Association of Pharmaceutical Scientists, Annual Meeting and Exposition, Salt Lake City, UT (2003).
28. H.W. Bosch, E.R. Cooper, S.L. McGurk, U. S. Patent No. 6,428,814 (2002).
29. S.B. Ruddy, W.M. Eickhoff, G. Liversidge, M.E. Roberts, U. S. Patent No. 5,593,657 (1997).
30. The United States Pharmacopeia (USP 26/NF 21), The United States Pharmacopeial Convention, Rockville, MD (2003) 1471.
31. Taxotere® (docetaxel), Sanofi-Aventis, in *PDR*® *Electronic Library*, Thomson Micromedex, Greenwood Village, CO (2006).
32. E. Merisko-Liversidge, G.G. Liversidge, E.R. Cooper, *Eur. J. Pharm. Sci.* **18** (2003) 113.
33. J.P. Donnelly, J.W. Mouton, N.M.A. Blijlevens, A. Smiets, P.E. Verweij, B.E. Depauw, American Society for Microbiology, 41st Interscience Conference on Antimicrobial Agents and Chemotherapy, Chicago, IL (2001).
34. L.R. Cantilena, Jr., unpublished results.
35. G.G. Liversidge, E.R. Cooper, J.M. Shaw, G.L. McIntire, U. S. Patent No. 5,318,767 (1994).
36. G. Wolf, U.S. Patent No. 5,496,536 (1996).
37. A.J. Hickey, in *Modern Pharmaceutics*, G.S. Banker, C.T. Rhodes (eds.). Marcel Dekker, New York (2002) 480.
38. Pulmicort Turbuhaler 200 mcg (budesonide inhalation powder), AstraZeneca LP, in *Physicians' Desk Reference*®, *58 Edition*, Thomson PDR, Montvale, NJ (2004) 642.
39. Flovent® Rotadisk® (fluticasone propionate inhalation powder), GlaxoSmithKline, in *Physicians' Desk Reference*®, *58 Edition*, Thomson PDR, Montvale, NJ (2004) 1507.
40. D.A. Edwards, J. Hanes, G. Caponetti, J. Hrkach, A. Ben-Jebria, M.L. Eskew, J. Mintzes, D. Deaver, N. Lotan, R. Langer, *Science* **276** (1997) 1868.
41. H.W. Bosch, American Association of Pharmaceutical Scientists, Annual Meeting and Exposition, New Orleans, LA (1999).
42. K.D. Ostrander, H.W. Bosch, D.M. Marcera, *Eur. J. Pharm. Biopharm.* **48** (1999) 207.
43. H.W. Bosch, D.M. Bondanza, R.A. Caspar, D.L. Gardner, F.A. Leith, International Society for Aerosols in Medicine Focus Symposium, Tours, France (1996).
44. C. Lipinski, *Am. Pharm. Rev.* **5** (2002) 82.

Transport, Deposition, and Removal of Fine Particles: Biomedical Applications

Goodarz Ahmadi and John B. McLaughlin

Introduction

In many biomedical, environmental, and industrial applications, small particle transport, deposition, and resuspension play a critical role. Pollutant transport and deposition in the respiratory passages, cardiovascular flows, pollutant transport in buildings and in cities, fluidized bed combustors, and fuel spray in internal combustion engines are but a few examples. Understanding motions of small particles suspended in a gas or liquid has received considerable attention in the past few decades due to its significance in numerous scientific and industrial applications.

Natural and man-made aerosols and colloids consist of a variety of solid and liquid particles suspended in a gas or liquid. Understanding the kinetics of particle dispersion and deposition in different passages has attracted considerable attention due to its importance in numerous industrial processes.[63] Reviews of the earlier experimental and modeling works on aerosols and particle transport and deposition processes were provided by Levich,[171] Fuchs,[109] Mercer,[198] Twomey,[267] Hinds,[134] Spurny,[250] Seinfeld,[241] and Vincent,[269] among others.

Extensive reviews of turbulent diffusion were provided by Levich[171] and Hinze.[135] Tchen[260] was the first investigator who modified the Basset-Boussinesq-Oseen (BBO) equation and applied it to study motions of small particles in a turbulent flow. Corrsin and Lumley[65] pointed out some inconsistencies of Tchen's modifications. Csanady[69] showed that the inertia effect on particle dispersion in the atmosphere is negligible, but the crossing trajectory effect is appreciable. Ahmadi[1] and Ahmadi and Goldschmidt[2] studied the effect of the Basset term on the particle diffusivity. Maxey and Riley[187] obtained a corrected version of the BBO equation, which includes the Faxen[102] correction for unsteady spatially varying Stokes flows.

Experimental studies of diffusion and dispersion of particles in turbulent flows were carried out by a number of researchers. A detailed experimental study on turbulent diffusivity was reported by Snyder and Lumley.[246] Calabrese and Middleman[33] and, Arnason[12] measured the dispersion of particles in a fully developed turbulent pipe flow. Extensive collections of data for the variation of relative mass diffusivity with particle relaxation time were presented by Goldschmidt et al.[115] and Arnason.[12] Experimental studies of the flow structure and dispersion from particle sources upwind of hills and near complex terrain were performed by Snyder and Britter.[245]

Computer simulation techniques were used by Ahmadi[1] and Riley and Patterson[225] for analyzing dispersion of small suspended particles in turbulent flows. Further progress was reported by Reeks,[222] Reeks and Mckee,[223] Rizk and Elghobashi,[226] and Maxey.[186] Wang and Stock[271,272] studied the dispersion of heavy particles under similar conditions. Direct simulation of particle dispersion in a decaying isotropic turbulence was performed by Elghobashi and Trusdell.[90] An extensive review of earlier works on turbulent diffusion of finite size particles was provided by Hinze.[135] Diffusion of fluid particles in simple shear flows was analyzed by Riley and Corrsin.[224] Tavoularis and Corrsin[259] used analytical and experimental methods for evaluating the heat diffusivity tensor in a turbulent simple shear flow with a constant mean temperature gradient field. These studies showed that the diffusivity tensor in a turbulent shear flow field is not only nondiagonal, but it is also nonsymmetric. Rouhiainen and Stachiewicz[227] and Lee and Durst[168] discussed the effect of the lift force. Ounis and Ahmadi[206,207,209] provided simulations and theoretical models for diffusivity of particles in isotropic and uniformly sheared turbulent flows under microgravity condition. Accordingly, the shear-induced lift force could increase the particle dispersion rate across a shear field.

Friedlander and Johnstone[107] developed a theory for particle deposition in turbulent flows using the concept of "free flight." Additional works in this direction were reported by Davies,[74] Sehmel,[240] Lane and Stukel,[164] Yaglom and Kader,[280] Wood,[278,279] and Fernandez de la Mora and Friedlander,[103] among others. Cleaver and Yates[55] suggested an inertial deposition mechanism during the "turbulent burst" as the key deposition mechanism. Progress along this line was reported by Fichman et al.,[104] and Fan and Ahmadi.[97–101] Accordingly, a sublayer-based model for the neutral and charged particle deposition rates on smooth and rough surfaces in turbulent flows was developed. Sample model predictions are reproduced in Figs. 1 and 2, which show the deposition velocity versus particle relaxation time in wall units. Particle diameter is also shown in the figure for clarity. Here E is the imposed electric field. Figures 1 and 2 shows good agreement with the experimental data and the computer simulations of Li and Ahmadi[173] and He and Ahmadi,[130] and semiempirical model predictions.

Liu and Agarwal[178] performed an experimental study on deposition of aerosol particles in turbulent pipe flows. McCoy and Hanratty,[191] Wood,[278] and Papavergos and Hedley[214] reported several collections of available data on wall deposition rates. Kvasnak et al.[161,163] reported their experimental data for the deposition rate of glass beads, various dust components, and glass fibers in a horizontal duct flow. Wood,[278,279] Hidy,[133] and Papavergos and Hedley[214] reviewed the available methods for evaluating the deposition velocity in turbulent duct flows and discussed different deposition mechanisms.

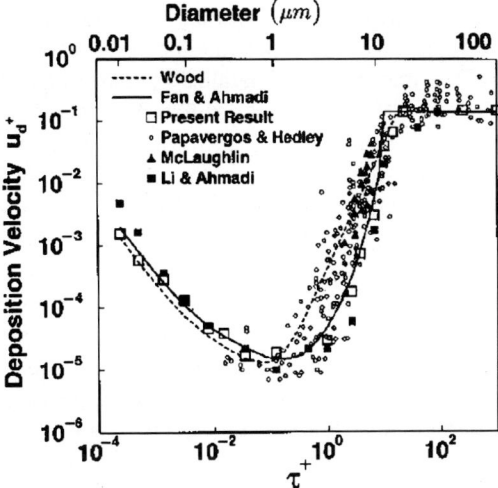

Fig. 1 Comparison of computer simulations of He and Ahmadi[130] for deposition of spherical particles in duct flows with experimental data collected by Papavergos and Hedley[214] and earlier simulation as well as model predictions of Fan and Ahmadi[97,98] and Wood[278,279]

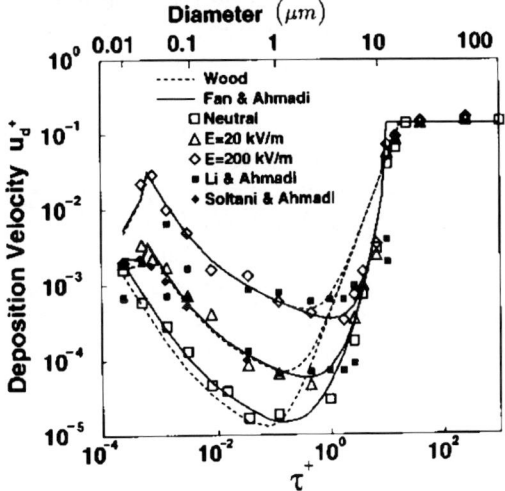

Fig. 2 Comparison of computer simulations of He and Ahmadi[130] for deposition of charged particles in duct flows with earlier simulation and model predictions of Fan and Ahmadi[97,98] and Wood[278]

Computer simulation of particle dispersion near surfaces in turbulent flows has attracted considerable attention in recent years. Using the direct simulation of turbulent flows, McLaughlin,[192] Ounis et al.,[210–212] Soltani et al.,[248,249] and Zhang and Ahmadi[284] performed a number of numerical simulations concerning deposition of particles in turbulent channel flows. The direct simulation methodology, while being exact, is a computationally intensive procedure. Li and Ahmadi[172,173] developed an approximate method for generating turbulent flow fields in complex geometry regions, and analyzed the dispersion and wall deposition of particles in turbulent air streams. The method was generalized to cover particle deposition rate on rough surfaces by Li and Ahmadi.[174] In addition, Li et al.[176] made use of an advanced (rate-dependent and thermodynamically consistent) turbulence model, and analyzed particle deposition rates in complex turbulent flows, in the presence and absence of gravity. Recently, He and Ahmadi[129,130] reported the results of their computer simulation for deposition of particle with thermophoresis and electrophoresis. Earlier Crowe,[67,68] Jurewicz and Stock,[143] and Drust et al.[87] developed computer models for analyzing two-phase gas-particle flows. Several computational two-fluid (Eulerian) models for turbulent two-phase flows were developed in the literature. Good examples were reported in the work of Elghobashi and Abou-Arab[89] and Chen and Wood.[43] Recently, Cao and Ahmadi[35–37] described a computational model for turbulent two-phase and granular flows that accounts for the phasic fluctuation energies and its interactions.

In the past decade, transport and deposition of nonspherical aerosol particles in laminar flows have received considerable attention. Gallily and his coworkers (Gallily and Eisner[111] and Krushkal and Gallily[157]) conducted a series of theoretical and experimental studies on motions of ellipsoidal particles. Asgharian and Yu,[14,15] and Asgharian and Ahmadi[13] studied the deposition of fibers in human lungs. Schamberger et al.[237] and Gradon et al.[121] considered the deposition of fibrous particles on a filter element. Recently, Fan and Ahmadi[99–101] analyzed the dispersion and deposition of ellipsoidal particles in turbulent flows. Experimental data for nonspherical particle deposition was reported by Kvasnak et al.,[163] Shapiro and Goldenberg,[242] and Kvasnak and Ahmadi.[161]

Direct numerical simulation studies of particle deposition rate in turbulent channel flows were performed by McLaughlin[192] and Ounis et al.,[211,212] Brooke et al.,[29] Squires and Eaton,[251] Soltani and Ahmadi,[247,249] and Soltani et al.[248] The effect of lift on particle deposition was studied by Wang et al.[273] A review of earlier works on direct numerical simulation was provided by McLaughlin.[196] A sample instantaneous velocity vector field in a section across a turbulent channel flow is shown in Fig. 3. The main flow is in the direction normal to the section shown. This figure shows that the velocity vary randomly across the section, and that there are certain structures, particularly, near the walls. Li et al.,[173,176] Ahmadi and Smith,[4] Ahmadi and Chen,[5] and He and Ahmadi[129] used anisotropic rate-dependent turbulence models for particle transport and dispersion analysis in complex regions. Fan and Ahmadi,[98–101] Kvasnak and Ahmadi,[162] Soltani and Ahmadi,[249] and Zhang et al.[285] performed simulations of elongated particles and fibers transport and deposition in turbulent flows. Comparison of DNS simulation of Zhang et al.[285] with the earlier experimental data and model predictions is shown in Fig. 4. It is seen that the DNS simulation is in good agreement with the experimental data and earlier model predictions.

Studies of aerosol transport and deposition under microgravity conditions including the g-jitter excitation are relatively scarce. Recently, Rogers and coworkers (Groszmann et al.[123]) considered the effect of inertia and gravity on particle

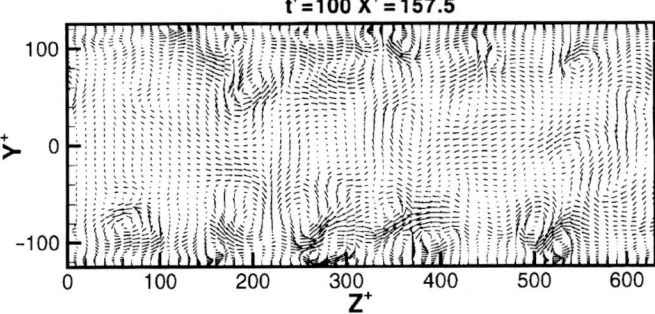

Fig. 3 Sample instantaneous velocity vector field across a duct as predicted by the direct numerical simulations (DNS)

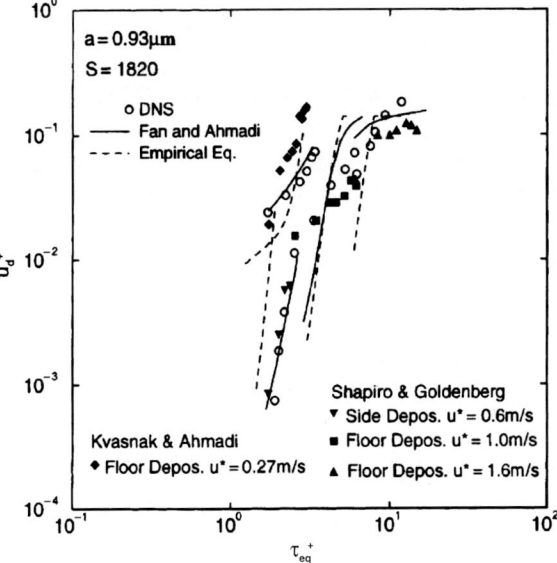

Fig. 4 Comparison of direct numerical simulations of Zhang et al.[285] for deposition of ellipsoidal particles in duct flows with earlier experimental data and semiempirical models

dispersion. Marshall[185] reported his study of the effect of charge on particulate clouds. Eaton[88] studied the attenuation of gas turbulence by fine particles, and Druzhinin and Elghobashi[84–86] and Ahmed and Elghobashi[6] reported the effect of gravity on sheared turbulence laden with bubbles or droplets. Trinh[266] described the acoustic streaming process in microgravity. Collins[61] studied the effect of particles on preferential particle concentration in an isotropic turbulence. G-jitter excitation was studied by Ellison et al.[91–93] and by Drolet and Vinals.[83]

Use of a virtual impactor and an electrical classifier for generating narrow size fibers was discussed by Chen et al.[44] and more recently by Baron et al.,[22,24,25] Baron,[23] and Wang et al.[274]

In this chapter, the fundamentals of particle transport, deposition, and removal are described, and sample computational results are presented and discussed. This is followed by a discussion of some biomedical and environmental applications.

Fundamentals of Particle Transport and Deposition Processes

Definition: Aerosols are suspensions of solid or liquid particles in a gas. Dust, smoke, mists, fog, haze, and smog are various forms of common aerosols. Colloids are suspensions of solid particles in a liquid. Colloidal particles are typically in the range of nanometers to few microns. The rheological properties of colloidal suspensions are strongly affected by the double-layer forces. Emulsions are suspensions of droplets of liquids in another immiscible liquid.

The range of diameters of common aerosol particles is between 0.01 and 40 μm. The lower limit of 10 nm roughly corresponds to the transition from molecule to particle. Particles larger than 40 μm normally do not remain suspended in air for a sufficient amount of time. Noting that the mean free path for air is about 0.07 μm and visible light has a wavelength band of 0.4–0.7 μm, the mechanical and optical behaviors of particles are significantly affected by their size. Particles greater than 5–10 μm are usually removed by the upper respiratory system. Particles smaller than 5 μm, however, can penetrate deep into the lung and become a health hazard. The important relevant dimensionless groups relevant to the motion of aerosols are listed in Table 1.

Note that the kinematic viscosity, speed of sound, and gas mean free path are related. That is,

$$v = 0.5 c^f \lambda. \tag{1}$$

The mean free path of the gas is given as

$$\lambda = \frac{1}{\sqrt{2}\pi n d_m^2} = \frac{kT}{\sqrt{2}\pi d_m^2 P}. \tag{2}$$

Table 1 Dimensionless groups

Knudsen number	$Kn = \dfrac{2\lambda}{d}$
Mach number	$M = \dfrac{|v^p - v^f|}{c^f}$
Schmidt number	$Sc = \dfrac{v}{D} = \dfrac{n^f \lambda d^2}{4}$
Reynolds number	$Re = \dfrac{|v^p - v^f| d}{v} = \dfrac{4M}{Kn}$

λ, Mean free path; d, Particle diameter; v^p, Particle velocity; v^f, Fluid (air) velocity; v, Kinematic viscosity; D, Diffusivity; c^f, Speed of sound; n, Number density; f, fluid; p, particle

Here, n is the gas number density, d_m is the gas molecule (collisional) diameter, $k = 1.38 \times 10^{-23}$ J/K is the Boltzmann constant, P is pressure, and T is temperature. For air, $d_m = 0.361$ nm and

$$\lambda\,(\mu m) = \frac{23.1T}{P}, \; P \text{ is in Pa, and } T \text{ in K}. \tag{3}$$

Hydrodynamic Forces

Drag Force and Drag Coefficient

A particle suspended in a fluid is subjected to hydrodynamic forces. For low Reynolds' number, the Stokes drag force on a spherical particle is given by

$$F_D = 3\pi\mu U d, \tag{4}$$

where d is the particle diameter, μ is the coefficient of viscosity, and U is the relative velocity of the fluid with respect to the particle. Equation (4) may be restated as

$$C_D = \frac{F_D}{\frac{1}{2}\rho U^2 A} = \frac{24}{Re}. \tag{5}$$

In (5), ρ is the fluid (air) density, $A = \pi d^2/4$ is the cross-sectional area of the spherical particle, and

$$Re = \frac{\rho U d}{\mu} \tag{6}$$

is the Reynolds number. The Stokes drag is applicable to the creeping flow regime (Stokes regime) with small Reynolds numbers ($Re < 0.5$). At higher Reynolds numbers, the flow and the drag coefficient deviates from (5). Oseen included the inertial effect approximately and developed a correction to the Stokes drag given as

$$C_D = \frac{24[1 + 3\,Re/16]}{Re}. \tag{7}$$

For $1 < Re < 1{,}000$, which is referred to as the transition regime, the following expressions may be used:[56]

$$C_D = \frac{24[1 + 0.15\,Re^{0.687}]}{Re}. \tag{8}$$

For $10^3 < Re < 2.5 \times 10^5$, the drag coefficient is roughly constant ($C_D = 0.4$). This regime is referred to as the Newton regime. At $Re \approx 2.5 \times 10^5$, the drag coefficient decreases sharply due to the transient from laminar to turbulent boundary layer around the sphere. That causes the separation point to shift downstream.

Predictions of various models for drag coefficient with the trend of the experimental data are shown in Fig. 5.

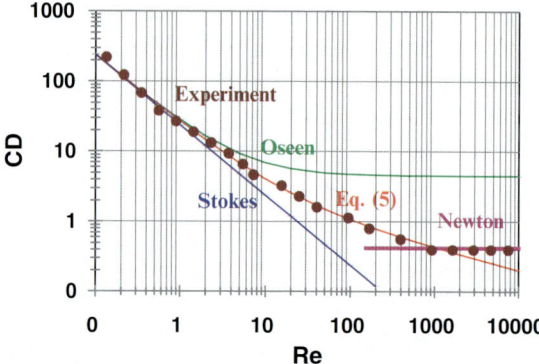

Fig. 5 Predictions of various models for drag coefficient for a spherical particle

Wall Effects on Drag Coefficient

For a particle moving near a wall, the drag force varies with distance of the particle from the surface. Brenner[27] analyzed the drag acting on a particle moving toward a wall under the creeping flow condition. To the first order, the drag coefficient is given as

$$C_D = \frac{24}{Re}\left(1 + \frac{d}{2h}\right), \tag{9}$$

where h is the distance of the particle center from the wall. For a particle moving parallel to the wall, the Stokes drag force needs to be modified. For large distances from the wall, Faxen[102] found that

$$C_D = \frac{24}{Re}\left[1 - \frac{9}{16}\left(\frac{d}{2h}\right) + \frac{1}{8}\left(\frac{d}{2h}\right)^3 - \frac{45}{256}\left(\frac{d}{2h}\right)^4 - \frac{1}{16}\left(\frac{d}{2h}\right)^5\right]^{-1}. \tag{10}$$

Cunningham Correction Factor

For very small particles, when the particle size becomes comparable with the gas mean free path, slip occurs and the expression for drag must be modified accordingly. Cunningham obtained the needed correction to the Stokes drag force:

$$F_D = \frac{3\pi\mu U d}{C_c}, \tag{11}$$

where the Cunningham correction factor C_c is given by

$$C_c = 1 + \frac{2\lambda}{d}\left[1.257 + 0.4 e^{-1.1 d/2\lambda}\right]. \tag{12}$$

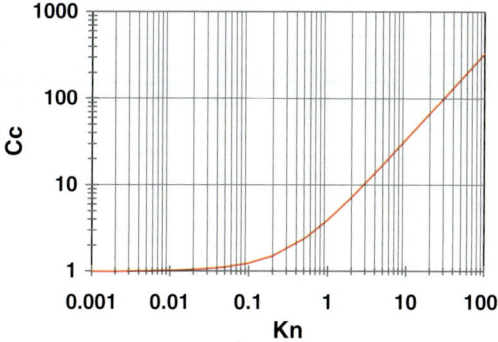

Fig. 6 Variation of Cunningham correction with Knudsen number

Table 2 Variations of C_c with d for $\lambda = 0.07\,\mu m$

Diameter (μm)	C_c
10	1.018
1	1.176
0.1	3.015
0.01	23.775
0.001	232.54

Here, λ denotes the molecular mean free path in the gas. Note that $C_c \geq 1$ for all values of d and λ. Figure 6 shows the variation of Cunningham correction factor with Knudsen number. It is seen that C_c is about 1 for $Kn < 0.1$ and increases sharply as Kn increases beyond 0.5. Table 2 illustrates the variation of Cunningham correction factor with particle diameter in air under normal pressure and temperature conditions with $\lambda = 0.07\,\mu m$. Equation (12) is applicable to a wide range of $Kn = \lambda/d \leq 1,000$ that covers slip, transition, and part of free molecular flows. The particle Reynolds number and Mach number (based on relative velocity), however, should be small.

Droplets

Drag force for liquid droplets at small Reynolds numbers is given as

$$F_D = 3\pi\mu^f U d \frac{1 + 2\mu^f/3\mu^p}{1 + \mu^f/\mu^p}, \qquad (13)$$

where the superscripts f and p refer to the continuous fluid and discrete particles (droplets, bubbles), respectively.

Table 3 Correction coefficient

Cluster shape	Correction	Cluster shape	Correction	Cluster shape	Correction
oo	$K = 1.12$	oooo	$K = 1.32$	oo oo	$K = 1.17$
ooo	$K = 1.27$	ooooo	$K = 1.45$	o o o o o	$K = 1.19$
o o o	$K = 1.16$	oooooo	$K = 1.57$	oo oo oo	$K = 1.17$
oooooo o o	$K = 1.64$	ooooooo	$K = 1.73$		

Nonspherical Particles

For nonspherical (chains or fibers) particles, Stokes' drag law must be modified, i.e.,

$$F_D = 3\pi\mu U d_e K, \tag{14}$$

where d_e is the diameter of a sphere having the same volume as the chain or fiber. That is,

$$d_e = \left(\frac{6}{\pi} \text{Volume}\right)^{1/3} \tag{15}$$

and K is a correction factor.

For a cluster of n spheres, $d_e = n^{1/3}d$. For tightly packed clusters, $k \leq 1.25$. Some other values of K are listed in Table 3.

Ellipsoidal Particles

For particles that are ellipsoids of revolution, the drag force is given by

$$F_D = 6\pi\mu U a K', \tag{16}$$

where a is the equatorial semiaxis of the ellipsoids and K' is a shape factor.

For the motion of a prolate ellipsoid along the polar axis as shown in Fig. 7a,

$$K' = \frac{\frac{4}{3}(\beta^2 - 1)}{\frac{(2\beta^2 - 1)}{(\beta^2 - 1)^{1/2}} \ln[\beta + (\beta^2 - 1)^{1/2}] - \beta}, \quad \beta = \frac{b}{a}, \tag{17}$$

where β is the ratio of the major axis b to the minor axis a.

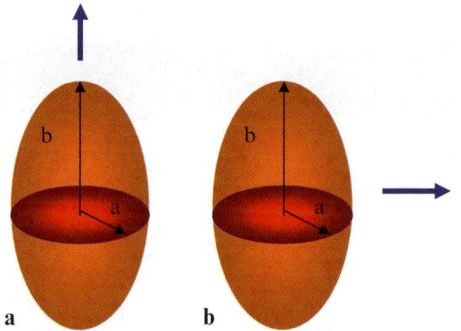

Fig. 7 Motions of prolate ellipsoids in a viscous fluid

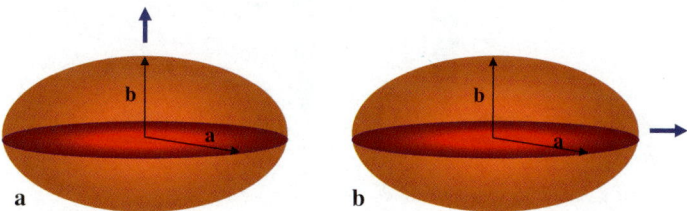

Fig. 8 Motions of oblate ellipsoids in a viscous fluid

For the motion of a prolate ellipsoid of revolution transverse to the polar axis, as shown in Fig. 7b

$$K' = \frac{\frac{8}{3}(\beta^2 - 1)}{\frac{(2\beta^2 - 3)}{(\beta^2 - 1)^{1/2}} \ln[\beta + (\beta^2 - 1)^{1/2}] + \beta}, \quad \beta = \frac{b}{a}. \tag{18}$$

Similarly for the motion of an oblate ellipsoid of revolution along the polar axis as shown in Fig. 8a,

$$K' = \frac{\frac{4}{3}(\beta^2 - 1)}{\frac{\beta(\beta^2 - 2)}{(\beta^2 - 1)^{1/2}} \tan^{-1}[(\beta^2 - 1)^{1/2}] + \beta}, \quad \beta = \frac{a}{b}. \tag{19}$$

For the motion of an oblate ellipsoid transverse to the polar axis as shown in Fig. 8b,

$$K' = \frac{\frac{8}{3}(\beta^2 - 1)}{\frac{\beta(3\beta^2 - 2)}{(\beta^2 - 1)^{1/2}} \tan^{-1}[(\beta^2 - 1)^{1/2}] - \beta}, \quad \beta = \frac{a}{b}. \tag{20}$$

By taking the limit as $\beta \to \infty$ in (16)–(20), the drag force on thin disks and needles may be obtained. These are as follows:

Thin disks of radius "a":

For motions perpendicular to the plane of the disk as shown in Fig. 9a

$$F_D = 16\mu a U. \tag{21}$$

Fig. 9 Motions of a thin disk in a viscous fluid

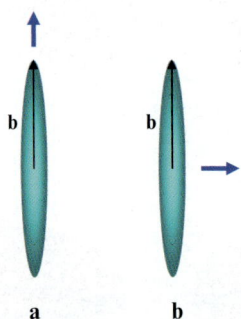

Fig. 10 Motions of a needle in a viscous fluid

For motions along the plane of the disk as shown in Fig. 9b

$$F_D = 32\mu a U/3. \tag{22}$$

Ellipsoidal needle of length $2b$:

For motions along the needle as shown in Fig. 10a

$$F_D = \frac{4\pi\mu U b}{\ln 2\beta}, \quad \beta = \frac{b}{a}. \tag{23}$$

For sideway motions of the needle as shown in Fig. 10b

$$F_D = \frac{8\pi\mu U b}{\ln 2\beta}. \tag{24}$$

Cylindrical needle:

For a cylindrical needle with a very large ratio of length to radius, moving transverse to its axis as shown in Fig. 11, the drag per unit length is given as

$$F_D = \frac{4\pi\mu U}{(2.002 - \ln R_e)}, \tag{25}$$

where $R_e = 2aU/\nu$ and a is the radius.

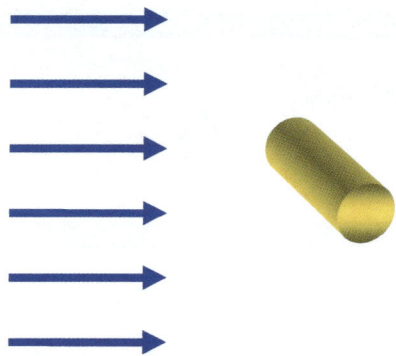

Fig. 11 Flow around a cylindrical needle

Particle Shape Factor

The ratio of the resistance of a given particle to that of a spherical particle having the same volume is called the dynamic shape factor of the particle, K. The radius of an equal volume sphere is referred to as the equivalent radius r_e. Clearly,

$$r_e = \alpha \beta^{1/3} \quad \text{for prolate spheroids,} \tag{26}$$

$$r_e = \alpha \beta^{-1/3} \quad \text{for oblate spheroids.} \tag{27}$$

Hence,

$$K = K' \beta^{1/3} \quad \text{for prolate ellipsoids,} \tag{28}$$

$$K = K' \beta^{-1/3} \quad \text{for oblate ellipsoids.} \tag{29}$$

The Stokes (sedimentation radius) of a particle is the radius of a sphere with the same density, which is settling with the terminal velocity of the particle in a quiescent fluid. Values of shape factors for a number of particles are available.[133,170]

Aerosol Particle Motion

Equation of Motion

Consider an aerosol particle in fluid flow as shown in Fig. 12. The equation of motion of a spherical aerosol particle of mass m and diameter d is given as

$$m \frac{du^p}{dt} = \frac{3\pi \mu d}{C_c}(u^f - u^p) + mg. \tag{30}$$

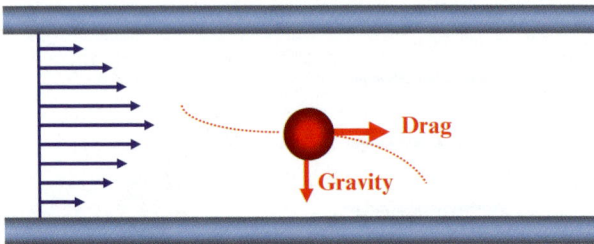

Fig. 12 Schematics of an aerosol motion in a gas flow

Here, u^p is the particle velocity, u^f is the fluid velocity, g is the acceleration due to gravity, and the buoyancy effect in air is neglected. Here, it is assumed that the particle is away from walls, and the Stokes drag is assumed.

Dividing (30) by $3\pi\mu d/C_c$ and rearranging, we find

$$\tau \frac{du^p}{dt} = (u^f - u^p) + \tau g, \qquad (31)$$

where the particle response (relaxation) time is defined as

$$\tau = \frac{mC_c}{3\pi\mu d} = \frac{d^2 \rho^p C_c}{18\mu} = \frac{Sd^2 C_c}{18\nu}, \qquad (32)$$

where $m = \pi d^3 \rho^p/6$, ν is the kinematic viscosity of the fluid, and $S = \rho^p/\rho^f$ is the density ratio. In practice, for non-Brownian particles, $C_c \approx 1$ and

$$\tau \approx \frac{d^2 \rho^p}{18\mu}. \qquad (33)$$

Terminal Velocity

For a particle starting from rest, the solution to (31) is given as

$$u^p = (u^f + \tau g)(1 - e^{-t/\tau}), \qquad (34)$$

where u^f is assumed to be a constant vector. For $u^f = 0$ and large t, the terminal velocity of particle u^t is given by

$$u^t = \tau g = \frac{\rho^p d^2 g C_c}{18\mu}. \qquad (35)$$

For different diameters, the values of relaxation time and terminal velocity of a particle with density of water are listed in Table 4. It is seen that for particles in the

Table 4 Relaxation time τ for a unit density particle in air ($p = 1$ atm, $T = 20\,^\circ\text{C}$)

Diameter (μm)	$u^t = \tau g$	v	Stop distance $u_o = 1$ m/s	Stop distance (mm) $u_o = 10$ m/s
0.05	0.39 μm/s	4×10^{-8}	0.04 m	4×10^{-4}
0.1	0.93 μm/s	9.15×10^{-8}	0.092 m	9.15×10^{-4}
0.5	10.1 μm/s	1.03×10^{-6}	1.03 m	0.0103
1	35 μm/s	3.57×10^{-6}	3.6 m	0.0357
5	0.77 mm/s	7.86×10^{-5}	78.6 m	0.786
10	3.03 mm/s	3.09×10^{-4}	309 m	3.09
50	7.47 cm/s	7.62×10^{-3}	7.62 mm	76.2

nanometer size ranges, the relaxation times and terminal velocities are extremely small. These values, however, increase rapidly as particle size increases.

Stopping Distance

In the absence of gravity and fluid flow, for a particle with an initial velocity of u_0^p, the solution to (31) is given by

$$x^p = u_0^p \tau (1 - e^{-t/\tau}), \qquad (36)$$

$$u^p = u_0 e^{-t/\tau}, \qquad (37)$$

where x^p is the position of the particle. As $t \to \infty$, $u^p \to 0$ and

$$x^p = u_0^p \tau \qquad (38)$$

is known as the stopping distance of the particle for a given initial velocity. For initial velocities of 1 and 10 m/s, the values of stop distance for various particles are listed in Table 4.

Particle Path

For constant fluid velocity, integrating (34), the position of the particle is given by

$$x^p = x_0^p + u_0^p \tau (1 - e^{-t/\tau}) + (u^f + \tau g)[t - \tau(1 - e^{-t/\tau})]. \qquad (39)$$

Here, x_0^p is the initial position of the particle. For a particle starting from rest, when the fluid velocity is in x-direction and gravity is in the negative y-direction, (39) reduces to

$$x^p / u^f \tau = [t/\tau - (1 - e^{-t/\tau})], \qquad (40)$$

$$y^p / u^f \tau = -\alpha [t/\tau - (1 - e^{-t/\tau})], \qquad (41)$$

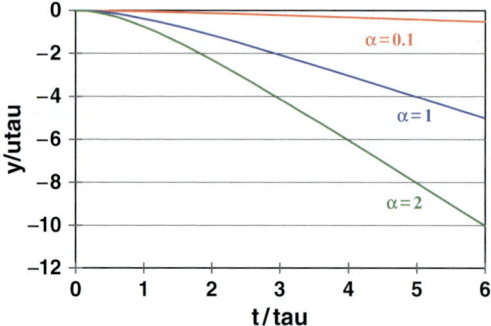

Fig. 13 Variations of the particle vertical position with time

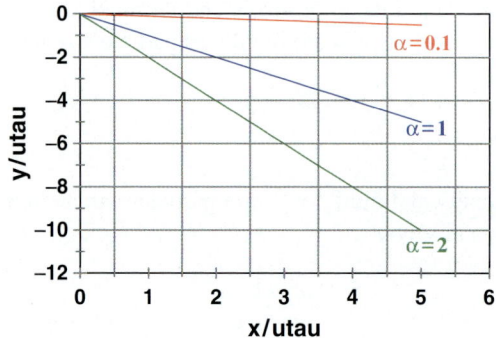

Fig. 14 Sample particle trajectories

where the ratio of the terminal velocity to the fluid velocity α is given by

$$\alpha = \frac{\tau g}{u^f \tau}. \tag{42}$$

Figure 13 shows the variation of vertical position of the particle with time.
From (40) and (41), it follows that

$$y^p = -\alpha \, x^p. \tag{43}$$

That is the particle paths are straight lines. Figure 14 shows sample particle trajectories.

Buoyancy Effects

For small particles in liquids, the buoyancy effect must be included. Thus, (30) is replaced by

$$(m+m^{\rm a})\frac{du^{\rm p}}{dt} = \frac{3\pi\mu d}{C_{\rm c}}(u^{\rm f} - u^{\rm p}) + (m - m^{\rm f})g, \qquad (44)$$

where $m^{\rm f}$ is the mass of the equivalent volume of fluid given as

$$m^{\rm f} = \frac{\pi d^3 \rho^{\rm f}}{6} \qquad (45)$$

and $m^{\rm a}$ is the apparent mass with $\rho^{\rm f}$ being the fluid density. For spherical particles,

$$m^{\rm a} = \frac{1}{2} m^{\rm f}. \qquad (46)$$

Keeping the same definition for particle relaxation time as given by (32), (31) may be restated as

$$\left(1 + \frac{1}{2S}\right)\tau \frac{du^{\rm p}}{dt} = (u^{\rm f} - u^{\rm p}) + \tau g\left(1 - \frac{1}{S}\right). \qquad (47)$$

The expression for the terminal velocity then becomes

$$u^{\rm t} = \tau g\left(1 - \frac{1}{S}\right) = \frac{\rho^{\rm p} d^2 g C_{\rm c}}{18\mu}\left(1 - \frac{\rho^{\rm f}}{\rho^{\rm p}}\right). \qquad (48)$$

Note that the Basset force and the memory effects are neglected in this analysis.

Lift Force

Small particles in a shear field as shown in Fig. 15 experience a lift force perpendicular to the direction of flow. The shear lift originates from the inertia effects in the viscous flow around the particle and is fundamentally different from aerodynamic lift force. The expression for the inertia shear lift was first obtained by Saffman.[235,236] That is,

$$F_{\rm L(Saff)} = 1.615 \, \rho v^{1/2} \, d^2 (u^{\rm f} - u^{\rm p}) \left|\frac{du^{\rm f}}{dy}\right|^{1/2} {\rm sgn}\left(\frac{du^{\rm f}}{dy}\right). \qquad (49)$$

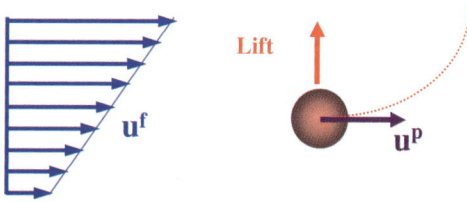

Fig. 15 Schematics of a particle in a shear flow

Here, u^f is the fluid velocity at the location of mass center of the particle, u^p is the particle velocity, $\dot{\gamma} = du^f/dt$ is the shear rate, d is the particle diameter, and ρ and ν are the fluid density and viscosity. Note that F_L is in the positive y-direction if $u^f > u^p$.

Equation (49) is subjected to the following constraints:

$$R_{es} = \frac{|u^f - u^p|d}{\nu} \ll 1, \quad R_{e\Omega} = \frac{\Omega d^2}{\nu} \ll 1 \tag{50}$$

$$R_{eG} = \frac{\dot{\gamma} d^2}{\nu} \ll 1, \quad \varepsilon = \frac{R_{eG}^{1/2}}{R_{es}} \gg 1. \tag{51}$$

Here, Ω is the rotational speed of the sphere. Dandy and Dwyer[70] found that the Saffman lift force is approximately valid at larger R_{es} and small ε. McLaughlin[193,194] and Cherukat and McLaughlin[52] and Cherukat et al.[53] evaluated a new expression for the shear lift force, which showed that the lift force decreases as ε decreases. On the basis of these studies, Mei[195] suggested the following empirical fit to the results of Dandy and Dwyer and McLaughlin. For large ε and R_{es},

$$\frac{F_L}{F_{L(\text{Saff})}} = \begin{cases} (1 - 0.3314\alpha^{1/2})\exp(-R_{es}/10) + 0.3314\alpha^{1/2} & \text{for } R_{es} \leq 40 \\ 0.0524(\alpha R_{es})^{1/2} & \text{for } R_{es} > 40 \end{cases}, \tag{52}$$

where

$$\alpha = \frac{\dot{\gamma} d}{2|u^f - u^p|} = \frac{R_{es}\varepsilon^2}{2} = \frac{R_{eG}}{2R_{es}}. \tag{53}$$

For $0.1 \leq \varepsilon \leq 20$

$$\frac{F_L}{F_{L(\text{Saff})}} = 0.3\{1 + \tan h[2.5 \log_{10}(\varepsilon + 0.191)]\}\{0.667 + \tan[6(\varepsilon - 0.32)]\}. \tag{54}$$

For large and small ε, McLaughlin obtained the following expressions

$$\frac{F_L}{F_{L(\text{Saff})}} = \begin{cases} 1 - 0.287\varepsilon^{-2} & \text{for } \varepsilon \gg 1 \\ -140\varepsilon^5 \ln(\varepsilon^{-2}) & \text{for } \varepsilon \ll 1 \end{cases}. \tag{55}$$

Note the change in sign of the lift force for small values of ε.

McLaughlin[194] included the effects of presence of the wall in his analysis of the lift force. The results for particles in a shear field but not too close to the wall were given in tabulated forms. Cherukat et al.[53] analyzed the lift force acting on spherical particles near a wall as shown in Fig. 16. Accordingly

$$F_{L(C-L)} = \rho \nu^2 d^2 I_L/4, \tag{56}$$

where

$$V = u^p - u^f = u^p - \dot{\gamma} l \tag{57}$$

and for nonrotating spheres,

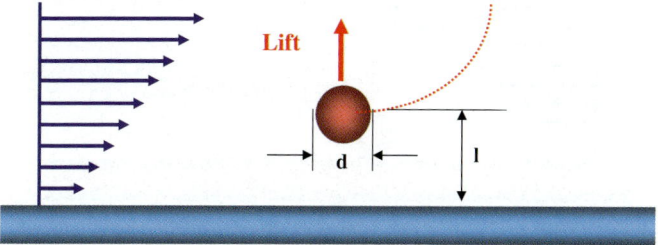

Fig. 16 Schematics of a particle near a wall in a shear flow

$$I_L = (1.7716 + 0.216K - 0.7292K^2 + 0.4854K^3)$$
$$- (3.2397/K + 1.145 + 2.084K - 0.9059K^2)\Lambda_G$$
$$+ (2.0069 + 1.0575 - 2.4007K^2 + 1.3174K^3)\Lambda_G^2. \quad (58)$$

For rotating (freely) spheres,

$$I_L = (1.7631s + 0.3561K - 1.1837K^2 + 0.845163K^2)$$
$$- (3.24139/K + 2.6760 + 0.8248K - 0.4616K^2)\Lambda_G$$
$$+ (1.8081 + 0.879585K - 1.9009K^2 + 0.98149K^3)\Lambda_G^2. \quad (59)$$

Here,

$$K = \frac{d}{2l}, \quad \Lambda_G = \frac{\dot{\gamma}d}{2v}. \quad (60)$$

Lift Force on a Particle Touching a Plane

Leighton and Acrivos[169] obtained the expression for the lift on the spherical particles resting on a plane substrate as shown in Fig. 17. They found that

$$F_{L(L-A)} = 0.576 \, \rho d^4 \, \dot{\gamma}^2, \quad (61)$$

which is always point away from the wall. Note that the Saffman expression given by (49) may be restated as

$$F_{L(Saff)} = 0.807 \, \rho v^{1/2} \, d^3 \, \dot{\gamma}^{3/2}. \quad (62)$$

Equation (66) with I_L given by (58) reduces to (61) for $K = 1$, $\Lambda_G = -1$.
For small particles in turbulent flows, using

$$u^+ = y^+, \quad u^+ = \frac{u}{u^*}, \quad y^+ = \frac{yu^*}{v}, \quad \dot{\gamma} = \frac{u^{*2}}{v}, \quad (63)$$

where u^* is the shear velocity, (61) and (62) become

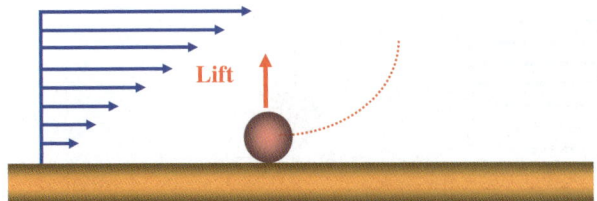

Fig. 17 Schematics of a sphere resting on a wall in a shear flow

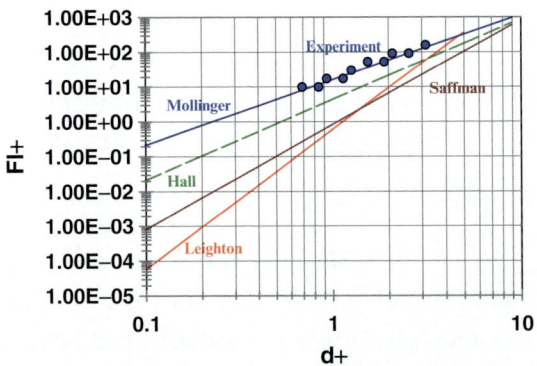

Fig. 18 Comparison of model predictions with the experimental data

$$F^+_{L(L-A)} = 0.576 d^{+4}, \tag{64}$$

$$F^+_{L(Saff)} = 0.807 d^{+3}, \tag{65}$$

where

$$F^+_L = \frac{F_L}{\rho v^2}, \quad d^+ = \frac{du^*}{v}. \tag{66}$$

Experimental studies of lift force were performed for generally larger particles in the range of 100 to several hundred micrometers. Hall[126] found that

$$F^+_{L(Hall)} = 4.21 d^{+2.31} \quad \text{for } d^+ > 1.5. \tag{67}$$

Mollinger and Nieuwstadt[200] found that

$$F^+_{L(MN)} = 15.57 d^{+1.87} \quad \text{for } 0.15 < d^+ < 1. \tag{68}$$

Figure 18 compares the model predictions with the experimental data of Hall. It is seen that the experimental data is generally much higher than the theoretical models.

Particle Equation of Motion (BBO Equation)

We have so far described drag and lift forces acting on a suspended particle. There are, however, additional hydrodynamic forces, such as Basset history, Faxen correction, and virtual mass effects that act on the particles. Some of these forces could become important especially for the particles suspended in a liquid. The general equation of motion of a small spherical particle suspended in fluid as obtained by Maxey and Riley[187] is given as

$$m_p \frac{du_i^p}{dt} = m_f \frac{Du_i^f}{Dt} - \frac{1}{2} m_f \frac{d}{dt}\left(u_i^p - u_i^f - \frac{1}{10} a^2 \nabla^2 u_i^f\right)$$
$$- 6\pi \mu a \left[(u_i^p - u_i^f) - \frac{1}{6} a^2 \nabla^2 u_i^f\right] + (m_p - m_f) g_i(t) + L_i + F_i, \quad (69)$$

where $m_p = (4\pi/3) a^3 \rho_p$, $m_F = (4\pi/3) a^3 \rho_f$, a is the radius of the spherical particle, v_i is the particle velocity, $u_i^f [x(t),t]$ is the fluid velocity at the particle location, μ is the viscosity, ν is the kinematic viscosity, ρ_p is the particle density, ρ_f is the fluid density, $g_i(t)$ is the acceleration of gravity, and $x(t)$ is the location of the particle. Saffman's[235,236] lift force L_i and other forces (Brownian, gravity, etc.) F_i are added to (69) for completeness.

Brownian Motion

When a small particle is suspended in a fluid, it is subjected to the impact gas or liquid molecules. For ultrafine (nano) particles, the instantaneous momentum imparted to the particle varies at random, which causes the particle to move on an erratic path now known as Brownian motion. Figure 19 illustrates the Brownian motion process.

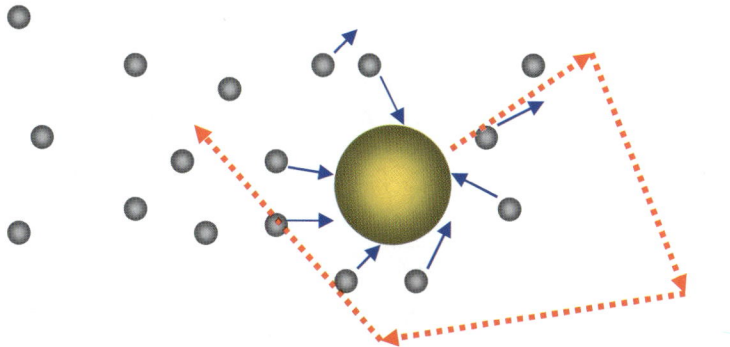

Fig. 19 Schematics of Brownian motion process

The Brownian motion of a small particle in a stationary fluid in x-direction is governed by the following Langevin equation,

$$\frac{du}{dt} + \beta u = n(t), \tag{70}$$

where u is the velocity of the particle,

$$\beta = 3\pi\mu d/C_c m = 1/\tau \tag{71}$$

and $n(t)$ is a white noise excitation due to the impact of fluid molecules on the particle. The intensity of noise is specified by its spectral intensity given as

$$S_{nn} = \frac{2kT\beta}{\pi m}, \tag{72}$$

where $k = 1.38 \times 10^{-16}$ erg/K is the Boltzmann constant and T is the temperature. It should be emphasized that the Brownian motion occurs in three dimensions, and (70) applies only to the x-component of the motion.

For the stochastic equation given by (70), using the standard linear system analysis, it follows that

$$S_{uu}(\omega) = |H(\omega)|^2 S_{nn}(\omega), \tag{73}$$

where $S_{uu}(\omega)$ is the power spectrum of the velocity of the Brownian particle, and $H(\omega)$ is the system function given by

$$H(\omega) = \frac{1}{i\omega + \beta}. \tag{74}$$

Hence,

$$S_{uu}(\omega) = \frac{2kT\beta/\pi m}{\omega^2 + \beta^2}. \tag{75}$$

The autocorrelation of the particle velocity field is defined as

$$R(\tau) = \overline{u(+\tau)u(t)}. \tag{76}$$

Here, a bar on the top of an expression implies the expected value. The autocorrelation is the inverse Fourier transform of the power spectrum function, i.e.,

$$R_{uu}(\tau) = \frac{1}{2}\int_{-\infty}^{+\infty} e^{i\omega\tau} S_{uu}(\omega) d\omega. \tag{77}$$

Hence,

$$S_{uu}(\omega) = \frac{1}{\pi}\int_{-\infty}^{+\infty} e^{-i\omega\tau} R_{uu}(\tau) d\tau. \tag{78}$$

From (75) and (77) it follows that

$$R_{uu}(\tau) = \frac{kT}{m} e^{-\beta|\tau|}. \tag{79}$$

The mass diffusivity is defined as

$$D = \frac{1}{2}\frac{d}{dt}\overline{x^2}(t) \quad \text{for large } t, \tag{80}$$

where $x(t)$ is the position of particle given by

$$x(t) = \int_0^t u(t_1)\,dt_1. \tag{81}$$

Using (81), one finds that

$$\overline{x^2(t)} = \int_0^t \int_0^t R_{uu}(\tau_1 - \tau_2)\,d\tau_1 d\tau_2. \tag{82}$$

Changing variables, after some algebra it follows that

$$\overline{x^2(t)} = 2\int_0^t (t - \tau) R_{uu}(\tau)\,d\tau. \tag{83}$$

Thus,

$$D = \int_0^\infty R_{uu}(\tau)\,d\tau. \tag{84}$$

Using expression (79) in (84), we find that

$$D = \frac{kT}{\beta m} = \frac{kTC_c}{3\pi\mu d}. \tag{85}$$

Fokker–Planck Approach

An alternative approach is to make use of the Fokker–Planck equation associated with the Langevin equation given by (70). That is,

$$\frac{\partial f}{\partial t} - \frac{\partial}{\partial u}(\beta\, uf) = \frac{kT\beta}{m}\frac{\partial^2 f}{\partial u^2}. \tag{86}$$

Here, f is the probability density of the velocity of the Brownian particle. The stationary solution to the Fokker–Planck equation given by (86) is given as

$$f = \frac{1}{\sqrt{2\pi kT/m}}\exp\left\{-\frac{mu^2}{2kT}\right\} \tag{87}$$

with $\overline{mu^2} = kT$.

Brownian Motion in a Force Field

Consider the following Langevin equation:

$$\ddot{x} + \beta \dot{x} - \frac{F(x)}{m} = n(t), \tag{88}$$

where

$$F(x) = -\frac{\partial V(x)}{\partial x} \tag{89}$$

is a conservative force field. The corresponding Fokker–Planck equation for the transition probability density function for the Brownian particle velocity and position is given as

$$\frac{\partial f}{\partial t} = -\frac{\partial (\dot{x} f)}{\partial x} + \frac{\partial}{\partial \dot{x}} \left[\left(\beta \dot{x} - \frac{1}{m} F(x) \right) f \right] + \frac{kT\beta}{m} \frac{\partial^2 f}{\partial \dot{x}^2}. \tag{90}$$

The stationary solution to (90) is given by

$$f = C_0 \exp\left\{ -\frac{m}{kT} \left[\frac{\dot{x}^2}{2} - \int_0^x \frac{F(x_1) dx}{m} \right] \right\}. \tag{91}$$

Using (89), we find that

$$f = C_0 \exp\left\{ -\frac{1}{kT} \left[\frac{m\dot{x}^2}{2} + V(x) \right] \right\}. \tag{92}$$

For a gravitational force field,

$$V(x) = mg(x - x_0) \tag{93}$$

and

$$f = C_0 \exp\left(-\frac{m\dot{x}^2}{2kT} \right) \exp\left(-\frac{mg(x - x_0)}{kT} \right). \tag{94}$$

Computer Simulation Procedure

As noted before, the Brownian force $n(t)$ may be modeled as a white noise stochastic process. White noise is a zero mean Gaussian random process with a constant power spectrum given in (72). Thus,

$$\overline{n(t)} = 0 \quad \overline{n(t_1)n(t_2)} = 2\pi S_{nn} \delta(t_1 - t_2). \tag{95}$$

The following procedure was used by Ounis et al.,[209] Ounis and Ahmadi,[209] and Li and Ahmadi,[173–175] among others.

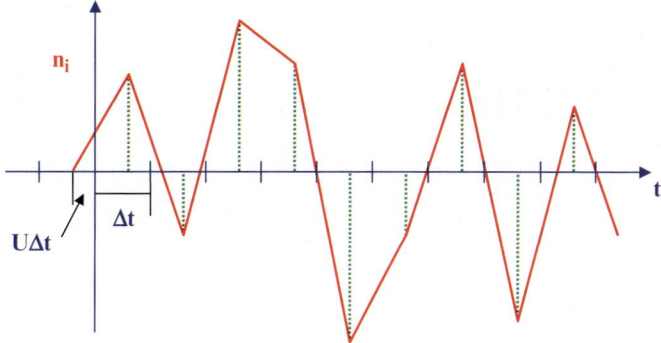

Fig. 20 Schematic of a numerically simulated Brownian excitation

- Choose a time step Δt (The time step should be much smaller than the particle relaxation time).
- Generate a sequence of uniform random numbers U_i (between 0 and 1).
- Transform pairs of uniform random numbers to pairs of unit variance zero mean Gaussian random numbers. This can be done using the following transformations:

$$G_1 = \sqrt{-2\ln U_1} \cos 2\pi U_2 \qquad (96)$$

$$G_2 = \sqrt{-2\ln U_1} \sin 2\pi U_2. \qquad (97)$$

- The amplitude of the Brownian force then is given by

$$n(t_i) = G_i \sqrt{\frac{\pi S_{nn}}{\Delta t}}. \qquad (98)$$

- The entire generated sample of Brownian force needs to be shifted by $U \Delta t$, where U is a uniform random number between zero and one.

Figure 20 shows the schematics of a numerically generated Brownian excitation.

Particle Dispersion and Deposition in a Viscous Sublayer

Ounis et al.[210,211] and Shams et al.[243] studied dispersion and deposition of nano and microparticles in turbulent boundary layer flows. A sample simulated Brownian force for a 0.01-μm particle is shown in Fig. 21. Here, the wall units with ν/u^* and ν/u^{*2} being, respectively, the length and the time scales are used. Note that the relevant scales and the wall layer including the viscous sublayer are controlled by kinematic viscosity ν and shear velocity u^*. The random nature of Brownian force is clearly seen from Fig. 21.

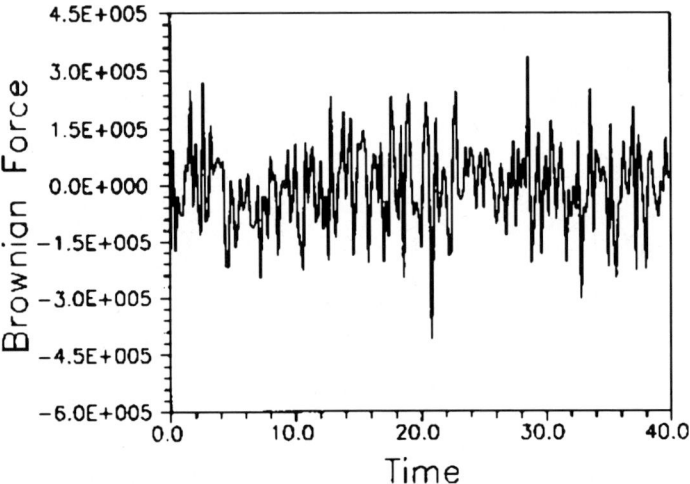

Fig. 21 Sample-simulated Brownian force

Using the definition of particle diffusivity, D, as given by (80), the variance of the particle position is given by

$$\overline{x^2}(t) = 2Dt. \tag{99}$$

Thus, for a given diffusivity, the variance of the spreading rate of particles may be evaluated from (99).

To verify the Brownian dynamic simulation procedure, Ounis et al.[209] studied that special case of a point source in a uniform flow with $U^+ = U/u^* = 1$. For different particle diameters, Fig. 22 displays the time variation of their simulated root mean square particle position. Here, for each particle size, 500 sample trajectories were evaluated, compiled, and statistically analyzed. The corresponding exact solutions given by (99) are also shown in this figure for comparison. It is seen that small nanometer-sized particles spread much faster by the action of the Brownian motion when compared with the larger micrometer-sized particles. Figure 22 also shows that the Brownian dynamic simulation results for the mean square displacement are in good agreement with the exact solutions.

Ounis et al.[209] performed a series of Lagrangian simulation studies for dispersion and deposition of particles emitted from a point source in the viscous sublayer of a turbulent near-wall flow. Figures 23 to 25 show time variation of particle trajectory statistics for different diameters, for the case that the point source is at a distance of 0.5 wall units away from the wall. In these simulations, it is assumed that when particles touch the wall they will stick to it. At every time step, the particle ordinates are statistically analyzed, and the mean, standard deviation, and the sample minimum and maximum were evaluated. The points at which the minimum curve touches the wall identify the locations of a deposited particle. Figure 23 shows that 0.05-μm particles have a narrow distribution, and in the duration of 40 wall units, none of

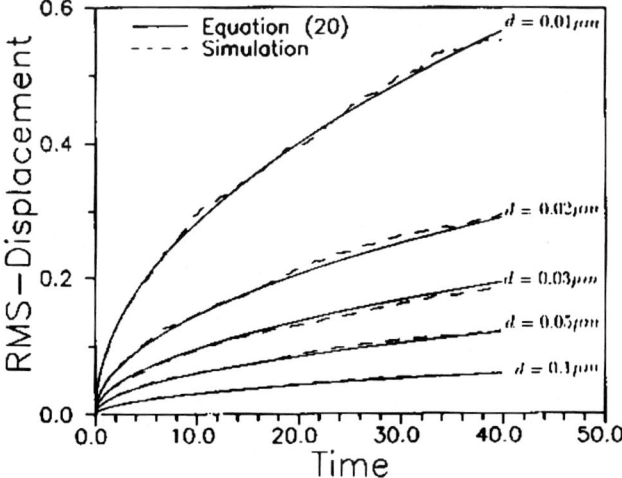

Fig. 22 Sample-simulated root-mean-square displacement for different particles

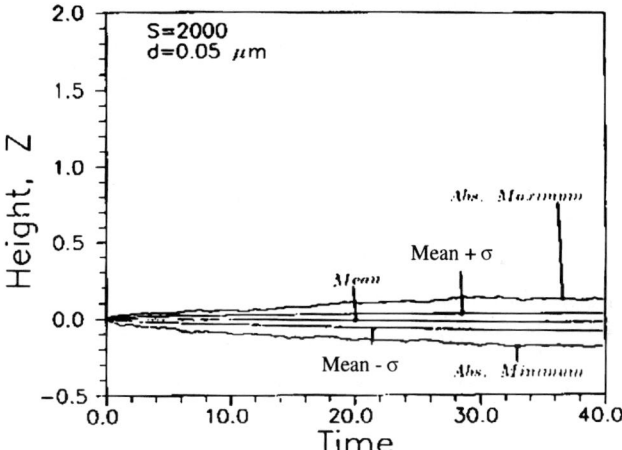

Fig. 23 Simulated trajectory statistics for 0.05 μm particles

these particles are deposited on the wall. As the particle diameter becomes smaller, their spreading due to Brownian diffusion increases and a number of particles reach the wall. For example, Fig. 7 shows that five 0.03-μm particles are deposited on the wall in the duration of 40 wall units, while Fig. 25 indicates that hundred and ninety 0.01-μm particles (out a sample of 500 particles) are deposited on the wall. Figures 23 to 25 further show that the Brownian diffusion of particles is strongly affected by their size. This is because the power spectral intensity of Brownian force in inversely proportional to the square of diameter.

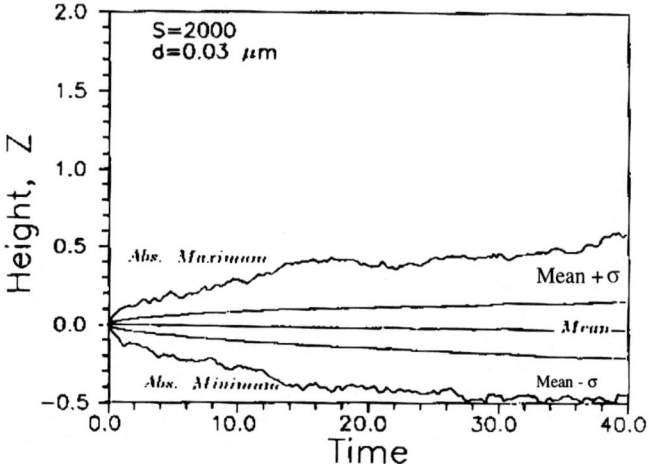

Fig. 24 Simulated trajectory statistics for 0.03 μm particles

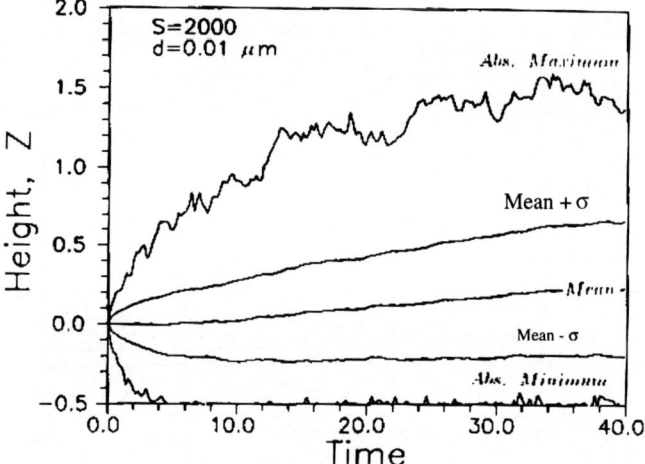

Fig. 25 Simulated trajectory statistics for 0.01 μm particles

Figure 26 shows variations of the number of deposited particles, N_t, with time for a point source at a distance of $z_o = 0.5$ wall units from the wall. The solid lines in this figure are the exact solution for a diffusion model given as

$$N_t = N_o erfc\left(\frac{z_o}{\sqrt{4Dt}}\right). \tag{100}$$

It is seen that the Brownian dynamic simulation results and the diffusion equation analysis are in good agreement for the range of particle diameters studied. Figure 26

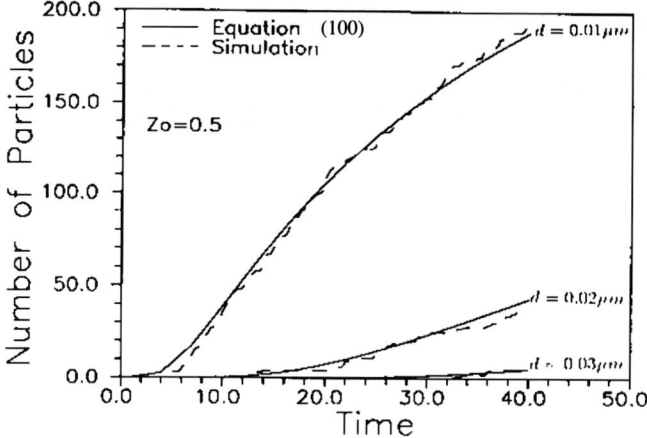

Fig. 26 Comparison of the simulated number of deposited particles with the diffusion model given by (100)

also shows that as the particle diameter decreases, the number of deposited particles increases sharply. Additional results (not shown here) indicate that the deposition rate decreases as the distance of the source from the wall increases. Figures 22 to 26 show that the Brownian motion process is a significant mechanism for nanoparticle diffusion and wall deposition.

Particle Adhesion

Small particles adhere to surfaces or to each other due to London–van der Waals, capillary, and electrostatic forces. The fundamental nature of these surface forces is described in this section.

London–van der Waals Force

The London–van der Waals force, which is generally attractive in nature, is a short-range force and decays rapidly to zero away from a surface. The origin of the London–van der Waals force lies in the instantaneous dipole generated by the fluctuation of electron cloud surrounding the nucleus of electrically neutral atoms. For a spherical particle of diameter d near a flat surface, the interaction energy is given by,

$$\phi = -\frac{A}{12}\left[\frac{1}{x} + \frac{1}{1+x} + 2\ln\frac{x}{1+x}\right], \qquad (101)$$

Table 5 Comparison of van der Waals force for 1-μm particles with the drag force

Particle	Surface	$F_v \times 10^8$ N (in air)	$\frac{F_v}{3\pi\mu dU}$	$F_v \times 10^8$ N (in water)	$\frac{F_v}{3\pi\mu dU}$
Polystyrene	Polystynene	1.2–1.8	70–100	0.2	12
Si	Si	13.6–14.4	800–850	7	410
Cu	Cu	17	1,000	9.8	580
Ag	Ag	18	1,060	15.5	910

Velocity of $U = 1$ m/s and a separation of $z_o = 4$ Å have been assumed

where $x = z/d$ and z is the distance of the sphere from the surface and A is the Hamaker constant. As the particle approaches the surface,

$$\phi \approx -\frac{Ad}{12z} \text{ as } z \to 0. \tag{102}$$

Thus, the energy becomes infinite for $z = 0$. Hence, the surface acts as a perfect sink for aerosol diffusion. The range of operation of the van der Waals force may be estimated by comparing the thermal energy with ϕ. Values of Hamaker constant A are in the range of 10^{-20} to 10^{-19} J. Thus,

$$z \leq \frac{Ad}{12kT} \simeq 0.2d \text{ for } A \simeq 10^{-20}. \tag{103}$$

In Table 5, values for van der Waals force for a number of materials are listed and the values of van der Waals force are compared with the Stokes drag force acting on a particle that is moving with a velocity of 1 m/s in air and in water. It is seen that the van der Waals force in air is comparatively larger than that in water. More importantly, van der Waals force is, generally, much larger that the drag force. The ratio of the van der Waals force to drag force in water is also less than that in air.

Values of van der Waals, surface tension, added mass, drag, and Basset forces acting on a sphere of different sizes moving with a velocity of 10 m/s are shown in Table 6. For a particle in air, it is seen that the surface tension force is order of magnitudes larger than the other forces, which is followed by the van der Waals force. Among the hydrodynamic forces, drag is the dominating force and the virtual mass and the Basset forces are negligibly small. In water, the surface tension force is absent and the drag force acting on large particles becomes comparable with the van der Waals adhesion force.

van der Waals Force Near a Surface

The van der Waals force for a sphere near a surface as shown in Fig. 27 is given by

$$F = \frac{A_{132}d}{12z_o^2}, \tag{104}$$

Table 6 Variation of forces (N) versus particle diameter, d (μm), for a flow velocity of $U_0 = 10$ m/s

Diameter	van der Waals	Surface tension	Added mass	Drag/lift	Basset
	$F_v \sim A_{123} \frac{d}{12 z_0^2}$	$F_{st} \sim 2\pi \gamma d$	$F_{am} \sim \rho d^3 \frac{dV}{dt}$	$F_D \sim \rho^f d^2 V^2$	$F_B \sim \frac{\mu d^2 V}{\sqrt{vt}}$
Air					
d(μm)	F_v	F_{st}	F_{am}	F_D	F_B
0.2	3×10^{-8}	9×10^{-5}	10^{-18}	10^{-12}	4×10^{-15}
2	3×10^{-7}	9×10^{-4}	10^{-15}	10^{-10}	4×10^{-13}
20	3×10^{-6}	9×10^{-3}	10^{-12}	10^{-8}	4×10^{-11}
Water					
d(μm)	F_v	F_{st}	F_{am}	F_D	F_B
0.2	2×10^{-9}	–	8×10^{-16}	8×10^{-10}	10^{-12}
2	2×10^{-8}	–	8×10^{-13}	8×10^{-8}	10^{-10}
20	2×10^{-7}	–	8×10^{-10}	8×10^{-6}	10^{-8}

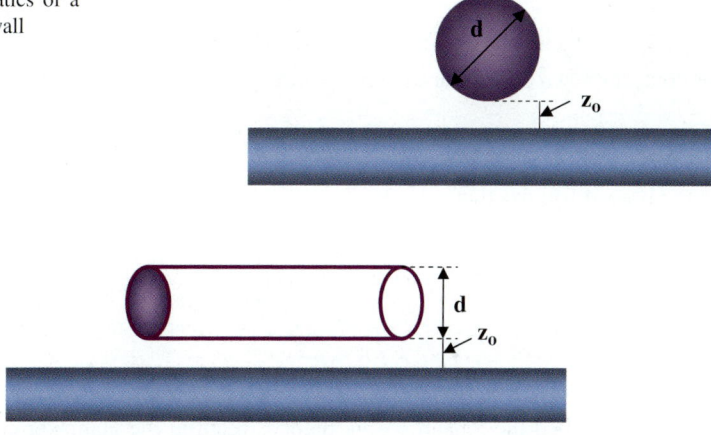

Fig. 27 Schematics of a particle near a wall

Fig. 28 Schematics of a cylindrical particle near a wall

where A_{132} is the Hamaker constant for substances "1" and "2" in presence of medium "3," and z_o is the separation distance. For a particle attached to a wall, z_o is about 4–10 Å. Typically $z_o = 4$ Å is used.

For cylinder–planar surface contacts shown in Fig. 28,

$$\frac{F}{\text{length}} = \frac{A_{132} d^{1/2}}{16 z_o^2}. \tag{105}$$

For two planar surfaces shown in Fig. 29,

$$\frac{F}{\text{area}} = \frac{A_{132}}{6\pi z_o^3}. \tag{106}$$

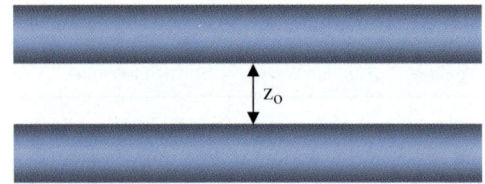

Fig. 29 Schematics of a two-planar surface at a separation distance of z_o

Hamaker Constants for Dissimilar Materials

For two dissimilar materials, the Hamaker constant may be estimated in terms of Hamaker constant of each material. That is,

$$A_{12} \approx \sqrt{A_{11}A_{22}} \tag{107}$$

or alternatively

$$A_{12} = \frac{2A_{11}A_{22}}{A_{11}+A_{22}}. \tag{108}$$

For contact of two dissimilar materials in the presence of a third media,

$$A_{132} = A_{12} + A_{33} - A_{13} - A_{23}. \tag{109}$$

From (108), it follows that

$$A_{131} = A_{11} + A_{33} - 2A_{13} = \frac{(A_{11}-A_{33})^2}{A_{11}+A_{33}} \simeq \left(\sqrt{A_{11}} - \sqrt{A_{33}}\right)^2 \tag{110}$$

or

$$A_{132} \simeq \left(\sqrt{A_{11}} - \sqrt{A_{33}}\right)\left(\sqrt{A_{22}} - \sqrt{A_{33}}\right). \tag{111}$$

Lifshitz developed the "macroscopic theory" relating the Hamaker constant to dielectric constants of the materials. Accordingly,

$$A_{132} = \frac{3}{4\pi} h\bar{\omega}_{132}. \tag{112}$$

Values of $h\bar{\omega}_{132}$ were given in tables for a number of materials.

London–van der Waals Surface Energy Between Particles

The London–van der Waals surface energy and force between two spherical particles of diameters d_1 and d_2 as shown in Fig. 30 were evaluated by Hamaker.[127] The corresponding surface energy is given as

Fig. 30 Schematics of contact of two dissimilar spheres

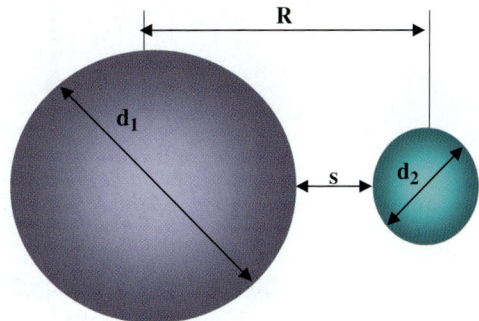

$$\Phi = -\frac{A}{6}\left[\frac{d_1 d_2/2}{R^2 - \left(\frac{d_1+d_2}{2}\right)^2} + \frac{d_1 d_2/2}{R^2 - \left(\frac{d_1-d_2}{2}\right)^2} + \ln\frac{R^2 - \left(\frac{d_1+d_2}{2}\right)^2}{R^2 - \left(\frac{d_1-d_2}{2}\right)^2}\right], \quad (113)$$

where

$$R = \frac{d_1 + d_2}{2} + s \quad (114)$$

is the distance between particles' center and s is the separation distance between surfaces.

For equal size particles, $d_1 = d_2 = d$, $r = d + s$, and

$$\Phi = -\frac{A}{6}\left[\frac{d^2}{2r^2} + \frac{d^2}{2(r^2 - d^2)} + \ln\left(1 - \frac{d^2}{r^2}\right)\right]. \quad (115)$$

As noted before, A is typically of the order of 10^{-19} to 10^{-21} J and depends on the properties of particles (of composition 1) and suspending medium (composition 2). Accordingly, the effective Hamaker constant is given by

$$A_{121} = A_{11} + A_{22} - A_{12} \approx \sqrt{(A_{11} - A_{22})}. \quad (116)$$

Particle Adhesion and Detachment Models

Figure 31 shows the schematic of a particle of diameter d attached to a flat surface. Here, P is the external force exerted on the particle, a is the contact radius, and F_{ad} is the adhesion force. The classical Hertz contact theory provides for the elastic deformation of bodies in contact, but neglects the adhesion force. Several models for particle adhesion to flat surfaces were developed in the past that improves the Hertz model by including the effect of adhesion (van der Waals) force.

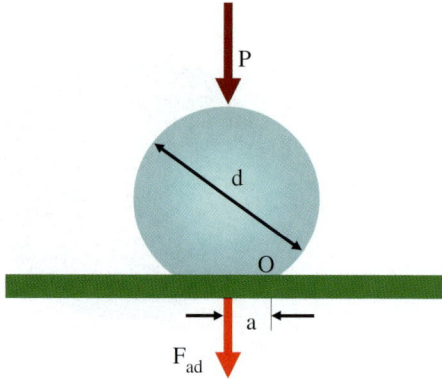

Fig. 31 Schematics of a spherical particle in contact with a plane

JKR Model

Johnson–Kandall–Roberts[141] developed a model (The JKR Model) that included the effect of adhesion force on the deformation of an elastic sphere in contact with an elastic half space. Accordingly, the contact radius is given as

$$a^3 = \frac{d}{2K}\left[P + \frac{3}{2}W_A\pi d + \sqrt{3\pi W_A dP + \left(\frac{3\pi W_A d}{2}\right)^2}\right]. \tag{117}$$

Here, W_A is the thermodynamic work of adhesion, and K is the composite Young's modulus given as

$$K = \frac{4}{3}\left[\frac{1-\upsilon_1^2}{E_1} + \frac{1-\upsilon_2^2}{E_2}\right]^{-1}. \tag{118}$$

In (118), E is the elastic modulus, υ is the Poisson ratio, and subscripts 1 and 2 refer to the materials of the sphere and substrate.

In the absence of surface forces, $W_A = 0$, and (117) reduced to the classical Hertz model. That is,

$$a^3 = \frac{dP}{2K}. \tag{119}$$

Pull-Off Force

The JKR model predicts that the force needed to remove the particle (the pull-off force) is given as

$$F_{po}^{JKR} = \frac{3}{4}\pi W_A d. \tag{120}$$

Contact Radius at Zero Force

The contact radius at zero external force may be obtained by setting $P = 0$ in (117). That is,

$$a_0 = \left(\frac{3\pi W_A d^2}{2K}\right)^{1/3}. \tag{121}$$

Contact Radius at Separation

The contact radius at the separation is obtained by setting $P = -F_{po}^{JKR}$ in (117). The corresponding contact radius is given by

$$a = \left(\frac{3\pi W_A d^2}{8K}\right)^{1/3} = \frac{a_0}{4^{1/3}}. \tag{122}$$

DMT Model

Derjaguin–Muller–Toporov[77] assumed that there is Hertz deformation and developed another model that included the effect of adhesion force. According to the DMT model, the pull-off force is given as

$$F_{Po}^{DMT} = \pi W_A d, \quad \left(F_{Po}^{DMT} = \frac{4}{3}F_{Po}^{JKR}\right). \tag{123}$$

Contact Radius at Zero Force

The contact radius at zero external force is given as

$$a_0 = \left(\frac{\pi W_A d^2}{2K}\right)^{1/3} \text{ (Hertz contact radius under adhesion force).} \tag{124}$$

Contact Radius at Separation

The DMT model predicts that the contact radius at the separation is zero. That is,

$$a = 0 \quad \text{(at separation)}. \tag{125}$$

Maugis–Pollock

While the JKR and the DMT models assume elastic deformation, there are experimental data that suggests that, in many cases, plastic deformation occurs. Maugis and Pollock[181] developed a model that included the plastic deformation effects. Accordingly, the relationship between the contact radius and external force is given as

$$P + \pi W_A d = \pi a^2 H, \tag{126}$$

where H is hardness and

$$H = 3Y \tag{127}$$

with Y being the yield strength.

Note that variations of contact radius with particle diameter at equilibrium, which is in the absence of external force, for elastic and plastic deformation are different. That is,

$$a_0 \sim d^{2/3} \text{ (elastic)}, \quad a_0 \sim d^{1/2} \text{ (plastic)}. \tag{128}$$

Thermodynamic Work of Adhesion

The thermodynamic work of adhesion (van der Waals surface energy per unit area) is given as

$$W_A = \frac{A}{12\pi z_0^2}, \tag{129}$$

where A is the Hamaker constant and z_0 is the minimum separation distance.

Nondimensional Forms

Nondimensional form of the relationship between contact radius and the external force and the corresponding moment are described in the section.

JKR Model

Equation (117) in nondimensional form may be restated as

$$a^{*3} = 1 - P^* + \sqrt{1 - 2P^*}, \tag{130}$$

where the nondimensional external force and contact radius are defined as

$$P^* = -\frac{P}{\frac{3}{2}\pi W_A d}, \quad a^* = \frac{a}{\left(\frac{3\pi W_A d^2}{4K}\right)^{1/3}}. \tag{131}$$

Variation of the nondimensional contact radius with the nondimensional force is shown in Fig. 32. Note that for $P^* = 0$, (130) and Fig. 32 show that $a_0^* = 1.26$.

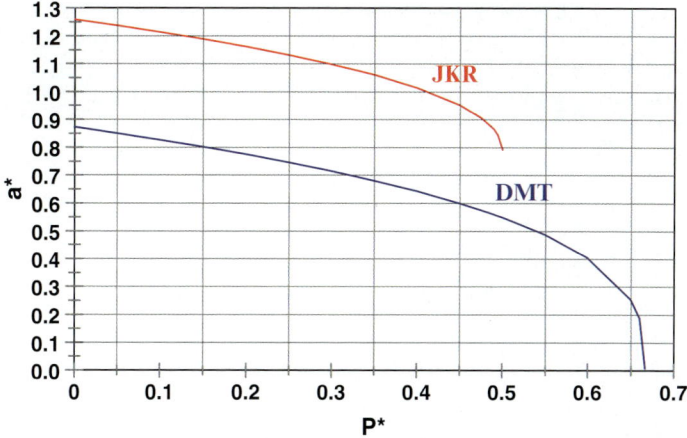

Fig. 32 Variations of contact radius with the exerted force

The corresponding resistance moment about point O in Fig. 31 as a function of nondimensional force is given as

$$M^{*JKR} = P^*a^* = P^*(1 - P^* + \sqrt{1 - 2P^*})^{1/3}. \tag{132}$$

Figure 33 shows the variation of the resistance moment as predicted by the JKR model. The corresponding maximum resistance moment then is given by

$$M^{*JKR}_{max} = 0.42 \tag{133}$$

and

$$P^*_{max} = F^{*JKR}_{po} = \frac{F^{JKR}}{\frac{3\pi}{2}W_A d} = 0.5. \tag{134}$$

The resistance moment at P^* is $M^{*JKR} = 0.397$. Also

$$P^*_{max} a^*_0 = 0.63. \tag{135}$$

DMT

For DMT Model, the approximate expression for the contact radius is given as

$$a^3 \approx \frac{d}{2K}(P + \pi W_A d) \tag{136}$$

or

$$a^{*3} = \left(\frac{a}{3\pi W_A d^2/4K}\right)^3 = -P^* + \frac{2}{3}. \tag{137}$$

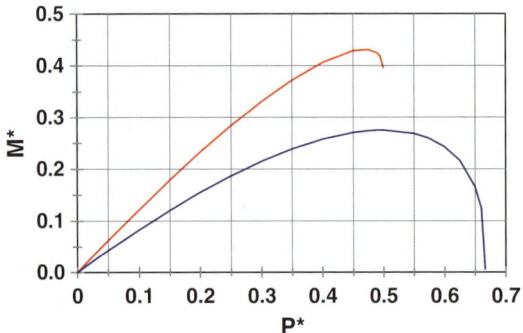

Fig. 33 Variations of resistance moment with the exerted force

Variation of the nondimensional contact radius with the nondimensional force as predicted by the DMT model is shown in Fig. 32 and is compared with the JKR model. Note that for $P^* = 0$, (137) and Fig. 32 show that $a_0^* = 0.874$.

The corresponding resistance moment as a function of nondimensional force as predicted by the DMT model is given as

$$M^{*\text{DMT}} = P^*(2/3 - P^*)^{1/3}. \tag{138}$$

The variation of the resistance moment as predicted by the DMT model is also shown in Fig. 33. The corresponding maximum resistance moment is

$$M_{\max}^{*\text{DMT}} = 0.28. \tag{139}$$

Note also that the maximum force (the pull-off force) is given by

$$P_{\max}^* = F_{po}^{*\text{DMT}} = \frac{F^{\text{DMT}}}{\frac{3\pi}{2}W_A d} = \frac{2}{3} \tag{140}$$

and

$$P_{\max}^* a_0^* = 0.58. \tag{141}$$

Comparing (133) and (139), it is seen that the JKR model predicts a larger resistance moment. That is,

$$M_{\max}^{*\text{JKR}} = 0.42 = 1.5 M_{\max}^{*\text{DMT}}, \ M_{\max}^{*\text{DMT}} = 0.28. \tag{142}$$

The resistance moment predicted by the JKR and the DMT models in dimensional form is given as

$$M_{\max}^{\text{JKR}} = 2.63 \frac{W_A^{4/3} d^{5/3}}{K^{1/3}}, \ M_{\max}^{\text{DMT}} = 1.83 \frac{W_A^{4/3} d^{5/3}}{K^{1/3}}. \tag{143}$$

Applications

Applications of particle transport, deposition, and removal are described in this section.

Particle Transport and Deposition in Respiratory Tracts

Introduction

Particle deposition in the nasal and lung airways has been a subject of great interest due to the health risk of particulate matter (PM) pollutants. Recently, there has been additional interest on the subject by pharmaceutical companies in connection with the targeted inhalation therapeutic drug delivery. PM deposition in nose and lung occurs by several mechanisms. Large particles deposit due to impaction, interception, turbulence, eddy impaction, and gravitational sedimentation. Particles in the nanometer-size range, however, deposit due to turbulence dispersion and Brownian motion.

Effects of airflow and turbulence on particle deposition in the lung were examined by Chan et al.[41] using the airflow measurements in hollow casts and airway bifurcation models. Heyder et al.[132] measured total and regional aerosol depositions through the mouth and the nose. Experimental studies for inspiratory particle depositions in single and double bifurcation airways were reported by Johnston and Schroter,[142] Kim and Iglesias,[155] Kim et al.,[154] and Kim and Fisher.[153]

Schiller et al. (1986) analyzed factors affecting ultrafine aerosol deposition in the human airway. Deposition of ultrafine particles in replicate cast models of the human nasal cavity is measured by Cohen et al.[57] and Swift et al.,[256,257] among others. Cohen and Asgharian[58] used experimental data and obtained an empirical expression for the deposition efficiency of ultra fine particles by the diffusion mechanism. Recently, Asgharian et al.[17] described a realistic model for lung deposition.

The fluid dynamics of the human larynx and upper tracheobronchial airways is studied by Martonen et al.[182,183] Balashazy and Hofmann,[19,20] and Balashazy[18] analyzed the particle trajectories in a three-dimensional bronchial airway bifurcation. Asgharian and Anjilvel[16] studied the inertial and gravitational depositions in a square cross-sectional bifurcating passage. Li and Ahmadi[177] studied particle transport and deposition in the first lung bifurcation and included the turbulence dispersion effects. Geometric factors for the quantification of particle deposition patterns in bifurcation airways were studied by Balashazy et al.[21] and Comer et al.[62] Airflow particle deposition in triple lung bifurcation was studied by Zhang et al.[285] and Mazaheri and Ahmadi.[188,189] Recently, Tian et al.[262] studied the airflow and deposition pattern in symmetric and asymmetric lung bifurcation models.

Despite a number of studies on particle deposition in human lung, the Brownian diffusion and turbulent dispersion effects were generally ignored in the earlier computational models. Martonen et al.[183] pointed out that the flow disturbances from the laryngeal jet could propagate into the trachea and main bronchus. In this section, computational modeling of inspiratory particle transport, dispersion, and deposition in the human nose and lung is described, and sample simulation results are presented and discussed. Attention was given to the effects of particle size on the deposition rates and comparison with the available experimental data.

Multibifurcation Airways

Weibel[275] and Raabe et al.[228] have shown that the structure of human bronchial airways can be approximated as a network of repeatedly bifurcating tubes. The bifurcations are generally asymmetric. In most computer models, however, symmetric conditions are assumed.

Recently, Tian et al.[262] have performed a series of computer simulations and compared the particle deposition for symmetric and asymmetric cases. Figure 34 shows the sample mesh used in the computations. Figure 35 shows the mean velocity contours in a plane across the trachea and the left and right bronchus of an adult person. Two different flow rates were considered in these analyses. For 15 l/min breathing rate, the flow was nearly laminar. For 60 l/min breathing rate, however, the flow was in turbulent state of motion. For the turbulence case, the stress transport model was used for the flow analysis and the instantaneous turbulence fluctuations were evaluated and were included in the particle trajectory analysis. Figure 35 shows that the asymmetric bifurcation leads to regions with high velocities.

To analyze the particle capture efficiency of the upper lung, particles of different diameters were initially released with a uniform distribution at the trachea inlet and their trajectories were analyzed. Figure 36 shows the locations at which the particles of different size were captured. It is seen that the relatively large 30-μm particles are mainly deposited on the carina by impaction mechanisms. On the other hand, the 10-nm particles have a more uniform distribution pattern. These small particles are deposited mainly by the diffusion process on the entire passage surface. Very few 1-μm particles are captured by the first lung bifurcation, since for this size range,

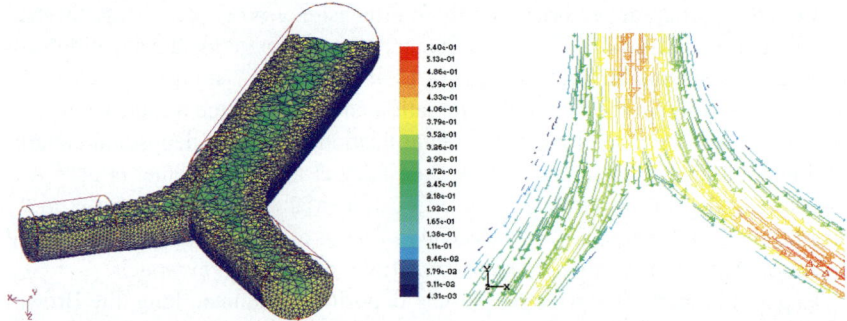

Fig. 34 A sample mesh that was used in the computation and a sample velocity vector field near carina

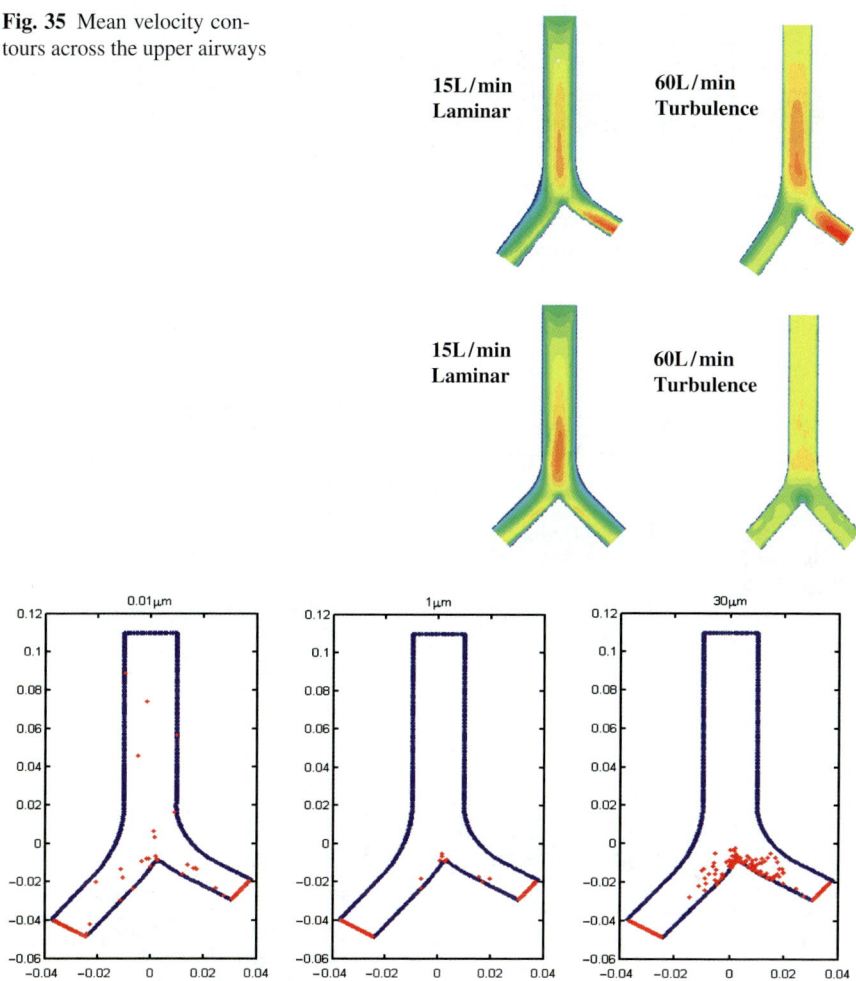

Fig. 35 Mean velocity contours across the upper airways

Fig. 36 Deposition pattern for different size particles

the diffusion is negligible and the inertia is small. These trends of behavior were also observed by Li and Ahmadi,[177] among others.

Variation of capture efficiencies of the symmetric and asymmetric first lung bifurcation as predicted by Tian and Ahmadi (2007) is compared with the earlier simulations and experimental data for the symmetric case in Fig. 37. It is seen that the model predictions are comparable to the experimental data. This figure also shows that the capture efficiency increases sharply as particle Stokes number increases beyond 0.05. Furthermore, the total capture efficiencies of the symmetric and asymmetric bifurcations do not differ to a noticeable extent.

Mazaheri and Ahmadi[188,190] studied the airflow and particle deposition pattern in a two-dimensional section of the top triple bifurcation of the lung accounting for

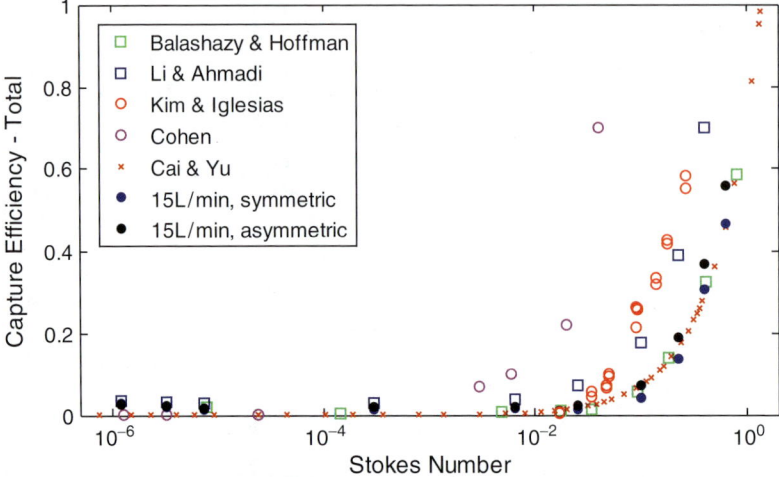

Fig. 37 Comparison of the capture efficiencies versus Stokes numbers

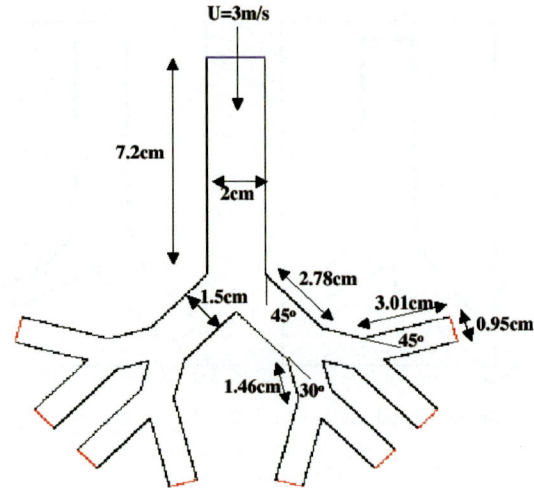

Fig. 38 Schematic of geometry of the triple bifurcation airway model (Mazaheri and Ahmadi[188])

the effects of turbulence. The geometric feature of the model is shown in Fig. 38, and a sample mesh is shown in Fig. 39. An unstructured grid of about 700,000 cells was used in these simulations.

Figure 40 shows the simulated velocity magnitude contours in the upper airways. This figure shows that the gas velocity in the trachea is about 3 m/s, and the airflow velocity magnitude decreases after each bifurcation. The boundary layer formation after each bifurcation can also be seen from this figure.

Sample simulated turbulence kinetic energy contours in the triple bifurcation model are shown in Fig. 41. This figure shows that the turbulent kinetic energy in

Fig. 39 Surface grid of a segment of the triple bifurcation airway

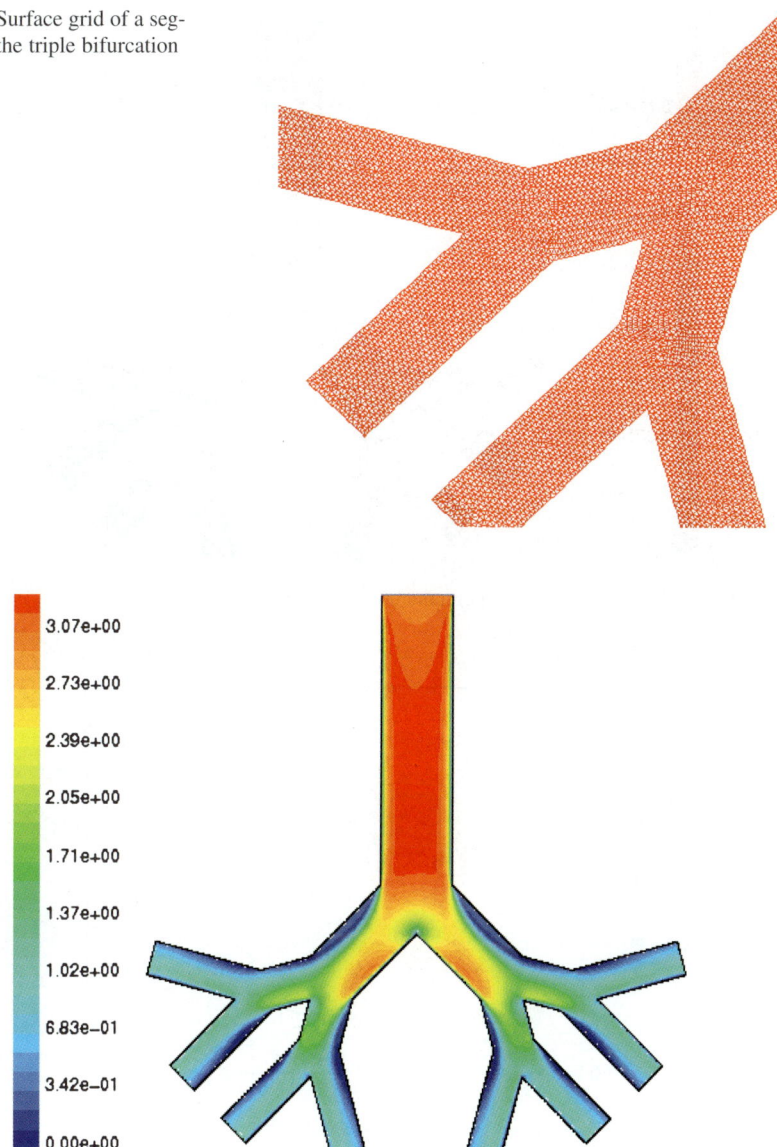

Fig. 40 Velocity magnitude contour plot of the multi bifurcation airway model

the trachea and in the left and right bronchus is about $1\,\text{m}^2/\text{s}^2$. It is also seen that the airflow turbulence may be present in the first few lung bifurcations.

Mazaheri and Ahmadi[188, 190] used an ensemble of 1,000 particles of different diameters and evaluated the corresponding capture efficiencies. Particles were initially released with a uniform distribution at the trachea inlet. Figure 42 shows the capture

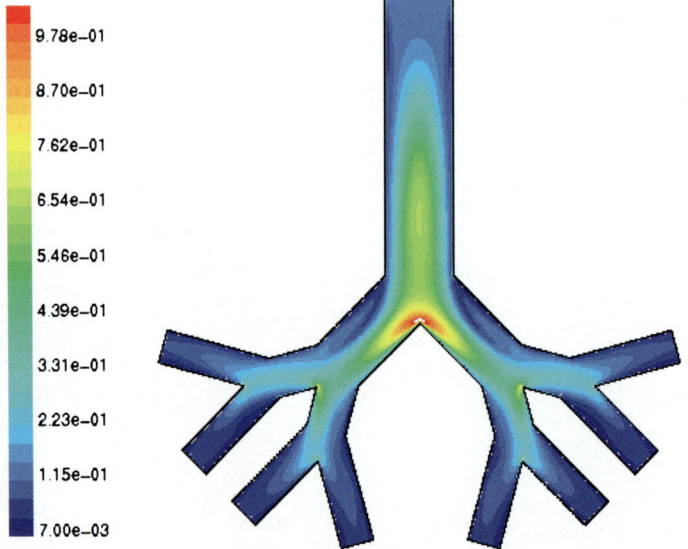

Fig. 41 Turbulence kinetic energy contours in the triple bifurcation airway

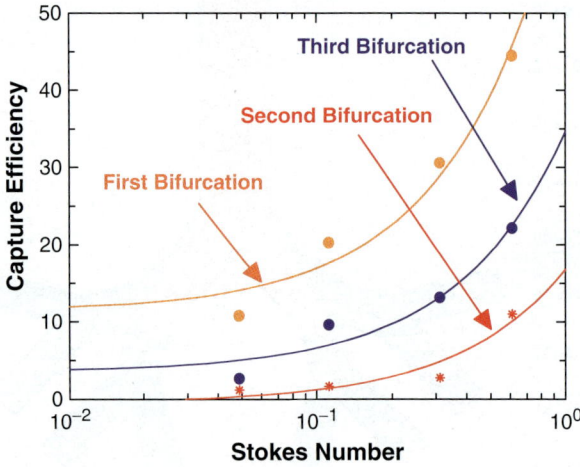

Fig. 42 Variation of capture efficiency with Stokes number

efficiency in different areas of the triple bifurcation airway versus the particle Stokes number, St. Here the effect of airflow turbulence was included in the analysis. It is seen that the capture efficiency increases as particle Stokes number increases. This figure also shows that the capture efficiency in the first bifurcation is highest due to particle impaction. It should be noted here that the particle Stokes number is evaluated on the basis of the trachea diameter and airflow velocity.

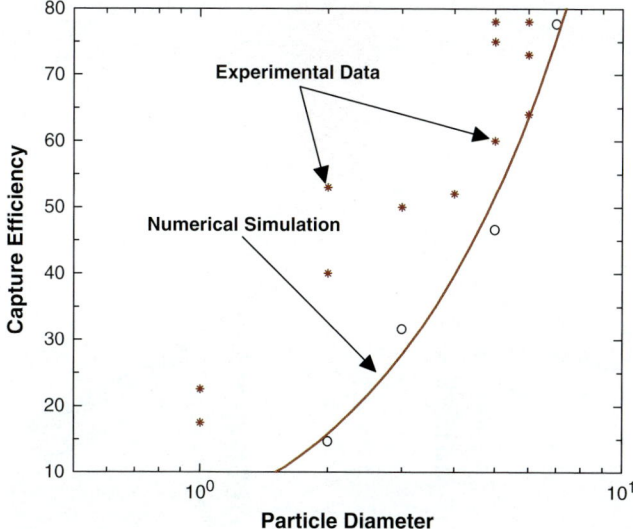

Fig. 43 Comparison of variation of total capture efficiencies with the experimental data collected by Hinds[134]

Figure 43 compares the simulated total capture efficiency of the triple bifurcation airway with the experimental data for different particle diameters. In this figure, solid lines denote the simulation results of Mazaheri and Ahmadi[188] and the stars represent the experimental data collected by Hinds.[134] This figure shows that total capture efficiency increases as particle diameter increases. It is also seen that the numerical simulation results are comparable with the experimental data. Mazaheri and Ahmadi also showed that when the effect of the turbulence in the lung was neglected, the particle capture efficiencies are reduced by about 40–60%.

Airflow and capture efficiency in a triple bifurcation airway was also studied by Zhang et al.,[285] where they used a three-dimensional model. They assumed laminar flow condition and used commercial software to study the airflow structure and particle transport pattern in the lung. They also showed that the particle depositions in the first bifurcation and first and second bifurcation are comparable with the earlier suggested correlations under laminar flow conditions.

A sample three-dimensional simulation of the triple bifurcation is shown in Fig. 44. Here, turbulence airflow condition is assumed. It is seen that the simulation results are comparable with the earlier simulation of Mazaheri and Ahmadi.[188,190]

Nasal Passages

Early measurements of airflow measurement in the nasal passages were reported by Swift and Proctor.[255] They used a miniature Pitot tube on a nasal cast of a cadaver

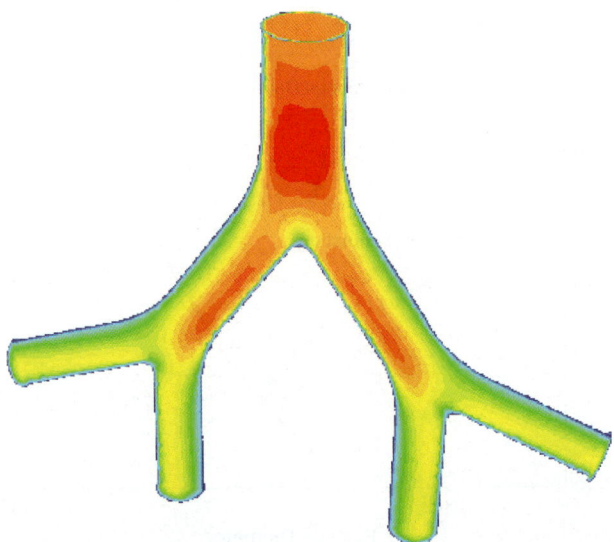

Fig. 44 Sample simulation results for air velocity condition in a three-dimensional model of the lung airways

and reported that the airflow was laminar at a breathing rate of 15 l/min and turbulence was detected downstream of the nasal valve for a breathing rate of 25 l/min. Airflow measurements in the nose using hot-wire and hot-film anemometers were reported by Schreck et al.[239] and Hahn et al.[125] They found that the airflow in the nose was laminar up to a breathing rate of 24 l/min. More recently, Kelly et al.[150] used particle image velocimetry (PIV) on a model fabricated from CT scans of the nasal passage for their airflow measurements. The common finding of the earlier experimental studies was that the flow regime inside of human nasal cavity for low-to-moderate breathing rates was laminar. In addition, a large portion of inspired airflow passes through the middle and inferior airways while a smaller fraction passes through olfactory and meatuses regions.

Computer simulations of airflow inside the human nasal passages were reported by Keyhani et al.[152] and Subramanian et al.,[254] among others. They constructed their computational model from MRI of a human subject and used the commercial software, FIDAP^{TM}, in their analysis. They also showed that a large part of the airflow passed through the middle and inferior airways.

Experimental study of deposition of particles in the human nasal passage was reported by Cheng et al.,[48,49] Swift et al.,[256–258] and Strong and Swift.[253] Cheng et al.,[48] Cheng,[47] and Martonen et al.[184] suggested empirical equations for the capture efficiency of the human nasal passage. Recently, Kelly et al.[151] measured the deposition of ultrafine particles in nasal airway replicas produced by a stereolithography machine.

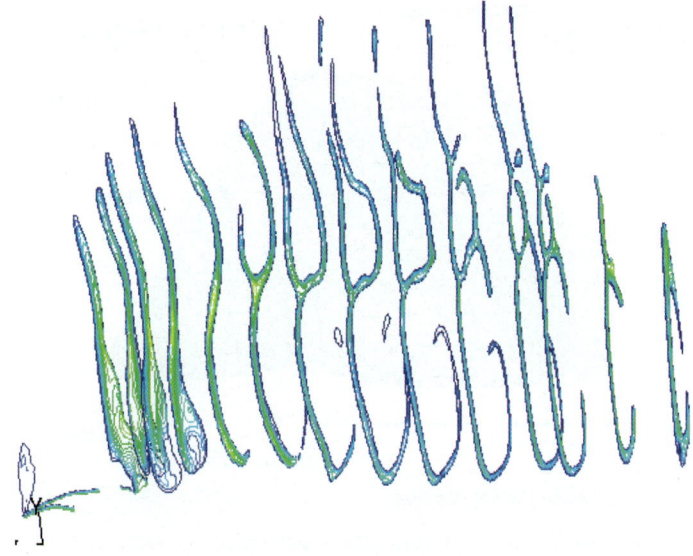

Fig. 45 Cross sections of the nose of a human male obtained from MRI and used in developing the computational model. The contours are velocity magnitudes

Computer simulation studies of transport and deposition of ultrafine particles in human nasal cavities were reported by Yu et al.[281] and Scherer et al.[244] In these simulations, a diffusion model for particle deposition was used. Recently, Zamankhan et al.[283] presented a series of simulations for particle deposition in the nasal passage of a human male using Lagrangian particle tracking approach.

Schematics of the cross section and the geometry of the nasal passage are shown in Figs. 45 and 46. The sections shown in Fig. 45 are obtained form MRI images of an anonymous male donor. These sections were used to construct the computational domain shown in Fig. 46. The computed velocity magnitude contours are also shown in Fig. 45. Various regions of the nasal passage are also identified in Fig. 46.

Figure 47 shows the unstructured computational mesh that was produced by Zamankhan et al.,[283] which included 965,000 tetrahedral elements and 250,000 computational points. When needed, in these simulations a finer computational mesh including boundary refinement was used.

The simulated velocity vector fields at various sections of the nasal passage are shown in Fig. 48. The corresponding velocity magnitude contours are plotted in Fig. 45. Detailed examination of the results shows that more than 70% of the flow passes through the inferior airway, the middle airway, and the region in between the septum side of the section. About 7% of the flow passes through the olfactory slit, 13% across the three meatuses areas, and around 10% across the superior airway. The corresponding pressure fields are shown in Fig. 49. It is seen that the pressure decreases from the nostril along the nose toward the nasopharynx region. Additional simulation results may be found in the work of Zamankhan et al.[283]

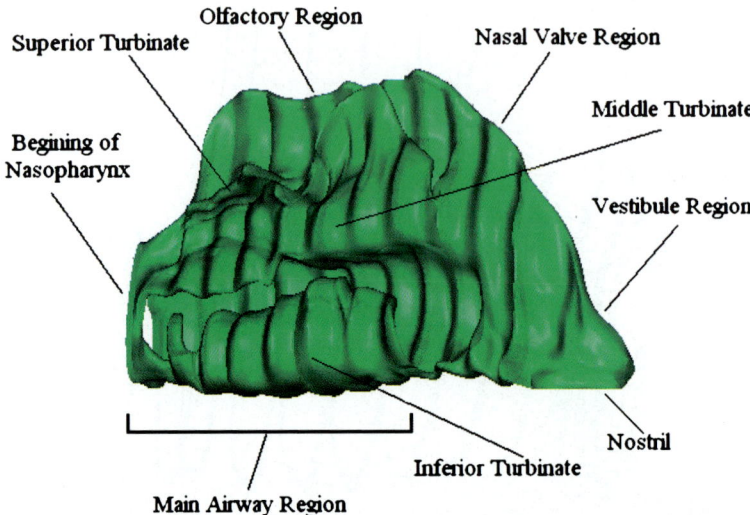

Fig. 46 The computational model of the nasal airway with various nose regions

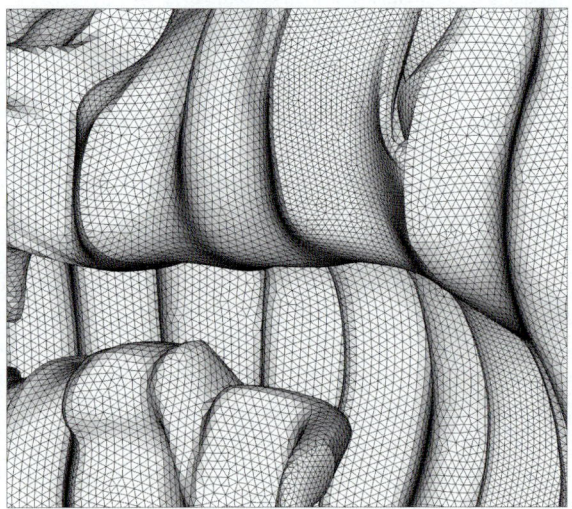

Fig. 47 Sample surface mesh used in the simulations (Zamankhan et al.[283])

Zamankhan et al.[283] have suggested an empirical expression for the nose friction factor. That is,

$$f = \frac{47.78}{Re}\left(1 + 0.127\, Re^{0.489}\right), \quad (144)$$

where the Reynolds number is defined as,

$$Re = \frac{u_m d}{\nu}. \quad (145)$$

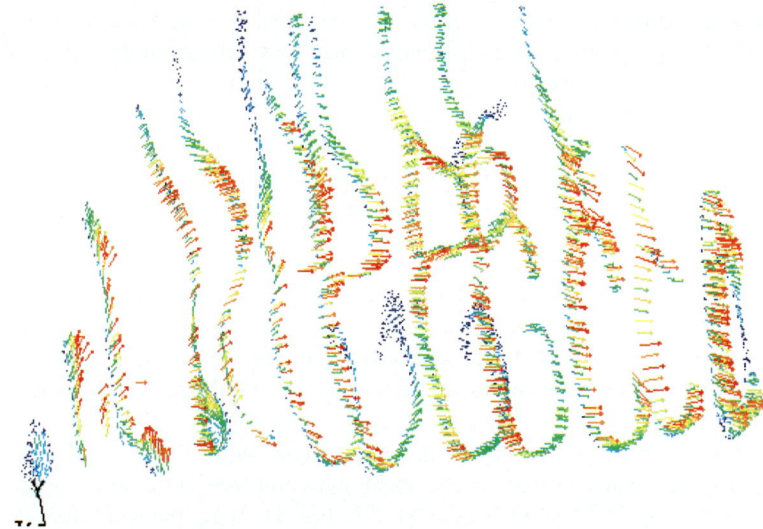

Fig. 48 Sample velocity vector field in different sections of the nasal airways

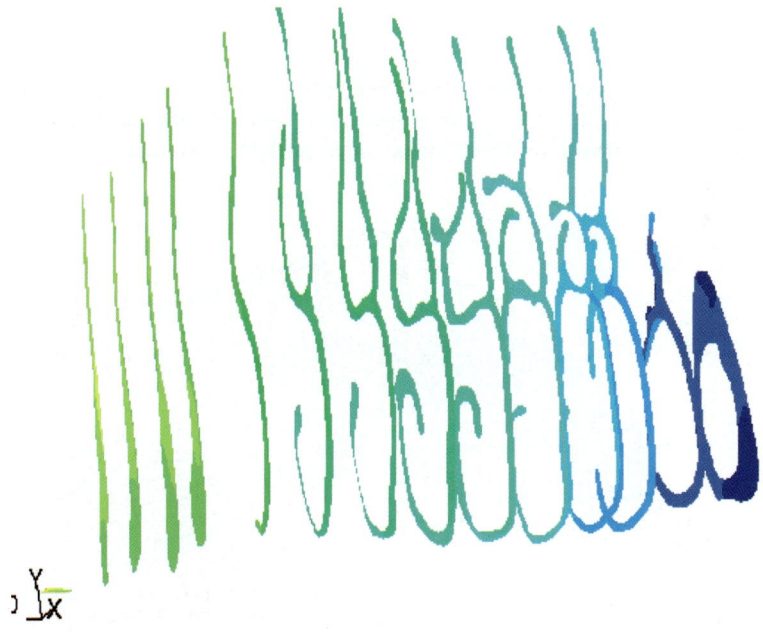

Fig. 49 Sample velocity magnitude contours in different section of the nasal airways

Here, v is the kinematic viscosity of air, d is the average hydrodynamic diameter of the nose coronal sections, and u_m is the average flow velocity at the nostril. With the friction factor given by (144), the mean pressure drop between the nostril and nasopharynx, Δp, is given as,

$$\frac{\Delta p}{\rho} = f \frac{L}{d} \frac{u_m^2}{2}. \qquad (146)$$

Here, L is the passage length, and ρ is the air density.

To study particle transport and deposition in the nasal passage, 700–1,200 particles of different sizes were introduced at the nostril with a uniform distribution, and their corresponding trajectories were analyzed. Figure 50 shows sample stream traces. It is seen that the stream traces are focused in certain areas of the nasal passage.

For a breathing rate of 4 l/min, the capture efficiency of the nose for different size particles smaller than 100 nm was evaluated and compared with the experimental data of Swift et al.[256,257] and Cheng et al.[50] in Fig. 51. Here, the solid lines correspond to the model prediction. While there are some scatters in the experimental data, this figure shows that the predicted capture efficiencies are in good agreement with the experimental data, particularly, for particles smaller than 20 nm The model prediction for particles larger than 20 nm is, however, somewhat higher than the average of the experimental data.

Zamankhan et al.[283] discussed the potential reason for the discrepancy and also performed a series of simulations with a more refined mesh. Accordingly, while the discrepancy can be reduced by using a finer mesh, some deviations remain due to the computational errors introduced by the linear interpolation used.

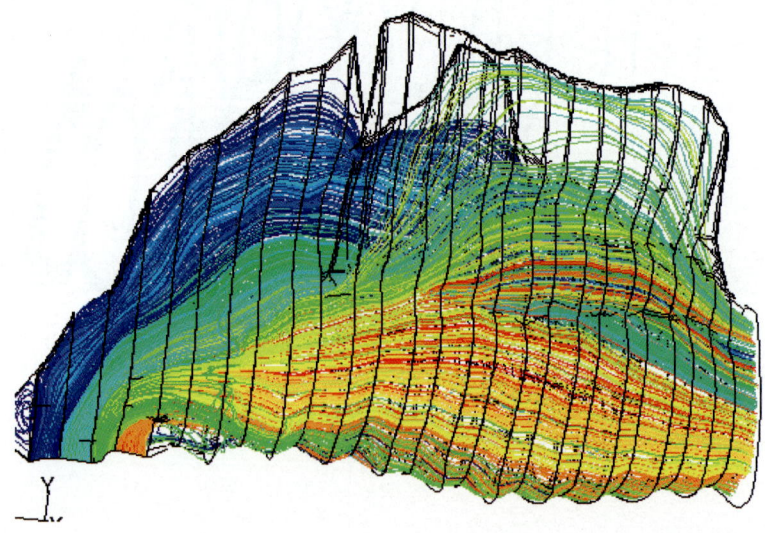

Fig. 50 Sample stream traced in the nasal airways

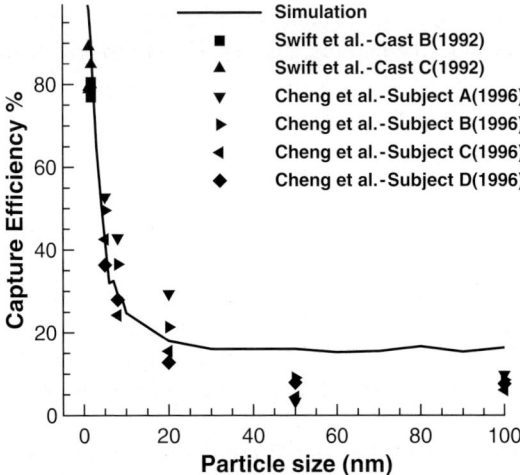

Fig. 51 Comparison of the simulation results for the nose capture efficiency with the experimental data for particle of different sizes

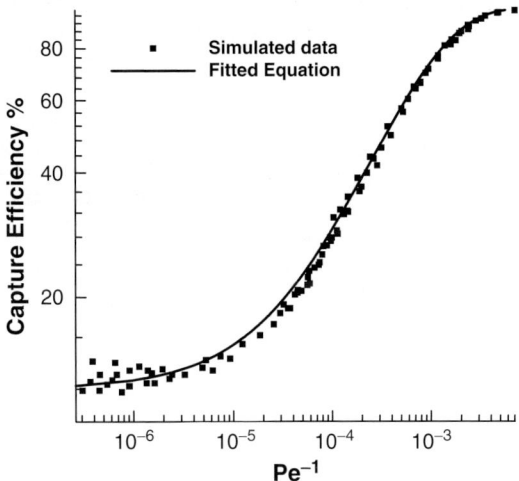

Fig. 52 Variations of nose capture efficiency with inverse Peclet number for particles smaller than 100 nm

Figure 52 shows the simulated capture efficiency results for particle smaller than 100 nm versus (inverse) Peclet number. Here, Peclet number is defined as

$$Pe = \frac{Q}{DL_s}, \qquad (147)$$

where L_s is the length of the nasal passage, Q is the flow rate, and D is the particle diffusivity. While there is some slight scatter for large values of Pe, it is seen that the simulation result for a range of parameters collapses to a single curve. This is as expected, and for ultrafine particle, the deposition process is dominated by the diffusion.

An empirical equation given by

$$\eta = 100(1 - 0.88\exp(-218 Pe^{-0.75})) \qquad (148)$$

may be fitted to the simulation results in Fig. 52. This figure shows that (148) provides a good fit to the simulation results. As the Peclet number increases, which is associated with larger particle size, some scatter appears in the simulation results. This suggests that for larger particles, in addition to the Peclet number, the capture efficiency could also depend on Stokes number. Zmankhan et al.[283] used a more refined mesh and found slightly different expression for the fit to the simulation results.

For breathing rates of 3.75, 7.5, and 15 l/min, the capture efficiency of the nose for particles in the size range of 200 nm to 10 μm were evaluated and results are shown in Fig. 53. It is seen that the capture efficiency increase as particle size increases. Furthermore, as the inspiratory flow rate increases, there is a marked increase in the capture efficiency of the nose.

The simulation results presented in Fig. 53 are reported in Fig. 54, versus Stokes number. It is seen that the simulation results collapse to a single curve. This observation suggests that the inertial impaction is the key mechanism for the deposition of large particles in the nasal passages.

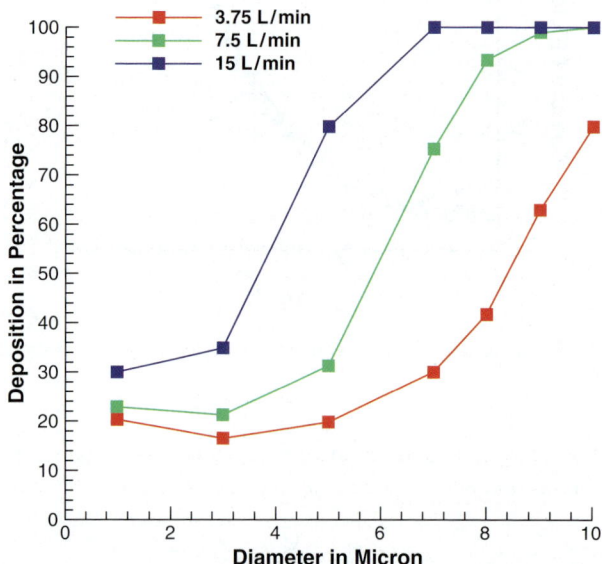

Fig. 53 Simulation results for the nose capture efficiency for particle of different sizes at different breathing rates

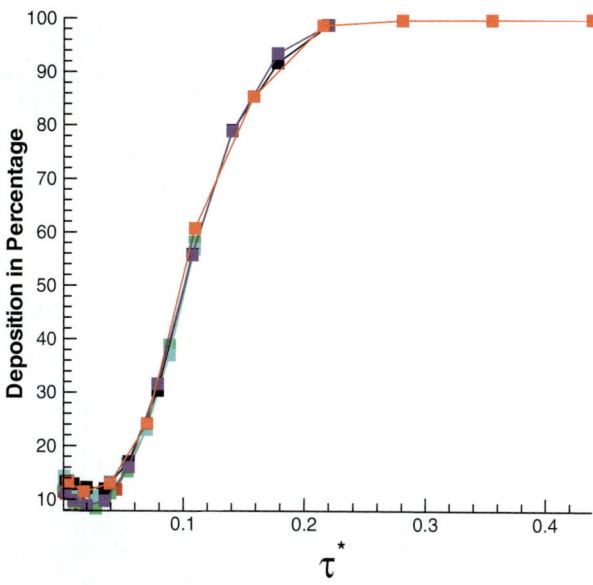

Fig. 54 Variations of nose capture efficiency with Stokes number for particles in the size range of 200 nm to 10 μm

Oral Airways

Cheng et al.[51] measured the regional deposition in a cast of human oral airways for particles in the size range of 1–30 μm for inspiratory flow rate of 15, 30, and 60 l/min. They found that the capture efficiency is a function of the Stokes number, which suggests that impaction was the dominant deposition mechanism. Stapleton et al.[252] studied the particle motion and deposition in a replica of the oral airways. They used the standard $k - \varepsilon$ turbulence model for airflow simulation and made use of the eddy life-time model of Gosman and Ioannides[120] for including the effect of turbulence on particle dispersion. Recently, Zhang et al.[288] simulated the airflow and the transport and deposition of microparticles in the oral airways. They used a low Reynolds number (LRN) $k - \omega$ turbulence model. They noted that the turbulence occurring after constriction in the oral airways for moderate and high-level breathing rates could enhance particle deposition in the trachea near the larynx. Zhang and Kleinstreuer[286] used a diffusion model and studied the transport and deposition of nanoparticles in the oral airway from the mouth to trachea and the upper tracheobronchial tree. They studied both steady and cyclic flow conditions. Zhang et al.[289] simulated the transport and deposition of nano and microparticles in the oral airways. Recently, Zahmatkesh et al.[282] analyzed the transport and deposition of particle in the human oral airways using a stress-transport turbulence model.

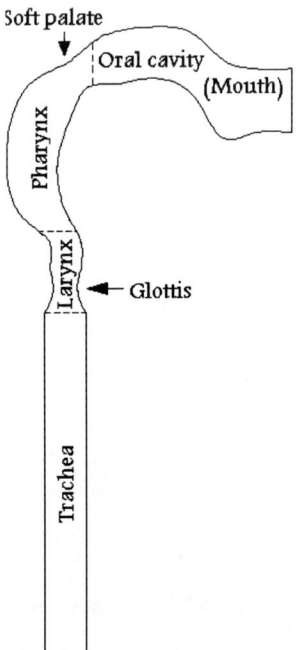

Fig. 55 Schematics of the oral airway

Schematics of the oral cavity from mouth to pharynx, larynx, and trachea are shown in Fig. 55. A computational grid of 3,36,000 structured cells was used by Zahmatkesh et al.[282] for the simulations of airflow conditions in the oral cavity.

Figure 56 shows the variations of mean airflow velocity field in the oral airways.[282] Here, the velocity distribution at the midsection oral airways and some cross-sectional views are shown. It is seen that the velocity distribution in the oral cavity and the lower half of the trachea are roughly uniform. The maximum velocity, however, shifts toward the outer bend in the pharynx and larynx. This is perhaps due to the centrifugal force generated when the passage bends. Figure 56 also shows that the sudden cross sectional changes in the pharynx and downstream of the glottis may lead to recirculation regions. It is also seen that a laryngeal jet is generated after the glottis.

Zahmatkesh et al.[282] presented their simulation results for variation of Reynolds stresses and have shown that the flow is strongly anisotropic. Their simulation results for the capture efficiency of the oral airway versus Stokes number are reproduced in Fig. 57 and are compared with the empirical correlation of Cheng et al.[51] It is seen that the simulation results are in reasonable agreement with the correlation of Cheng et al.[51] In addition, the capture efficiency of the oral cavity increases sharply as Stokes number increases.

The influence of airflow turbulence on particle deposition fraction was also studied by Zahmatkesh et al.[282] Their simulation results for deposition fractions in the absence of turbulence fluctuations are also shown in Fig. 57. It is seen that turbulence

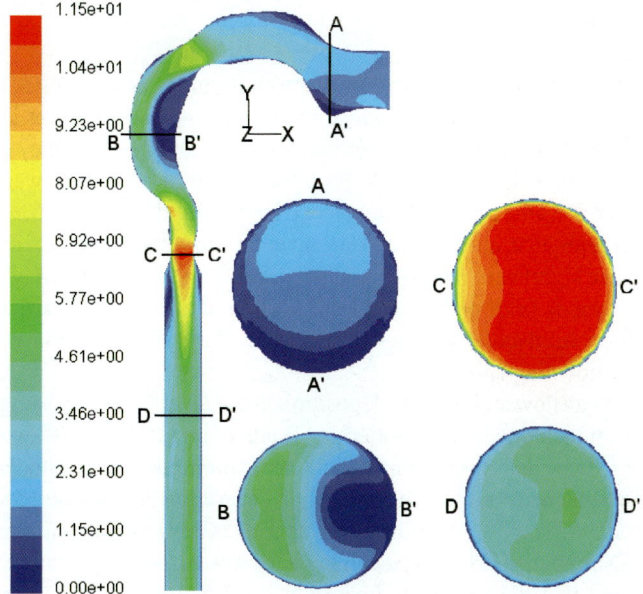

Fig. 56 Velocity contours at a midplane and some sections of the oral airways (velocity magnitudes are in m/s)

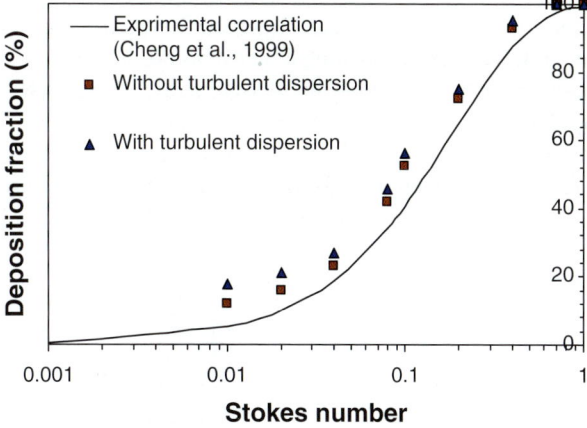

Fig. 57 Comparisons of particle capture efficiency in the human upper oral airway with experimental data (Zahmatkesh et al.[282])

enhances particle deposition, which is due to the increased fluctuation motions that lead to higher dispersion rate of particle trajectories. The enhancement of the capture efficiency, however, decreases as the Stokes number increases. This is because the larger particles are mainly deposition of the impact and are less affected by the airflow fluctuations.

Alveolar Cavities

Alveolar cavities are the lowest portion of the lung airways where gas exchange with blood takes place. Alveoli are formed in clusters and each alveolus consists of a tubular alveolar duct leading into a terminal alveolar cavity. Before they terminate, the alveolar ducts branch several times. If all the alveoli in both lungs were laid out flat, the total surface area available for gas exchange would be about $75\,m^2$.

Inhaled particles that are deposited on alveolar surfaces reduce their effective area, could cause significant damages to the surrounding tissues, and may even lead to cancer and serious heart problems. In this section, the process of particulate pollutant deposition in alveolar cavities is reviewed. Particular attention is given to computational modeling approach.

Simulation of airflow and particle deposition in the alveolar region using a Monte Carlo simulation technique was reported by Tsuda et al. (1994a,b). They noted the significance of Brownian motion on particle deposition rate. Darquenne and Paiva (1996) and Darquenne (2001) simulated a four-generation alveolar structure, using a 2D model with different orientation angles as well as a 3D model of a single alveolar structure. Darquenne (2002) extended her earlier work to a large class of particle sizes. She found that the gravity is important for 0.5-μm particles or larger. Recently, Chang and Ahmadi[42] reported a series of simulation for particle deposition in the alveolar passages of lung and included the effect of Brownian motion and the Cunningham slip correction to the drag in the particle equation of motion.

Chang and Ahmadi[42] used two alveolar models, a three-generation and a nine-generation model for the alveolar regions. Figure 58 shows the schematics of their three-geneation model. The corresponding dimensions are also shown on this figure. These dimensions are consistent with the morphometric data for the human pulmonary acinus reported by Haefeli-Bleuer and Weibel (1988).

For an inlet velocity of $U = 0.0075\,m/s$, Fig. 58 displays the velocity magnitude contours for a three-generation alveolar cavity. Airflow velocity appears to be symmetric at upper and lower branches of the alveolar structure. The velocity

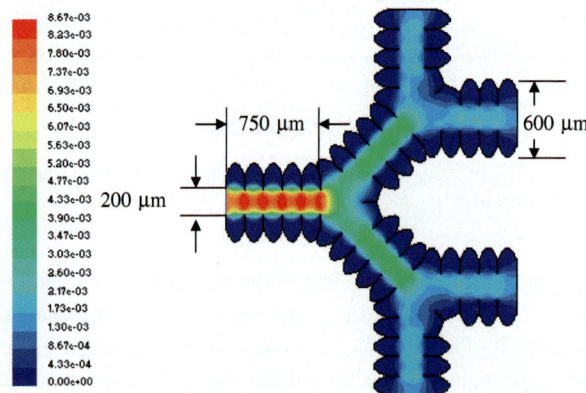

Fig. 58 Velocity magnitude contours in a three-generation alveolar cavity (Chang and Ahmadi[42])

Fig. 59 Sample particle trajectories in a three-generation alveolar cavity (Chang and Ahmadi[42])

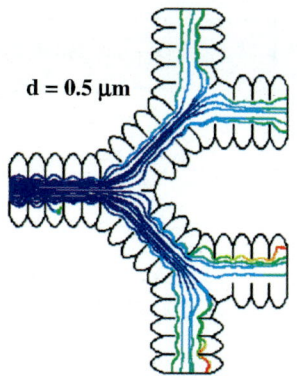

magnitude is relatively high in the central channel in the first alveolar duct, and it deceases as the flow passes through successive alveolar duct and alveoli. In alveoli region, the flow velocity is very small compared with the flow in the alveolar duct.

Sample particle trajectories for 0.5-μm particles in a three-generation alveolar structure are shown in Fig. 59. Here, the color corresponds to the particle residence time. For 0.5-μm particles, it is seen that the Brownian motion and gravity do not have noticeable effects on the particle trajectories, and these particles follow the airflow in the alveolar duct. Chang and Ahmadi[42] showed that for smaller particle, the Brownian motion effect becomes important, while for larger particle gravitational sedimentation effect is significant.

An empirical equation for the capture efficiency (deposition fraction) in the alveolar region as a percentage of the amount of particles that reaches the alveolar region, DF_{ALC}, was obtained by Chang and Ahmadi.[42] That is,

$$DF_{ALC} = \frac{DF_{AL}}{1 - (DF_{HA} + DF_{TB})}, \tag{149}$$

where DF_{AL}, DF_{HA}, and DF_{TB} are, respectively, the capture efficiency of the alveolar region, the head airway, and the tracheobronchial region. These are given as

$$DF_{AL} = \left(\frac{0.0155}{d_P}\right)$$
$$\times \left[\exp\left(-0.416(\ln d_p + 2.84)^2\right) + 19.11\exp(-0.482(\ln d_p - 1.362)^2))\right], \tag{150}$$

$$DF_{HA} = IF\left(\frac{1}{1 + \exp(6.84 + 1.183\ln d_p)} + \frac{1}{1 + \exp(0.924 - 1.885\ln d_p)}\right), \tag{151}$$

$$DF_{TB} = \left(\frac{0.00352}{d_P}\right)$$
$$\times \left[\exp(-0.234(\ln d_P + 3.40)^2) + 63.9\exp(-0.819(\ln d_P - 1.61)^2)\right]. \tag{152}$$

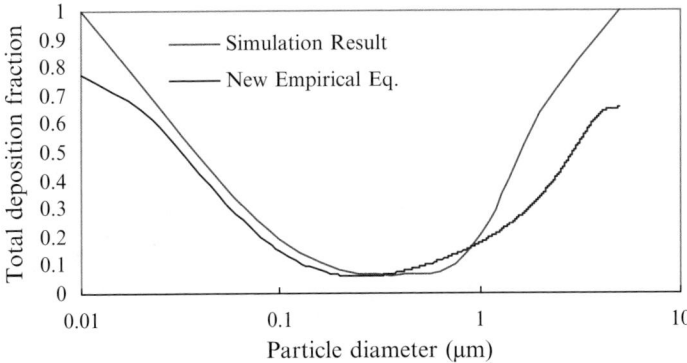

Fig. 60 Comparison of the simulation results for capture efficiency of the alveolar cavity with the empirical equation

Here, d_P is particle size in mm and IF is the inhalable fraction defined as

$$\text{IF} = 1 - 0.5\left(1 - \frac{1}{1 + 0.00076\, d_P^{2.8}}\right). \tag{153}$$

Figure 60 compares the simulation results of Chang and Ahmadi[42] for the particle capture efficiency of the three-generation alveolar cavity with the prediction of (149). This figure shows that the simulation results are in general agreement with the empirical model predictions. Both simulation and empirical equations show a V-shaped variation with the highest deposition efficiency occurring for very small or very large particle sizes. The largest discrepancies also occur for the large and small particle cases.

Blood Flow

In this section, some of the characteristics of blood flow are described with emphasis on the fact that blood is a suspension of particulates. First, the types of cells present in human blood are described, and the flow of blood through the human vasculature is discussed with emphasis on the multiscale nature of the relevant phenomena. As an example of bioengineering research dealing with the vascular system, some of the transport-related aspects of the onset of atherosclerois are discussed. Finally, the use of DNS to model suspensions of red blood cells is reviewed, and prospects for multiscale models of transport phenomena associated with blood are discussed.

Much of the material in this section is based on reviews by Goldsmith and Skalak,[119] Skalak et al.,[230] Fung,[110] Ku,[158] Berger and Jou,[26] Buerk,[32] and Kamm.[145] Pozrikidis[219] discusses modeling of the red blood cell membrane and related topics. Alberts et al.[10] is a useful source for the molecular biology of blood cells.

Composition of Human Blood

Human blood is a suspension of cells in blood plasma. Three types of cells are present in blood: red blood cells, white blood cells, and platelets. The concentrations of these cells in blood are 4.5 to 5.5×10^9 red cells/ml, 2.5 to 5.0×10^8 white cells/ml, and 5 to 10×10^6 platelets/ml (Goldsmith and Skalak[119] and Fung[110]).

This section focuses on red cells since most of the particulate volume in human blood is occupied by red cells. Red cells ("erythrocytes") are biconcave disks under equilibrium conditions in blood plasma as shown in Fig. 61. The diameter of the disk is between 6 and 8 μm. Red cells lack a nucleus and are largely filled with the protein hemoglobin, which binds O_2 molecules and several other biologically important molecules. The red cells transport O_2 molecules from the lung to tissues throughout the body. They also transport waste products, such as CO_2, away from the tissues.

The "hematocrit" is the volume percent of red cells in blood. A typical value of the hematocrit is 45%. The volume fractions of the white cells and platelets are much smaller; red cells occupy more than 99% of the particulate volume in blood. From a rheological standpoint, blood can be viewed as a concentrated suspension of red cells in plasma. The effective viscosity of blood is surprisingly small since suspensions of solid particles at comparable volume fractions behave like pastes.

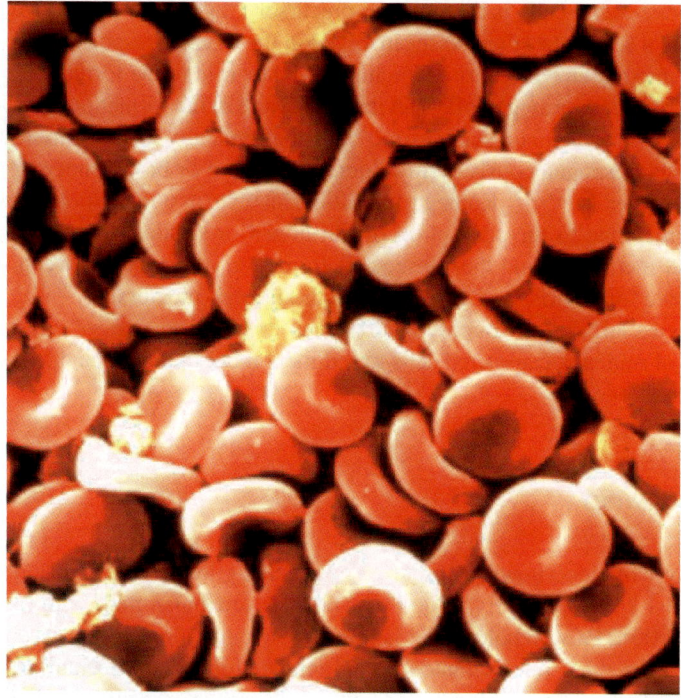

Fig. 61 A microphotograph showing red blood cells (Adapted from www.sciencemuseum.org.uk)

The low effective viscosity of blood is largely due to the elastic properties of the red cell membrane.

White cells ("leucocytes") protect the body against various kinds of infections. Unlike red cells, they cross the walls of small blood vessels to perform certain tasks. There are several types of white cells. Neutrophils constitute about two-thirds of all white cells. Their main function is to provide protection against bacteria and other causes of inflammations. Eosinophils deal with parasitic infections. Basophils deal with allergic and antigen response by releasing histamine. Various types of lymphocytes produce antibodies, coordinate immune responses, and kill cancer cells and viruses. Monocytes engulf foreign objects ("phagocytosis"). Once they enter tissues, monocytes are referred to as "macrophages." The sizes of most white cells lie in the range 8–15 µm. Monocytes are the largest white cells and are typically between 16 and 25 µm in size.

Platelets are small biconvex disks derived from the cortical cytoplasm of larger cells called "megakaryotes." They adhere to the endothelial cell lining of damaged blood vessels, where they help repair injuries and play a role in the process of blood clotting. Platelets also play a role in the formation of atherosclerotic plaques.

Blood cells are suspended in blood plasma. Blood plasma is a Newtonian fluid having a viscosity roughly equal to 0.0012 Pa s at 37 °C. Blood plasma is more than 90% water. The most prevalent solute is protein, but lipids, glucose, amino acids, hormones, and other molecules are also present.

One of the blood plasma proteins that performs an important role in the interactions between red cells is the enzyme precursor ("zymogen") fibrinogen. When activated by a sequence of processes involved in the clotting response, fibrinogen is converted into the enzyme fibrin, which plays a key role in the formation of a clot. In its normal state, fibrinogen plays an important role in the formation of "rouleaux" of red cells. A rouleau is a stack of red cells that typically involves more than six cells. Rouleaux spontaneously form in quiescent blood or blood under sufficiently small shear rates.[95,118,205] Rouleau formation involves a competition between attractive forces due to the adsorption of fibrinogen or similar macromolecules on the red cell membrane and the subsequent formation of bridges between red cells and repulsive forces arising from anions on the red cell membranes.[30,31,139,140,197] Shear stresses larger than roughly 0.2 Pa break up all aggregates of red cells. The formation of rouleaux causes red cells to sediment more rapidly than dispersed cells in quiescent blood.

Blood Rheology

The rheology of blood flow is complex. The specific gravity of plasma and red cells is 1.03 and 1.10, respectively. Blood plasma is a Newtonian fluid with a viscosity equal to 0.00012 Pa s. Whole blood is a non-Newtonian fluid with a viscosity that depends on the shear rate, hematocrit, and temperature. Experimental data on the variation of the viscosity of blood as a function of shear rate for different hematocrits and at various temperatures were reported by Fung,[110] among others. These results show that the viscosity of blood decreases with shear rate and increases rapidly with hematocrit. While there has been disagreement about the existence of a yield stress for the blood, most data indicates that the normal human blood has a yield stress roughly equal to 0.005 Pa, which is almost independent of the temperature in the range 10–37 °C. There is general agreement that the blood yield stress is a strong function of hematocrit.

In large, straight arteries, where the shear rate is larger than about $100\,\mathrm{s}^{-1}$, blood can be approximated as a Newtonian fluid (Womersley[277] and Berger and Joe[26]). This is because the viscosity of blood becomes nearly constant for shear rates of this order or larger. There are, however, complications that can arise even in large arteries. For example, immediately downstream of bifurcations, regions of separated flow develop (see, for example, Berger and Jou,[26] and Kaazempur-Mofrad et al.[144]). In the unsteady, recirculating flows downstream of the bifurcation, local shear rates can be small in magnitude. Moreover, phenomena such as red cell depletion may occur in these regions.

Flow Behavior of Red Cells

Under equilibrium conditions, red cells exist as biconcave disks. In flowing blood, red cells may have significantly different shapes. The smallest blood vessels in the human body have diameters that are smaller than the largest linear dimension of an equilibrium red cell. To pass through such vessels, a red cell must undergo large deformations. This is made possible by the unusual characteristics of the red cell membrane. The red cell membrane permits very large shape deformations provided they do not significantly change the surface area of the cell. The elastic characteristics of the red cell membrane were modeled by Skalak et al.[229] (see also Goldsmith and Skalak[119]). The red cell membrane is comprised of a phospholipid bilayer that is attached to a protein network (the "cytoskeleton") that is inside the cell. The cell membrane strongly resists changes in area because of the properties of the phospholipid bilayer. The elastic response of the membrane is due to the elastic properties of the cytoskeleton. Evans and Skalak[94] introduced a modification of the Skalak et al. model that separates the effects of shear deformation and area changes.

The behavior of isolated red cells in laminar shear flows has been extensively studied. Deformation is observed at shear stresses larger than 0.1 Pa (Goldsmith[117] and Goldsmith and Marlow[118]). In laminar tube flow, isolated red cells migrate to the center of the tube. This effect is observed even when the inertial effects are negligible. Deformable drops also migrate to the center of a laminar tube flow.[116,146] Chaffey et al.[39] and Chaffey and Brenner[40] showed that this effect is caused by the interaction of the tube wall and the disturbance flow created by the drop. This interaction creates an inward radial flow close to the deformed drop. Hetsroni et al.[131] generalized the above analysis to Poiseuille flow. Leal[167] reviewed more recent studies of drop migration due to deformation as well as inertial migration.

The phenomenon of inward migration is associated with the "Fåhraeus effect" and the "Fåhraeus–Lindqvist effect."[95,96] The Fåhraeus effect refers to the fact that, in tubes with diameters smaller than 0.5 mm, the average concentration of red cells is smaller than in the feed reservoir or the collection reservoir. Moreover, the average concentration of red cells decreases with decreasing tube diameter. The region of fluid near the tube wall has a lower concentration of cells than the core region because of the tendency of red cells to migrate away from the wall. Since the fluid near the tube wall moves more slowly than the fluid in the core region, it has a larger residence time in the tube and, hence, the average concentration of red cells inside the tube is smaller than in the feed reservoir.

The Fåhraeus–Lindqvist effect is related to the Fåhraeus effect. In tubes having diameters smaller than 0.5 mm, when the shear rate is large enough that the apparent viscosity no longer increases with increasing flow rate, the apparent viscosity of blood decreases with decreasing tube diameter. This phenomenon is due to the fact that there exists a layer of cell-depleted liquid adjacent to the wall, and the ratio of the thickness of this layer to the tube radius increases as the tube radius decreases. Thus, since the suspending liquid has a lower viscosity than the suspension, the apparent viscosity decreases.

The results of Cokelet and Goldsmith[59] and Cokelet et al.[60] suggest that, in large blood vessels with high shear rate, if a cell-depleted zone exists at all, it is likely to be very small (in the order of a red cell diameter). The reason for this is that the wall shear rate in large arteries is typically $10^2\,\text{s}^{-1}$ or larger. The study by Cokelet and Goldsmith[59] demonstrated that small shear rates are needed to produce a significant cell-depleted zone; the shear rates in large arteries are at least an order of magnitude larger than the maximum shear rate at which cell depletion has been observed.

In regions immediately downstream of bifurcations in large arteries, separated flow regions can exist (Berger and Jou,[26] Kaazempur-Mofrad et al.[144]). Such regions are known to be associated with low shear rates. For example, the finite element solutions reported by Kaazempur-Mofrad et al. show large regions near the bifurcation in a human carotid artery where the shear stress is less than 1% of the value on straight sections of artery. Figure 62 shows discretization of a human carotid obtained and the distribution of shear stress in the artery based on a finite element. The figure was supplied to the authors by Professor C. Aidun of Georgia Institute of Technology. It is, therefore, conceivable that significant cell-depleted regions might exist in such regions. Although the diameter of large arteries, such as the carotid artery, is typically a few millimeters (see e.g., Kaasempur-Mofrad et al.[144]), the experiments reported by Cokelet et al.[60] demonstrated that cell depletion can occur in Couette flows with 1-mm gaps provided that the shear rate was sufficiently small ($2.2\,\text{s}^{-1}$ in their experiment). Furthermore, the cell-depleted region can be as large as 100 μm in thickness. Cokelet et al.[60] pointed out that "cell-depleted" does not mean "cell-free"; the concentration of red cells is small but not zero in the cell-depleted zone.

The aforementioned results pose a formidable challenge for simulations of flows downstream of arterial bifurcations. Normally, one assumes that blood can be treated as a Newtonian fluid in large arteries since the shear rates are large in long, straight

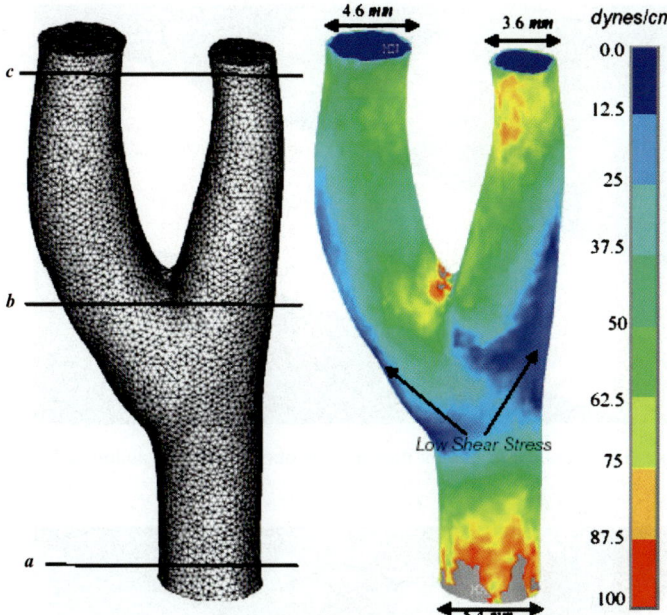

Fig. 62 A computational discretization of a human carotid artery bifurcation and the corresponding distribution of wall shear stress obtained from a finite element simulation (Courtesy of Prof. C. Aidun)

sections of arteries. In the recirculation zone, the shear rate is much smaller and one must consider the additional complication of the possible existence of a cell-depleted zone. Thus, one should expect non-Newtonian behavior and a variable hematocrit in such regions, and this calls into question the usefulness of finite element solutions that are based on the assumption of Newtonian behavior with spatially and temporally uniform properties.

Computer simulations of flow in the abdominal aorta were carried out by Artoli et al.,[11] and Ahmadi and Joseph[3]. Sample pressure and surface shear stress contours are shown in Figs. 63 and 64. The upper arteries connected to the abdominal aorta are also identified in Fig. 63. The computational domain actually included the inferior mesenteric artery and iliac bifurcation at the end of abdominal aorta. It is seen that there is noticeable pressure drop from the main abdominal aorta to the bifurcating arteries. The celiac and superior mesenteric arteries are at higher pressure compared with the left and right renal arteries. While not shown in the figures, the inferior mesenteric and iliac arteries are at lower pressure. Figure 63 also shows that the pressure decreases rapidly along the smaller arteries.

Figure 64 shows that the wall shear stress is relatively low in the abdominal aorta but is relatively high in the smaller arteries. The shear stress is relatively high in the renal arteries. It is also seen that the shear stress at the interface between the bifurcating arteries and the abdominal aorta is relatively high.

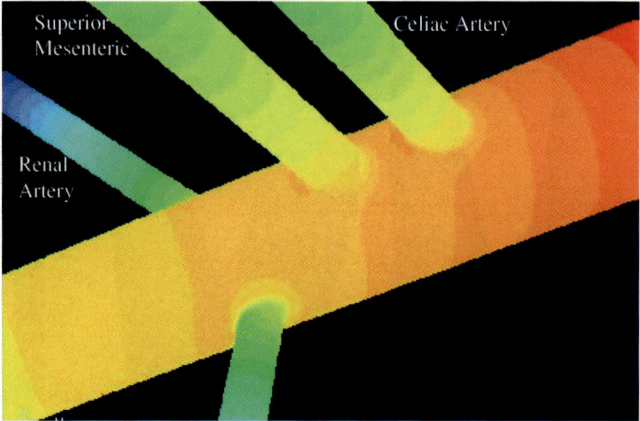

Fig. 63 Sample computational results for pressure contours in a human abdominal aorta (Ahmadi and Joseph[3])

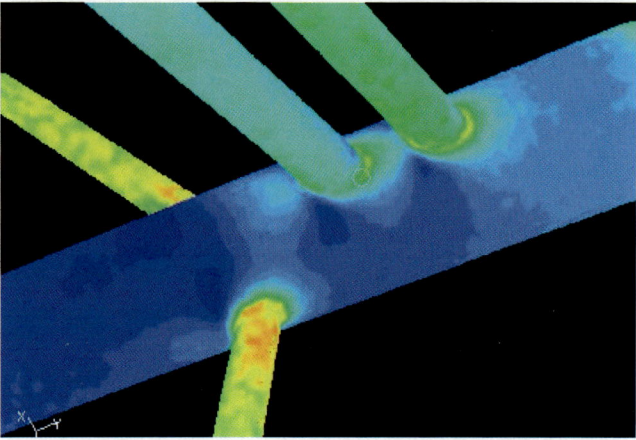

Fig. 64 Sample computational results for the shear stress contours in a human abdominal aorta (Ahmadi and Joseph[3])

Atherosclerosis

The flow near arterial bifurcations is of considerable medical interest since atherosclerosis typically develops in such regions.[159,233] Moreover, it is known from molecular studies[81,204] that there is a direct regulation of proatherosclerotic gene expression by mechanical forces. Complex interactions between hemodynamic forces, NO, and reactive species (e.g., $ONOO^{-1}$ and H_2O_2) play an important role in the pathogenesis of atherosclerosis.[122,128] The rate of production of NO in the endothelial cells lining the surfaces of blood vessels and the rate of scavenging of NO by hemoglobin in red cells is quite different in the straight ("uniform flow") and bifurcated or highly curved ("disturbed flow") sections of arteries due to differences in the hemodynamic stresses on the endothelial cells. This difference in biochemistry caused by differences in the hemodynamic stresses is important in the localization of atherosclerotic lesion formation. Although plausible, it has not been established that this explains the effect of the disturbed flow stress on the onset of atherosclerosis in artery bifurcations.

Motivated in part by the need to develop a better understanding of the onset of atherosclerotic lesions, many studies have been aimed at understanding the characteristics of the endothelial cells lining artery walls. The importance of endothelial response to hemodynamics was established by Caro and Nerem.[38] The book by Davies[75] provides a detailed discussion. Many in vitro experiments have demonstrated that mechanical stresses caused by blood flow are extremely important throughout the atherosclerotic disease process.[46,76,234]

In the relatively straight sections of arteries, the shear stress is relatively large (greater than 1 Pa). The endothelial cells lining the artery walls in these sections are elongated in the flow direction with relatively tight cell–cell junctions. These regions show a much higher resistance to atherosclerosis development compared to regions near bifurcations or regions where the curvature is large. In the latter region, flow separation and recirculating flow result in low average shear stress or high oscillatory shear index;[78,112,148,149,159,160,202–204] the oscillatory index is defined to be the ratio of the mean magnitude of the instantaneous local shear stress to the instantaneous local shear stress. Much is known about the correlation between the hemodynamic stress and endothelial cell response, but less is known about the underlying mechanism(s). In what follows, some recent progress in obtaining a better understanding of these mechanisms is briefly reviewed.

Cai and Harrison[34] showed that NO inhibits some of the earliest events in the onset of atherosclerotic lesion formation. These events include the downregulation of expression of the monocyte chemoattractant peptide-1 (MCP1) and the vascular cell adhesion molecule-1 (VCAM-1), prevention of propagation of lipid oxidant, and decreased platelet aggregation. It has also been shown that NO inhibits expression and activity of matrix metalloproteinases (MMPs), limiting cell migration, enhancing atherosclerotic plaque stability, and inhibiting cardiac hypertrophy.[9,46,232,268] The production of NO is important to inhibiting disease because almost all of the risk factors for atherosclerosis including hypercholesterolemia, diabetes, insulin resistance, hypertension, and cigarette smoking reduce endothelial production of NO (Cai and Harrison[34]). In many cases, decreased NO bioavailability is due to increased production of O_2^- ("superoxide") and other reactive oxide species (ROS). When NO reacts with superoxide, it produces the anion peroxynitrite ($ONOO^-$), which is a strong oxidant that can promote many of the proatherosclerotic events described earlier.[201]

For the aforementioned reasons, the balance of NO, O_2^-, and $ONOO^-$ plays a central role in the onset of atherosclerotic lesion formation. The process by which this balance is regulated depends on a combination of shear-induced NO production in the endothelial cells, the presence of heparin sulfate (HS) in the so-called glycocalyx (see later), scavenging of NO by hemoglobin in red cells, hematocrit distribution near the surfaces of the endothelial cells, and diffusive transport of NO, O_2^-, ONOO$^-$, and other ROS throughout the vasculature from the lumen to the glycocalyx (see later), endothelial cells, vessel wall, and the tissue beneath the endothelial cells.

The glycocalyx is a layer on the surface of the endothelial cells. It is also referred to as the "endothelial surface layer" or "ESL." It was first identified by Luft.[180] Vink and Duling[270] observed the glycocalyx in vivo in hamster muscle capillaries using bright field and fluorescence microscopy. They estimated that the glycocalyx was roughly 0.5-µm thick. The glycocalyx serves several different purposes. It acts as a transport barrier, a porous interface in the motion of red and white cells in capillaries, and as a mechanotransducer of fluid shear stress to the actin cortical cytoskeleton of the endothelial cell (Weinbaum et al., 2003).

Squire et al.[231] showed that the glycocalyx is a quasiperiodic array of bush-like structures consisting of glycoprotein fibers with extended glycan sidechains. Figure 65 is a sketch based on a more detailed model suggested by Weinbaum et al.[276] Weinbaum et al.[276] developed a mathematical model of the glycocalyx. The transport properties were modeled by a Brinkman medium with a Darcy permeability that was based on the structure sketched in Fig. 65. The mechanical properties of the glycocalyx were based on a computation of the flexural rigidity of the glycoprotein fibers.

The glycocalyx's properties as a transport barrier and mechanotransducer are of particular interest in understanding the onset of atherosclerosis. In straight sections of arteries, the glycocalyx serves to prevent macromolecules and platelets from accumulating on the endothelial cell membrane. In bifurcations and strongly curved sections of arteries, this protective function breaks down. At present, it is not possible to offer a definitive explanation for this phenomenon, but it seems possible that changes occur in the structure of the glycocalyx. Additionally, the lower shear stress may affect the production of certain molecular species including NO and ROS. A full understanding of the complex mechanisms will require a multiscale approach. The shear stress on the glycocalyx is determined by the flow in the artery. The de-

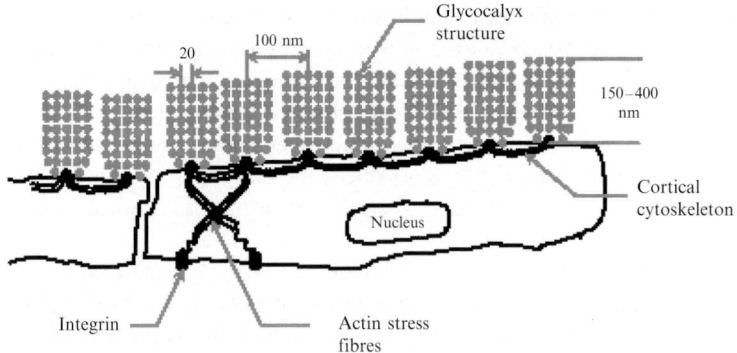

Fig. 65 A schematic representation of the endothelial cell showing the bush-like structure of the glycocalyx in the Weinbaum et al.[?] model

termination of the flow in a region of disturbed flow is complicated by the fact that the hematocrit distribution is a priori unknown, and must, therefore, be computed as part of the problem. Moreover, the hematocrit distribution is also needed to compute the concentrations of various molecular species, such as NO, since red cells scavenge these molecules efficiently.[32] Finally, the structure of the glycocalyx in regions of low or oscillatory shear stress is currently not fully understood. It seems likely that any approach to understanding these complicated, interrelated phenomena will require an iterative approach involving different models for different scales. One of the challenges is to compute the flow and hematocrit distribution in disturbed flow regions. A possible approach to this challenge is briefly discussed below.

DNS of Blood Flow

Several different DNS methods have been used to simulate blood flow. These include the finite element method (FEM), boundary element method (BEM), and the lattice Boltzmann method (LBM). For example, Kaazempur-Mofrad et al.[144] used the FEM to simulate blood flow in an artery. For rough calculation purposes, this may be acceptable. As noted above, however, one has to assume that the blood behaves as a Newtonian fluid to perform such simulations. Moreover, the simulations do not provide information about the spatial distribution of the hematocrit.

By contrast, the BEM that involves solving boundary integral equations with a collocation method[219,221] is restricted to small Reynolds numbers. The method takes advantage of the linearity of the Stokes equation and permits one to perform simulations for highly complex, time-dependent geometries. For example, Pozrikidis[218] used the BEM to simulate the motion of a red cell in a capillary vessel. Since red cells are comparable to or larger than the diameter of capillaries, considerable distortion occurs as was discussed earlier. The BEM permits highly accurate simulations of such processes. If, however, one wishes to investigate phenomena associated with the onset of atherosclerosis, the BEM is not suitable since the Reynolds numbers in large arteries are of the order of thousands.

The LBM is an alternative to the above approaches that may be useful as a tool for investigating flowing suspensions of red cells at realistic Reynolds numbers. The LBM originated from the method of lattice gas automata (LGA).[108] Both the LGA and LBM compute solutions of the Navier–Stokes equation using approaches that are quite different from conventional CFD methods. The LGA used "computational particles" on a lattice of regularly spaced lattice points. By requiring that particles could move only from one lattice point to an adjacent lattice point and imposing certain "collision rules" on the lattice points, it was possible to show that one could obtain approximate solutions of the Navier–Stokes equation by certain averaging methods. Although the LGA had certain computational advantages over competing CFD methods such as locality of the operations, which facilitated computations on massively parallel computers, the averaging step proved too costly for the method to be competitive.

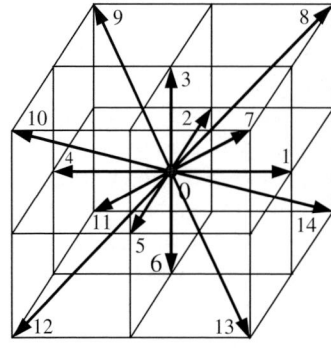

$$e_i = \begin{cases} (0,0,0), & i=0 \\ perm(\pm 1,0,0), & i=1,\dots,6 \\ perm(\pm 1, \pm 1, \pm 1), & i=7,\dots,14 \end{cases}$$

Fig. 66 The 15-velocity lattice

The LBM is similar to the LGA in that one performs simulations for populations of computational particles on a lattice. It differs from the LGA in that one computes the time evolution of "particle distribution functions." These particle distribution functions are a discretized version of the particle distribution function that is used in Boltzmann's kinetic theory of dilute gases. There are, however, several important differences. First, the Boltzmann distribution function is a function of three continuous spatial coordinates, three continuous velocity components, and time. In the LBM, the velocity space is truncated to a finite number of directions. One popular lattice uses 15 lattice velocities, including the rest state. The dimensionless velocity vectors are shown in Fig. 66. The length of the lattice vectors is chosen so that, in one time step, the population of particles having that velocity will propagate to the nearest lattice point along the direction of the lattice vector. If one denotes the distribution function for direction i by $f_i(x,t)$, the fluid density, ρ, and fluid velocity, u, are given by

$$\rho(x,t) = \sum_i f_i(x,t), \quad \rho(x,t)u(x,t) = \sum_i f_i(x,t)e_i. \tag{154}$$

In (1), the lattice velocity vector is denoted by e_i. The distribution functions satisfy a linear algebraic evolution equation that involves a collision term. The most popular version of the LBE is called the BGK formulation and it is based on the simplified model of the Boltzmann kinetic equation that was devised by Bhatnagar et al. (1954). In the BGK formulation, the LBE takes the following form:

$$f_i(x,t+1) = f_i(x,t) + \frac{1}{\tau}(f_i(x,t) - f_i^{eq}(x,t)). \tag{155}$$

Here, it is understood that the space and time variables are dimensionless and that the magnitude of the time step is unity. The symbol f_i^{eq} denotes the "equilibrium" distribution function, which is chosen so that, when the Chapman–Enskog procedure is applied to (2), one obtains the Navier–Stokes equation. The details of this procedure may be found in the review by Chen and Doolen.[45] Each time step involves two substeps: (1) a collision substep in which the collision term is added to the value of the distribution function on the current time step, (2) a propagation step in which the updated distribution function is assigned to the closest lattice point along the direction of the lattice vector for that distribution function. All of the operations involved in the solution of the LBE are local. This greatly facilitates implementation of LBM programs on parallel computers.

There are two general approaches that one might follow in trying to apply the LBM to flowing suspensions of red cells. One approach is to treat the red cell as a drop. Dupin et al.[82] used this approach. Since one is considering a suspension of drops, a two-fluid version of the LBM must be considered. Chen and Doolen[44] described some early work on developing two-fluid versions of the LBM. A drawback of this approach is that it is not clear how such an approach could correctly describe the equilibrium shape of a red cell. One would need to incorporate a membrane with the correct elastic properties into the LBM formulation.

Ding and Aidun[80] took an alternative approach to simulating red cell suspensions using a single-phase LBM. They treated the cells as rigid particles, but with the correct equilibrium shape. The fluid obeyed rigid, no-slip boundary conditions on the particle surfaces, and the motion of the particles was determined by numerically solving Newton's law using the distribution of stress on the particle surface to compute the force and torque on the particle. Figure 67 shows velocity profiles that were computed by Ding and Aidun[80] for different hematocrits. Their method may

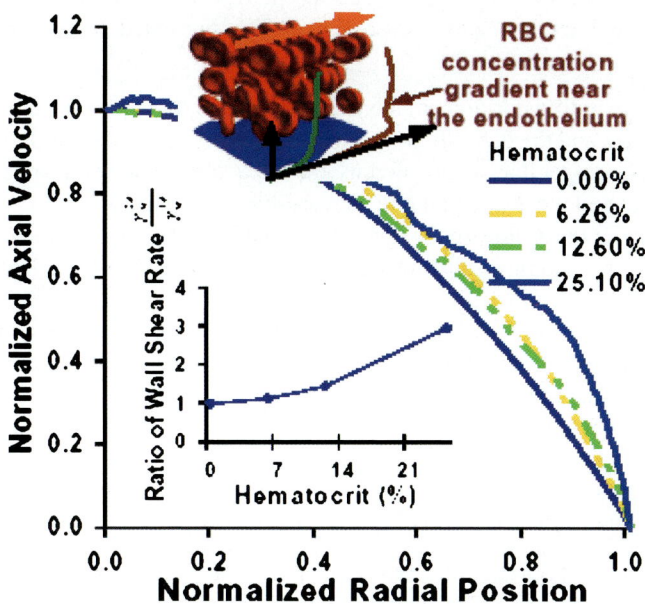

Fig. 67 Velocity profiles obtained by Ding and Aidun[80] for different hematocrits using a LBM simulation

also be criticized since red cells deform significantly at typical hematocrits under flow conditions that are typical of large arteries or veins. It seems feasible, however, to correct this problem. Aidun and Qi[7] discussed how the LBM can be used to simulate fluid interactions with deformable membranes. Furthermore, Ding and Aidun[79] described LBM techniques for simulating suspended particles in near contact. Aidun and Ding[8] used the latter technique to simulate the trajectories of pairs of particles sedimenting in a vertical channel and obtained excellent agreement with finite element solutions for the same problem.

Thus, it appears feasible to develop LBM simulations of deformable particles with the shape and mechanical characteristics of red cells. As yet, however, no such simulations have been published. The extent to which LBM simulations can simulate blood flow in realistic geometries for arteries remains to be seen.

Conclusions

In this chapter, fundamentals of particle transport, deposition, and removal were reviewed, and some of their biomedical applications were described. Particular attention was given to recent advances in computational modeling of nano and microparticle transport and deposition in human airways. Transport and deposition processes in lung bifurcations, nose and oral passages, as well as in alveolar cavities were discussed. Rheological properties of blood are also discussed, and sample simulation results are presented. The presented results showed the following:

- The molecular diffusion and turbulence are dominant for deposition of nanometer size particles and the deposition rate increases as particle diameter decreases.
- For particles larger than a few microns, the impaction is the dominant deposition mechanism and the deposition rate increases with the particle size.
- Turbulence in the airflow in the upper airway affects the capture efficiency.
- Computer simulation results are in good agreement with the available experimental data.
- Nonlinear behavior of the blood can be included in the computational model.
- The computational model provides some detailed information on the details of blood flow in arterial bifurcations.

Acknowledgments The financial support of National Science Foundation as part of the Combined Research and Curriculum Development Project is gratefully acknowledged. The authors also would like to thank Cyrus Aidun, Omid Abouali, Parsa Zamankhan, Lin Tian, Alireza Mazaheri, Ilsoo Chang and Rajosh Joseph for making their work available to us. Thanks is also given to CRCD project team members Cetin Cetinkaya, Suresh Dhaniyala, Jeffery Taylor, Stephen Doheny-Farina, Mark Glauser, Fa-Gung Fan, Ahmed Busnaina, Xinli Jia, David Schmidt and Kambiz Nazridoust for many helpful discussions.

References

1. Ahmadi, G. (1970). Analytical Prediction of Turbulent Dispersion of Finite Size Particle. Ph.D. Thesis, Purdue University.
2. Ahmadi, G. and Goldschmidt, V.W. (1971). Motion of Particle in a Turbulent Fluid—The Basset History Term. J. Appl. Mec. Trans. ASME, Vol. 38, pp. 561–563.
3. Ahmadi, G. and Joseph, R. (2007). Simulations of Pulsatile Blood Flow in the Abdominal Aorta with Newtonian and Non-Newtonian Models. Comput. Biol. Med. (submitted for publication).
4. Ahmadi, G. and Smith, D.H. (1998). Particle Transport and Deposition in a Hot-Gas Cleanup Pilot Plant. Aerosol Sci. Technol., Vol. 29, pp. 183–205.
5. Ahmadi, G. and Chen, Q. (1998). Dispersion and Deposition of Particles in a Turbulent Pipe Flow with Sudden Expansion. J. Aerosol Sci., Vol. 29, pp. 1097–1116.
6. Ahmed, A.M. and Elghobashi, S.E. (2000). On the mechanisms of modifying the structure of turbulent homogeneous shear flows by dispersed particles. Phys. Fluids, Vol. 12, pp. 2906–2930.
7. Aidun, C.K. and Qi, D. (1998). A New Method for Analysis of the Fluid Interaction with a Deformable Membrane. J. Stat. Phys., Vol. 90, pp. 145–158.
8. Aidun, C.K. and Ding, E.-J. (2003). Dynamics of Particle Sedimentation in a Vertical Channel: Period-Doubling Bifurcation and Chaotic State. Phys. Fluids, Vol. 15, pp. 1612–1621.
9. Akool, E-S., Kleinert, H., Hamada, F.M.A., Abdelwahab, M.H., Förstermann, U., Pfeilschifter, J., and Eberhardt, W. (2003). Nitric Oxide Increases the Decay of Matrix Metalloproteinase 9 mRNA by Inhibiting the Expression of mRNA-Stabilizing Factor HuR. Mol. Cell Biol., Vol. 23, pp. 4901–4916.
10. Alberts, B., Bray, D., Lewis, J., Raff, M., Roberts, K., and Watson, J.D. (1989). Molecular Biology of the Cell, Garland Publishing, New York/London.
11. Artoli, A.M., Hoekstra, A.G., and Sloot, P.M.A. (2006). Mesoscopic Simulations of Systolic Flow in the Human Abdominal Aorta. J. Biomech., Vol. 39, pp. 873–884.
12. Arnason, G. (1982). Measurements of Particle Dispersion in Turbulent Pipe Flow. Ph.D. Thesis, Washington State University, Pullman, WA.
13. Asgharian, B. and Ahmadi, G. (1998) Effect of Fiber Geometry on Deposition in Small Airways of the Lung. Aerosol Sci. Technol., Vol. 29, pp. 459–474.
14. Asgharian, B. and Yu, C.P. (1988). Deposition of Inhaled Fibrous Particles in the Human Lung. J. Aerosol Med., Vol. 1, pp. 37–50.
15. Asgharian, B. and Yu, C.P. (1989). Deposition of Fibers in the Rat Lung. J. Aerosol Sci., Vol. 20, pp. 355–366.
16. Asgharian, B. and Anjilvel, S. (1994). Inertial and Gravitational Deposition of Particles in a Square Cross Section Bifurcating Airway. Aerosol Sci. Technol., Vol. 20, pp. 177–193.
17. Asgharian, B., Price, O.T., and Hofmann, W. (2006). Prediction of Particle Deposition in the Human Lung Using Realistic Models of Lung Ventilation. J. Aerosol Sci., Vol. 37, pp. 1209–1221.
18. Balashazy, I. (1994). Simulation of Particle Trajectories in Bifurcating Tubes. J. Comput. Phys., Vol. 110, pp. 80–88.
19. Balashazy, I. and Hofmann, W. (1993). Particle Deposition in Airway Bifurcations: I. Inspiratory Flow. J. Aerosol Sci., Vol. 24, pp. 745–772.
20. Balashazy, I. and Hofmann, W. (1993). Particle Deposition in Airway Bifurcations: II. Expiratory Flow. J. Aerosol Sci., Vol. 24, pp. 773–786.
21. Balashazy, I., Hofmann, W., and Heistracher, T. (1999). Computation of Local Enhancement Factors for the Quantification of Particle Deposition Patterns in Airway Bifurcations. J. Aerosol Sci., Vol. 30, pp. 185–203.
22. Baron, P.A., Deye, G.J., and Fernback, J. (1994). Length Separation of Fibers. Aerosol Sci. Technol., Vol. 21, pp. 79–192.
23. Baron, P.A. (2001). Measurement of Airborne Fibers: A Review. Ind. Health, Vol. 39, pp. 39–50.

24. Baron, P.A., Sorensen, C.M., and Brockmann, J.B. (2001). Nonspherical Particle Measurements: Shape Factors, Fractals, and Fibers. In Aerosol Measurement, 2nd Ed, Edited by Baron, P.A. and Willeke, K., Wiley-Interscience, New York.
25. Baron, P.A., Deye, G., Aizenberg, V., and Castranova, V. (2002). Generation of Size-Selected Fibers for a Nose-Only Inhalation Toxicity Study. Ann. Occup. Hyg, Vol. 46, Suppl. 1, pp. 186–190.
26. Berger, S.A. and Jou, L.-D. (2000). Flows in Stenotic Vessels. Ann. Rev. Fluid Mech., Vol. 32, pp. 347–382.
27. Brenner, H. (1961). The Slow Motion of a Sphere through a Viscous Fluid Towards a Plane Surface. Chem. Eng. Sci., Vol. 16, pp. 242–251.
28. Bhatnagar, P.L., Gross, E.P., and Krook, M. (1954). A model for collision processes in gases. I: small amplitude processes in charged and neutral one-component systems. Phys Rev. Vol. 94, pp. 511–525.
29. Brooke, J.W., Kontomaris, K., Hanratty, T.J., and McLaughlin, J.B. (1992). Turbulent deposition and trapping of aerosols at a wall. Phys. Fluids A, Vol. 4, pp. 825–834.
30. Brooks, D.E. and Seaman, G.V.F. (1973). Effect of Neutral Polymers on Electrokinetic Potential of Cells and Other Charged Particles. I. Models for the Zeta Potential Increase. J. Colloid Interface Sci., Vol. 43, pp. 670–686.
31. Brooks, D.E. (1973). Effect of Neutral Polymers on Electrokinetic Potential of Cells and Other Charged Particles. II. Model for Effect of Adsorbed Polymer on Diffuse Double Layer. J. Colloid Interface Sci., Vol. 43, pp. 687–699.
32. Buerk, D.E. (2001). Can We Model Nitric Oxide Biotransport? A Survey of Mathematical Models for a Simple Diatomic Molecule with Surprisingly Complex Biological Activities. Ann. Rev. Biomed. Eng., Vol. 3, pp. 109–143.
33. Calabrese, R.V. and Middleman, S. (1979). The Dispersion of Discrete Particles in a Turbulent Fluid Field. AIChE J., Vol. 25, pp. 1025–1035.
34. Cai, H. and Harrison, D.G. (2000). Endothelial Dysfunction in Cardiovascular Diseases: The Role of Oxidant Stress. Circ. Res., Vol. 87, pp. 840–844.
35. Cao, J. and Ahmadi, G. (1995). Gas-Particle Two-Phase Turbulent Flow in a Vertical Duct. Int. J. Multiphase Flow, Vol. 21, pp. 1203–1228.
36. Cao, J. and Ahmadi, G. (1996). Gravity Granular Flows Down an Inclined Bumpy Chute. J. Fluid Mech., Vol. 316, pp. 197–221.
37. Cao, J. and Ahmadi, G. (2000). Gas-Particle Two-Phase Turbulent Flows in Horizontal and Vertical Ducts. Int. J. Eng. Sci., Vol. 38, pp. 1961–1981.
38. Caro, C.G. and Nerem, R.M. (1973). Common Carotid Artery Transport of 14C-4-Cholesterol Between Serum and Wall in the Perfused Dog. Circ. Res., Vol. 32, pp. 187–205.
39. Chaffey, C., Brenner, H., and Mason, S.G. (1965). Particle Motions in Sheared Suspensions, Part 18: Wall Migration (Theoretical). Rheol. Acta, Vol. 4, pp. 64–72.
40. Chaffey, C.E. and Brenner, H. (1967). A Second-Order Theory for Shear Deformation of Drops. J. Colloid Interface Sci., Vol. 24, pp. 258–269.
41. Chan, T.L. and Schreck, R.M. (1980). Effect of the Laryngeal Jet in the Human Trachea and Upper Bronchial Airways. J. Aerosol Sci., Vol. 11, pp. 447–459.
42. Chang, I-S. and Ahmadi, G. (2007). Pneumonic Alveolar Cavity Transport and Deposition during Inhalation. Comput. Biol. Med. (Submitted for publication).
43. Chen, C.P. and Wood, P.E. (1985). A turbulence Closure Model for Dilute Gas-Particle Flows. Can. J. Chem. Eng., Vol. 63, pp. 349–360.
44. Chen, B.T., Yeh, H.C., and Johnson, N.F. (1996). Design and use of a virtual impactor and an electrical classifier for generation of test fiber aerosols with narrow size distributions. J. Aerosol Sci., Vol. 27, pp. 83–94.
45. Chen, S. and Doolen, G.D. (1998). Lattice Boltzmann Method for Fluid Flows. Ann. Rev. Fluid Mech., Vol. 30, pp. 329–364.
46. Chen, H.H. and Wang, D.L. (2004). Nitric Oxide Inhibits Matrix Metalloproteinase-2 Expression via the Induction of Activating Transcription Factor 3 in Endothelial Cells. Mol. Pharmacol., Vol. 65, pp. 1130–1140.

47. Cheng, Y.S. (2003). Aerosol Deposition in the Extrathoracic Region. Aerosol Sci. Technol., Vol. 37, pp. 659–671.
48. Cheng, Y.S., Su, Y.F., Yeh, H.C., and Swift, D.L. (1993). Deposition of Thoron Progeny in Human Head Airways. Aerosol Sci. Technol., Vol. 18, pp. 359–375.
49. Cheng, Y.S., Yamada, Y., Yeh, H.C., and Swift, D.L. (1988) Diffusional Deposition of Ultrafine Aerosols in a Human Nasal Cast. J. Aerosol Sci., Vol. 19, pp. 741–751.
50. Cheng, Y.S., Yeh, H.C., Guilmette, R.A., Simpson, S.Q., Cheng, K.H., and Swift, D.L. (1996). Nasal Deposition of Ultrafine Particles in Human Volunteers and Its Relationship to Airway Geometry. Aerosol Sci. Technol., Vol. 25, pp. 274–291.
51. Cheng, Y.S., Zhou, Y., and Chen, B.T. (1999). Particle Deposition in a Cast of Human Oral Airways. Aerosol Sci. Technol., Vol. 31, pp. 286–300.
52. Cherukat, P. and McLaughlin, J.B. (1990). Wall-Induced Lift on a Sphere. Int. J. Multiphase Flow, Vol. 16, pp. 899–907.
53. Cherukat, P. and McLaughlin, J.B., and Graham, A.L. (1994). The inertial Lift on a Rigid Sphere Translating in a Linear Shear Flow Field. Int. J. Multiphase Flow, Vol. 20, pp. 339–353.
54. Chowdhury, S.J. and Ahmadi, G. (1992). Int. J. Nonlinear Mech., Vol. 27, pp. 705–718.
55. Cleaver, J.W. and Yates, B. (1975). A Sublayer Model for Deposition of Particles from Turbulent Flows. Chem. Eng. Sci., Vol. 30, pp. 983–992.
56. Clift, R., Grace, J.R., and Weber, M.E. (1978). Drops and Particles, Academic Press, New York.
57. Cohen, B.S., Sussman, R.G., and Lippmann, M. (1990). Ultrafine Particle Deposition in a Human Tracheobronchial Cast. Aerosol Sci. Technol., Vol. 12, pp. 1082–1091.
58. Cohen, B.S. and Asgharian, B. (1990). Deposition of Ultrafine Particles in the Upper Airways: An Empirical Analysis. J. Aerosol Sci., Vol. 21, pp. 789–797.
59. Cokelet, G.R. and Goldsmith, H.L. (1991). Decreased Hydrodynamic Resistance in the Two-Phase Flow of Blood through Small Vertical Tubes at Low Flow Rates. Circ. Res., Vol. 68, pp. 1–17.
60. Cokelet, G.R., Brown, J.R., Codd, S.L., and Seymour, J.D. (2005). Magnetic Resonance Microscopy Determined Velocity and Hematocrit Distributions in a Couette Viscometer. Biorheology, Vol. 42, pp. 385–399.
61. Collins, L.R. (2000). Reynolds Number Scaling of Preferential Concentration of Particles in Isotropic Turbulence. AIChE Annual Meeting, Los Angeles, CA.
62. Comer, J.K., Kleinstreuer, C., Hyun, S., and Kim, C.S. (2000). Aerosol Transport and Deposition in Sequentially Bifurcation Airways. ASME J. Biomech. Eng., Vol. 122, pp. 152–158.
63. Cooper, D.W. (1986). Particle Contamination and Microelectronics Manufacturing: An Introduction. Aerosol Sci. Technol., Vol. 5, pp. 287–299.
64. Corn, M. (1976). In Air Pollution, Edited by Stren, A.C., Academic Press, New York.
65. Corrsin, S. and Lumley, J.L. (1956). On the Equation of Motion for a Particle in Turbulent Fluid. Appl. Sci. Res., Vol. 6, pp. 114–116.
66. Cox, R. G. and Brenner, H. (1967). The Slow Motion of a Sphere through a Viscous Fluid Towards a Plane Surface. II. Small Gap Widths, Including Inertial Effects. Chem. Eng. Sci., Vol. 22, pp. 1753–1777.
67. Crowe, C.T. (1982). REVIEW – Numerical Models for Dilute Gas-Particle Flows. J. Fluid Eng. Trans. ASME, Vol. 104, pp. 297–303.
68. Crowe, C.T. (1986). Two-Fluid vs. Trajectory Model: Range of Applicability. Gas-Solid Flows, ASME FED, Vol. 35, pp. 91–96.
69. Csanady, G.T. (1963). Turbulent Diffusion of Heavy Particles in the Atmosphere. J. Atmos. Sci., Vol. 20, pp. 201–208.
70. Dandy, D.S. and Dwyer, H.A. (1990). A Sphere in Shear Flow at Finite Reynolds Number: Effect of Shear on Particle Lift, Drag and Heat Transfer. J. Fluid Mech., Vol. 216, pp. 381–410.
71. Darquenne, C. (2001). A realistic two-dimensional model of aerosol transport and deposition in alveolar zone of human lung. Journal of Aerosol Science, Vol. 32, pp. 1161–1174.

72. Darquenne, C. (2002). Heterogeneity of aerosol deposition in two-dimensional model human alveolar ducts. Journal of Aerosol Science, Vol. 33, pp. 1161–1174.
73. Darquennne, C. and Paive, M. (1996). Two- and three dimensional simulations of aerosol transport and deposition in alveolar zone of human lung. Journal of Applied Physiology, Vol. 80, pp. 1401–1414.
74. Davies, C.N. (1966). Aerosol Science, Academic Press, London.
75. Davies, P.F. (1988). Endothelial Cells, Hemodynamic Stress, and the Localization of Atherosclerosis. CRC Press, Boca Raton.
76. Davies, P.F. (1995). Flow-Mediated Endothelial Mechanotransduction. Physiol. Rev., Vol. 75, pp. 519–560.
77. Derjaguin, B.V., Muller, V.M., and Toporov, Y.P.T. (1975). Effect of Contact Deformation on the Adhesion of Particles. J. Colloid Interface Sci., Vol. 53, pp. 314–326.
78. Dewey, C.F., Jr., Bussolari, S.R., Gimbrone, M.A., Jr., and Davies, P.F. (1981). The Dynamic Response of Vascular Endothelial Cells to Fluid Shear Stress. J. Biomech. Eng., Vol. 103, pp. 177–185.
79. Ding, E.-J. and Aidun, C.K. (2003). Extension of the Lattice-Boltzmann Method for Direct Simulation of Suspended Particles Near Contact. J. Stat. Phys., Vol. 112, pp. 685–708.
80. Ding, E.-J. and Aidun, C.K. (2006). Cluster Size Distribution and Scaling for Spherical Particles and Red Blood Cells in Pressure-Driven Flows at Small Reynolds Number. Phys. Rev. Lett., Vol. 96, pp. 204502-1-4.
81. De Keulenaer, G.W. (1998). Oscillatory and Steady Laminar Shear Stress Differentially Affect Human Endothelial Redox State: Role of a Superoxide-Producing NADH Oxidase. Circ. Res., Vol. 82, 1094–1101.
82. DuPin, M.M., Halliday, I., and Care, C.M. (2006). A Multi-Component Lattice Boltzmann Scheme: Towards the Mesoscale Simulation of Blood Flow. Med. Eng. Phys., Vol. 28, pp. 13–18.
83. Drolet, F. and Vinals, J. (1998). Fluid Flow Induced by a Random Acceleration Field. Microgr. Sci. Technol., Vol. 11, pp. 64–68.
84. Druzhinin, O.A. and Elghobashi, S.E. (1999). A Lagrangian-Eulerian mapping solver for direct numerical simulation of a bubble-laden homogeneous turbulent shear flow using the two-fluid formulation. J. Comput. Phys., Vol. 154, pp. 174–196.
85. Druzhinin, O.A. and Elghobashi, S.E. (1999). On the decay rate of isotropic turbulence laden with microparticles. Phys. Fluids, Vol. 11, pp. 602–610.
86. Druzhinin, O.A. and Elghobashi, S.E. (2000). Direct numerical simulation of a three-dimensional spatially-developing bubble-laden mixing layer with two-way coupling. J. Fluid Mech.
87. Drust, F., Milojevic, D., and Schonung, B. (1984). Appl. Math. Model., Vol. 8, pp. 101–115.
88. Eaton, J.K. (2000). Attenuation of Gas Turbulence by a Nearly Stationary Dispersion of Solid Particles. Fifth Microgravity Fluid Physics and Transport Phenomena Conference, Cleveland.
89. Elghobashi, S. and Abou-Arab, T.W. (1983). A Two-Equation Turbulence Model for Two-Phase Flows. Phys. Fluids, Vol. 26, pp. 931–938.
90. Elghobashi, S. and Trusdell, G.C. (1992). Direct Simulation of Particle Dispersion in a Decaying Isotropic Turbulence. J. Fluid Mech., Vol. 242, pp. 655–700.
91. Ellison, J., Ahmadi, G., and Grodsinsky, C. (1995). Stochastic Model for Microgravity Excitation. ASCE J. Aerospace Eng., Vol. 8, pp. 100–106.
92. Ellison, J., Ahmadi, G., Regel, L., and Wilcox, W. (1995). Particle Motion in a Liquid Under g-Jitter Excitation. Microgr. Sci. Technol., Vol. 8, pp. 140–147.
93. Ellison, J., Ahmadi, G., and Grodsinsky, C. (1997). Stochastic Response of Passive Vibration Control Systems to g-jitter Excitation. Microgr. Sci. Technol., Vol. 10, pp. 2–12.
94. Evans, E.A. and Skalak, R. (1980). Mechanics and Thermodynamics of Biomembranes, CRC Press, Boca Raton.
95. Fåhraeus, R. (1929). The Suspension Stability of the Blood. Physiol. Rev., Vol. 9, pp. 241–274.

96. Fåhraeus, R. and Lindqvist, T. (1931). The Viscosity of Blood in Narrow Capillary Tubes. Am. J. Physiol., Vol. 96, pp. 562–568.
97. Fan, F.G. and Ahmadi, G. (1993). A Sublayer Model for Turbulent Deposition of Particles in Vertical Ducts with Smooth and Rough Surfaces. J. Aerosol Sci., Vol. 24, pp. 45–64.
98. Fan, F.G. and Ahmadi, G. (1994). On the Sublayer Model for Turbulent Deposition of Particles in Presence of Gravity and Electric Fields. Aerosol Sci. Technol., Vol. 21, pp. 49–71.
99. Fan, F.G. and Ahmadi, G. (1995a). Dispersion of Ellipsoidal Particles in an Isotropic Pseudo-Turbulent Flow Field. ASME J. Fluid Eng., Vol. 117, pp. 154–161.
100. Fan, F. and Ahmadi, G. (1995b). A Sublayer Model for Wall Deposition of Ellipsoidal Particles in Turbulent Stream. J. Aerosol Sci., Vol. 25, pp. 813–840.
101. Fan, F. and Ahmadi, G. (2000). Wall Deposition of Small Ellipsoids from Turbulent Air Flow – A Brownian Dynamics Simulation. J. Aerosol Sci., Vol. 31, pp. 1205–1229.
102. Faxen, H. (1923). Die Bewegung einer starren Kugel längs der Achse eines mit zäher Flüssigkeit gefüllten Rohres. Arkiv. Mat. Astron. Fys., Vol. 17, No. 27.
103. Fernandez de la Mora, J. and Friedlander, S.K. (1982). Aerosol and Gas Deposition to Fully Rough Surfaces: Filtration Model for Blade-Shaped Elements. Int. J. Heat Mass Transfer, Vol. 25, pp. 1725–1735.
104. Fichman, M., Gutfinger, C., and Pnueli, D. (1988). A Model for Turbulent Deposition of Aerosols. J. Aerosol Sci., Vol. 19, pp. 123–136.
105. FLUENT^{TM} User's Guide. (2001). Computational Fluid Dynamic Software, Version 6.0.12, Fluent, New Hampshire.
106. Friedlander, S.K. (1977). Smoke, Dust and Haze – Fundamentals of Aerosol Behaviour, Wiley, New York.
107. Friedlander, S.K. and Johnstone, H.F. (1957). Deposition of Suspended Particles from Turbulent Gas Streams. Ind. Eng. Chem., Vol. 49, pp. 1151–1156.
108. Frisch, U., Hasslacher, B., Pomeau, Y. (1986). Lattice-Gas Automata for the Navier-Stokes Equation. Phys. Rev. Lett., Vol. 56, 1505–1508.
109. Fuchs, N.A. (1964). The Mechanics of Aerosols, Pergamon Press, Oxford.
110. Fung, Y.C. (1993). Biomechanics, Mechanical Properties of Living Tissues, 2nd Ed, Springer-Verlag, Berlin.
111. Gallily, I. and Eisner, A.D. (1979). On the Orderly Nature of the Motion of Non-Spherical Aerosol Particles. I. Deposition from a Laminar Flow. J. Colloid Interface Sci., Vol. 68, pp. 320–337.
112. Giddens, D.P., Zarins, C.K., and Glagov, S. (1993). The Role of Fluid Mechanics in the Localization and Detection of Atherosclerosis. J. Biomech. Eng., Vol. 115, pp. 588–594.
113. Goldman, A.J., Cox, R.G., and Brenner, H. (1967a). Slow Viscous Motion of a Sphere Parallel to a Plane Wall. I. Motion Through a Quiescent Fluid. Chem. Eng. Sci., Vol. 22, pp. 637–651.
114. Goldman A.J., Cox, R.G., and Brenner, H. (1967b). Slow Viscous Motion of a Sphere Parallel to a Plane Wall. II. Couette Flow. Chem. Eng. Sci., Vol. 22, pp. 653–660.
115. Goldschmidt, V., Householder, M.K., Ahmadi, G., and Chuang, S.C. (1972). Turbulent Diffusion of Small Particles Suspended in Turbulent Jets. Prog. Heat Mass Transfer, Vol. 6, pp. 487–508.
116. Goldsmith, H.L. and Mason, S.G. (1962). The Flow of Suspensions through Tubes. I. Single Spheres, Rods, and Discs. J. Colloid Sci., Vol. 17, pp. 448–476.
117. Goldsmith, H.L. (1971). Deformation of Human Red Cells in Tube Flow. Biorheology, Vol. 7, pp. 235–242.
118. Goldsmith, H.L. and Marlow, J. (1972). Flow Behavior of Erythrocytes I. Rotation and Deformation in Dilute Suspensions. Proc. R. Soc. Lond. B, Vol. 182, pp. 351–384.
119. Goldsmith, H.L. and Skalak, R. (1975). Hemodynamics. Ann. Rev. Fluid Mech., Vol. 7, pp. 213–247.
120. Gosman, A.D. and Ioannides, E. (1983). Aspects of Computer Simulation of Liquid-Fueled Combustors. J. Energy, Vol. 7, pp. 482–490.
121. Gradon, L., Grzybowski, P., and Pilacinski, W. (1988). Analysis of Motion and Deposition of Fibrous Particles on a Single Filter Element. Chem. Eng. Sci., Vol. 43, pp. 1253–1259.

122. Griendling, K.K. and Harrison, D.G. (2001). Out, Damned Dot: Studies of the NADPH Oxidase in Atherosclerosis. J. Clin. Invest., Vol. 108, pp. 1423–1424.
123. Groszmann, D. E., et al. (1998). Decoupling the Roles of Inertia and Gravity on Particle Dispersion. Proceeding of the Fourth Microgravity Fluid Physics & Transport Phenomena Conference, NASA Lewis Research Center, Cleveland, OH, pp. 117–118.
124. Haefeli-Bleuer, B. and Weibel, E.R. (1988). Morphometry of the human pulmonary acinus. Anatomical record, Vol. 220, pp. 401–414.
125. Hahn, I., Scherer P.W., and Mozell M.M. (1993). Velocity Profiles Measured for Airflow through A Large-Scale Model of the Human Nasal Cavity. J. Appl. Physiol., Vol. 75, pp. 2273–2287.
126. Hall, D. (1988). Measurements of the Mean Force on a Particle Near a Boundary in Turbulent Flow. J. Fluid Mech., Vol. 187, pp. 451–466.
127. Hamaker, H.C. (1937). The London-van der Waals Attraction Between Spherical Particles. Physica, Vol. 4, pp. 1058–1072.
128. Harrison, D., Griendling, K.K., Landmesser, U., Hornig, B., and Dexler, H. (2003). Role of Oxidative Stress in Atherosclerosis. Am. J. Cardiol., Vol. 91, pp. 7A–11A.
129. He, C. and Ahmadi, G. (1998). Particle Deposition with Themophoresis in Laminar and Turbulent Duct Flows. Aerosol Sci. Technol., Vol. 29, pp. 525–546.
130. He, C. and Ahmadi, G. (1999). Particle Deposition in a Nearly Developed Turbulent Duct Flow with Electrophoresis. J. Aerosol Sci., Vol. 30, pp. 739–758.
131. Hetsroni, Haber, G.S., Brenner, H., and Greenstein, T. (1971). A Second-Order Theory for a Deformable Drop Suspended in a Long Conduit. In Progress in Heat and Mass Transfer, Vol. 6, Edited by Hetsroni, G., Pergamon, Oxford/New York, pp. 591–612.
132. Heyder, J., Gabhart, J., Rudolf, G., Schiller, C.F., and Stahlhofen, W. (1986). Deposition of Particles in the Human Respiratory Tract in the Size Range 0.005–15 μm. J. Aerosol Sci., Vol. 17, pp. 811–825.
133. Hidy, G. M. (1984). Aerosols, an Industrial and Environmental Science, Academic Press, New York.
134. Hinds, W.C. (1982). Aerosol Technology, Properties, Behavior, and Measurement of Airborne Particles, Wiley, New York.
135. Hinze, J. O. (1975). Turbulence, McGraw Hill, New York.
136. Hoppel, W.A. and Frick, G.M. (1986). Ion-aerosol attachment coefficients and the steady-state charge distribution on aerosol in a bipolar environment. Aerosol Sci. Technol., Vol. 5, pp. 1–21.
137. Hwang, N.H.C., Gross, D.R., and Patel, D.J. (Eds.) (1978). Quantitative Cardiovascular Studies: Clinical and Research Applications of Engineering Principles, University Park Press, Baltimore, pp. 289–351.
138. Ibrahim, A.H., Dunn, P.F., and Brach, R.M. (2003). Microparticle Detachment from Surfaces Exposed to Turbulent Air Flow: Controlled Experiments and Modeling. J. Aerosol Sci., Vol. 34, pp. 765–782.
139. Jan, K.-M. and Chien, S. (1973a). Role of Surface Electric Charge on Red Blood-Cell Interactions. J. Gen. Physiol., Vol. 61, pp. 638–654.
140. Jan, K.-M. and Chien, S. (1973b). Influence of Ionic Composition of Fluid Medium on Red-Cell Aggregration. J. Gen. Physiol., Vol. 61, pp. 655–668.
141. Johnson, K.L., Kendall, K., and Roberts, A.D. (1971). Surface Energy and the Contact of Elastic Solids. Proc. R. Soc. Lond. A, Vol. 324, pp. 301–313.
142. Johnson, J.R. and Schroter, R.C. (1979). Deposition of Particles in Model Airways. J. Appl. Physiol., Vol. 47, pp. 947–953.
143. Jurewicz, J.T. and Stock, D.E. (1976). ASME WAM, Paper No. 76-WA/PE-33.
144. Kaazempur-Mofrad, M.R., Isasi, A.G., Younis, H.F., Chan, R.C., Hinton, D.P., Sukhova, G., Lamuraglia, G.M., Lee, R.T., and Kamm, R.D. (2004). Characterization of the Atherosclerotic Carotid Bifurcation using MRI, Finite Element Modeling, and Histology. Ann. Biomed. Eng., Vol. 32, pp. 932–946.
145. Kamm, R.D. (2002). Cellular Fluid Mechanics. Ann. Rev. Fluid Mech., Vol. 34, pp. 211–232.

146. Karnis, A. and Mason, S.G. (1967). Particle Motions in Sheared Suspensions. J. Colloid Interface Sci., Vol. 24, pp. 161–169.
147. Karino, T. and Goldsmith, H.L. (1980). Disturbed Flow in Models of Branching Vessels. Trans. Am. Soc. Artif. Internal Organs, Vol. 26, pp. 500–506.
148. Karino, T. and Goldsmith, H.L. (1984). Role of Blood Cell-Wall Interactions in Thrombogenesis and Atherosclerosis: A Microrheological Study. Biorheology, Vol. 21, pp. 587–601.
149. Karino, T., Asakura, T., and Mabuchi, S. (1988). Role of Hemodynamic Factors in Atherogenesis. Adv. Exp. Med. Biol., Vol. 242, pp. 51–57.
150. Kelly, J.T., Prasad, A.K., and Wexler, A.S. (2000). Detailed Flow Patterns in The Nasal Cavity. J. Appl. Physiol., Vol. 89, pp. 323–337.
151. Kelly, J.T., Asgharian, B., Kimbell, J.S., Wong, B.A. (2004). Particle Deposition in Human Nasal Airway Replicas Manufactured by Different Methods. II. Ultrafine Particles. Aerosol Sci. Technol., Vol. 38, pp. 1072–1079.
152. Keyhani, K., Scherer, P.W., and Mozell, M.M. (1995) Numerical Simulation of Airflow in the Human Nasal Cavity. J. Biomech. Eng., Vol. 117, pp. 429–441.
153. Kim, C.S. and Fisher, D.M. (1999). Deposition of Aerosol Particles in Successively Bifurcating Airways Models. Aerosol Sci. Technol., Vol. 31, pp. 198–220.
154. Kim, C.S., Fisher, D.M., Lutz, D.J., and Gerrity, T.R. (1994). Particle Deposition in Bifurcating Airway Models with Varying Airway Geometry. J. Aerosol Sci., Vol. 25, pp. 567–581.
155. Kim, C.S. and Iglesias, A.J. (1989). Deposition of Inhaled Particles in Bifurcating Airways Models. I. Inspiratory Deposition. J. Aerosol Med., Vol. 2, pp. 1–14.
156. Kimbell, J.S. (2001). Computational Fluid Dynamics of the Extrathoracic Airways, Medical Applications of Computer Modelling: The Respiratory System, WIT Press, United Kingdom.
157. Krushkal, E.M. and Gallily, I. (1984). On the Orientation Distribution Function of Non-Spherical Aerosol Particles in a General Shear Flow. I. The Laminar Case. J. Colloid Interface Sci., Vol. 99, pp. 141–152.
158. Ku, D.N. (1997). Blood Flow in Arteries. Annu. Rev. Fluid Mech., Vol. 29, 399–434.
159. Ku, D.N., Giddens, D.P., Phillips, D.J., and Strandress, D.E., Jr. (1985a). Hemodynamics of the Normal Human Carotid Bifurcation: In Vitro and In Vivo Studies. Ultrasound Med. Biol., Vol. 11, pp. 13–26.
160. Ku, D.N., Giddens, D.P., Zarins, C.K., and Glagov, S. (1985b). Pulsatile Flow and Atherosclerosis in the Human Carotid Bifurcation. Positive Correlation between Plaque Location and Low Oscillating Shear Stress. Arterioscler. Thromb. Vasc. Biol., Vol. 5, pp. 293–302.
161. Kvasnak, W. and Ahmadi, G. (1995). Fibrous Particle Deposition in a Turbulent Channel Flow – An Experimental Study. Aerosol Sci. Technol., Vol. 23, pp. 641–652.
162. Kvasnak, W. and Ahmadi, G. (1996). Deposition of Ellipsoidal Particles in Turbulent Duct Flows. Chem. Eng. Sci., Vol. 51, pp. 5137–5148.
163. Kvasnak, W., Ahmadi, G., Bayer, R., and Gaynes, M.A. (1993). Experimental Investigation of Dust Particle Deposition in a Turbulent Channel Flow. J. Aerosol Sci., Vol. 24, pp. 795–815.
164. Lane, D.D. and Stukel, J.J. (1978). Aerosol Deposition on a Flat Plate. Aerosol Sci., Vol. 9, pp. 191–197.
165. Launder, B.E., Reece, G.J., Rodi, W. (1975). Progress in Development of a Reynolds-Stress Turbulence Closure, J. Fluid Mech., Vol. 68, pp. 537–566.
166. Lawless, P.A. (1996). Particle Charging Bounds, Symmetry Relations, and an Analytic Charging Rate Model for the Continuum Regime. J. Aerosol Sci., Vol. 27, pp. 191–215.
167. Leal, L.G. (1980). Particle Motions in a Viscous Fluid. Annu. Rev. Fluid Mech., Vol. 12, pp. 435–476.
168. Lee, S.L. and Durst, F. (1982). On the Motion of Particles in Turbulent Duct Flows. Int. J. Multiphase Flow, Vol. 8, pp. 125–146.
169. Leighton, D.T. and Acrivos, A. (1985). The Lift on a Small Sphere Touching a Plane in the Presence of a Simple Shear Flow. Z. Angew. Math. Phys., Vol. 36, pp. 174–178.
170. Lerman, A. (1979). Geochemical Processes, Wiley, New York.

171. Levich, V. (1962). Physicochemical Hydrodynamics, Prentice-Hall, Englewood Cliffs, NJ.
172. Li, A. and Ahmadi, G. (1992). Dispersion and Deposition of Spherical Particles from Point Sources in a Turbulent Channel Flow, Aerosol Sci. Technol., Vol. 16, pp. 209–226.
173. Li, A. and Ahmadi, G. (1993). Deposition of Aerosol on Surfaces in a Turbulent Channel Flow. Int. J. Eng. Sci., Vol. 31, pp. 435–451.
174. Li, A. and Ahmadi, G. (1993). Computer Simulation of Deposition of Aerosols in a Turbulent Channel Flow With Rough Walls. Aerosol Sci. Technol., Vol. 18, pp. 11–24.
175. Li, A. and Ahmadi, G. (1993). Aerosol Particle Deposition with Electrostatic Attraction in a Turbulent Channel Flow. J. Colloid Inteface Sci., Vol. 158, pp. 476–482.
176. Li, A., Ahmadi, G., Bayer, R., and Gaynes, M.A. (1994). Aerosol Particle Deposition in an Obstructed Turbulent Duct Flow. J. Aerosol Sci., Vol. 25, pp. 91–112.
177. Li, A. and Ahmadi, G. (1995). Computer Simulation of Particle Deposition in the Upper Tracheobronchial Tree. Aerosol Sci. Technol., Vol. 23, pp. 201–223.
178. Liu, B.Y.H. and Agarwal, J.K. (1974). J. Aerosol Sci., Vol. 5, pp. 145–155.
179. Liu, B.Y.H. and Kapadia, H. (1978). Combined Field and Diffusion Charging of Aerosol Particles in the Continuum Regime. J. Aerosol Sci., Vol. 9, pp. 227–242.
180. Luft, J.H. (1966). Fine Structures of Capillary and Endocapillary Layer as Revealed by Ruthenium Red. Fed. Proc., Vol. 25, pp. 1773–1783.
181. Maugis, D. and Pollock, H.M. (1984). Surface Forces, Deformation and Adherence at Metal Microcontact. Acta Met., Vol. 32, pp. 1323–1334.
182. Martonen, T.B., Yang, Y., and Xue, Z.Q. (1994). Effects of Carinal Ridge Shapes on Lung Airstreams. Aerosol Sci. Technol., Vol. 21, pp. 119–136.
183. Martonen, T.B., Zhang, Z., and Lessmann, R.C. (1993). Fluid Dynamics of the Human Larynx and Upper Tracheobronchial Airways. Aerosol Sci. Technol., Vol. 19, pp. 133–156.
184. Martonen, T.B., Zhang, Z., Yue, G., and Musante, C.J. (2003). Fine Particle Deposition within Human Nasal Airways. Inhal. Toxicol., Vol. 15, pp. 283–303.
185. Marshall, J.R. (1998). Electrostructural Phase Changes in Charged Particulate Clouds: Planetary and Astrophysical Implications. LPSC, Vol. 29, pp. 1132.
186. Maxey, M.R. (1987). The Gravitational Settling of Aerosol Particle in Homogeneous Turbulence and Random Flow Fields. J. Fluid Mech., Vol. 174, pp. 441–445.
187. Maxey, M.R. and Riley, J.J. (1983). Equation of Motion for a Small Rigid Sphere in a Nonuniform Flow. Phys. Fluid., Vol. 26, pp. 883–889.
188. Mazaheri, A.R. and Ahmadi, G. (2002). Inspiratory Particle Deposition in the Upper Three Airway bifurcation. 21st Annual Conference of the American Association for Aerosol Research, AAAR 2002, Charlotte, NC.
189. Mazaheri, A.R. and Ahmadi, G. (2003). Inspiratory Particle Deposition in the Upper Three Airway bifurcation. 21st Annual Conference of the American Association for Aerosol Research, AAAR 2002, Charlotte, NC.
190. Mazaheri, A.R. and Ahmadi, G. (2004). Modeling Inspiratory Particle Deposition. ASME Heat Transfer/Fluid Engineering Summer Conference, Charlotte, NC.
191. McCoy, D. D. and Hanratty, T.J. (1977). Rate of Deposition of Droplets in Annular Two-Phase Flow. Int. J. Multiphase Flow, Vol. 3, pp. 319–331.
192. McLaughlin, J.B. (1989). Aerosol Particle Deposition in Numerically Simulated Channel Flow. Phys. Fluids A, Vol. 7, pp. 1211–1224.
193. McLaughlin, J.B. (1991). Inertial Migration of a Small Sphere in Linear Shear Flows. J. Fluid Mech., Vol. 224, pp. 261–274.
194. McLaughlin, J.B. (1993). The Lift on a Small Sphere in Wall-Bounded Linear Shear Flows. J. Fluid. Mech., Vol. 246, pp. 249–265.
195. Mei, R. (1992). An Approximate Expression for the Shear Lift Force on a Spherical Particle at Finite Reynolds Number. Int. J. Multiphase Flow, Vol. 18, pp. 145–147.
196. McLaughlin, J.B. (1994). Numerical Computation of Particles-Turbulence Interaction. Int. J. Multiphase Flow, Vol. 20, pp. 211–232.
197. Meiselman, H.J., Merrill, E.W., Salzman, E., Gilliland, E.R., Pelletier, G.A. (1967). Effect of Dextran on Rheology of Human Blood – Low Shear Viscometry. J. Appl. Physiol., Vol. 22, pp. 480–486.

198. Mercer, T.T. (1973). Aerosol Technology in Hazard Evaluation of Airborne Particles, Academic Press, New York.
199. Modarress, D., Wuerer, J., and Elghobashi, S. (1984). Chem. Eng. Commun., Vol. 28, pp. 341–354.
200. Mollinger, A.M. and Nieuwstadt, F.T.M. (1996). Measurement of the Lift Force on a Particle Fixed to the Wall in the Viscous Sublayer of a Fully Developed Turbulent Boundary Layer. J. Fluid Mech., Vol. 216, pp. 285–306.
201. Mulivor, A.W. and Lipowsky, H.H. (2004). Inflammation- and Ischemia-Induced Shedding of Venular Glycocalyx. Am. J. Physiology Heart Circ. Physiol., Vol. 286, pp. H1672–H1680.
202. Nerem, R.M. (1985). Atherosclerosis: Hemodynamics, Vascular Geometry, and the Endothelium. Biorheology, Vol. 21, pp. 565–569.
203. Nerem, R.M. and Girard, P.R. (1990). Hemodynamic Influences on Vascular Endothelial Biology. Toxicol. Pathol., Vol. 18, pp. 572–582.
204. Nerem, R.M. (1993). Hemodynamics and Vascular Endothelial Biology. J. Cardiovasc. Pharmacol., Vol. 21, pp. S6–S10.
205. Norris, R. (1869). On the Laws and Principles Concerned in the Aggregation of Blood-Corpuscles Both Within and Without the Vessels. Proc. R. Soc. Lond., Vol. 17, pp. 429–436.
206. Ounis, H. and Ahmadi, G. (1989). Motions of Small Rigid Spheres in a Simulated Random Velocity Field. ASCE J. Eng. Mech., Vol. 115, pp. 2107–2121.
207. Ounis, H. and Ahmadi, G. (1990). Analysis of Dispersion of Small Spherical Particles in a Random Velocity Field. ASME J. Fluid Eng., Vol. 112, pp. 114–120.
208. Ounis, H. and Ahmadi, G. (1990). A Comparison of Brownian and Turbulent Diffusions. Aerosol Sci. Technol., Vol. 13, pp. 47–53.
209. Ounis, H. and Ahmadi, G. (1991). Motions of Small Particles in Simple Shear Flow Field under Microgravity Condition. Phys. Fluids A, Vol. 3, pp. 2559–2570.
210. Ounis, H., Ahmadi, G., and McLaughlin, J.B. (1991a). Brownian Diffusion of Submicron Particles in the Viscous Sublayer. J. Colloid Interface Sci., Vol. 143, pp. 266–277.
211. Ounis, H., Ahmadi, G., and McLaughlin, J.B. (1991b). Dispersion and Deposition of Brownian Pericles from Point Sources a Simulated Turbulent Channel flow. J. Colloid interface Sci., Vol. 147, pp. 233–250.
212. Ounis, H., Ahmadi, G., and McLaughlin, J.B. (1993). Brownian Particle Deposition in a Directly Simulated Turbulent Channel Flow. Phys. Fluids A, Vol. 5, pp. 1427–1432.
213. Owen, P.R. (1960). Aerodynamic Capture of Particles. Pergamon, Oxford, p. 8.
214. Papavergos, P.G. and Hedley, A.B. (1984). Particle Deposition Behavior from Turbulent Flow. Chem. Eng. Res. Des., Vol. 62, pp. 275–295.
215. Peters, M.H., Cooper, D.W, and Miller, R.J. (1989). J. Aerosol Sci., Vol. 20, pp. 123–136.
216. Pope, S.B. and Chen, Y.L. (1990). Phys. Fluids, Vol. 2, pp. 1437–1449.
217. Pope, S.B. (1991). Ann. Rev. Fluid Mech., Vol. 26, pp. 23–63.
218. Pozrikidis, C. (1990). The Axisymmetric Deformation of a Red Blood Cell in Uniaxial Straining Flow. J. Fluid Mech., Vol. 216, pp. 231–254.
219. Pozrikidis, C. (1992). The Axisymmetric Deformation of a Red Blood Cell in Uniaxial Straining Flow. J. Fluid Mech., Vol. 216, pp. 231–254.
220. Pozrikidis, C. (Ed.) (2003). Modeling and Simulation of Capsules and Biological Cells, Chapman and Hall/CRC Mathematical Biology and Medicine Series, Boca Raton.
221. Pozrikidis, C. (2002). A Practical Guide to Boundary Element Methods with the Software Library BEMLIB, Chapman and Hall/CRC Mathematical Biology and Medicine Series, Boca Raton.
222. Reeks, M.W. (1977). On the Dispersion of Small Particles Suspended in an Isotropic Turbulent Flow. J. Fluid Mech., Vol. 83, pp. 529–546.
223. Reeks, M.W. and Mckee, S. (1984). The Dispersive Effect of Basset History Forces on Particle Motion in a Turbulent Flow. Phys. Fluid, Vol. 27, pp. 1573–1582.
224. Riley, J.J. and Corrsin, S. (1974). The Relation of Turbulent Diffusivities to Lagrangian Velocity Statistics for the Simplest Shear Flow. J. Geophys. Res., Vol. 79, pp. 1768–1771.
225. Riley, J.J. and Patterson, G.S., Jr. (1974). Diffusion Experiments with Numerically Integrated Isotropic Turbulence. Phys. Fluid, Vol. 17, pp. 292–297.

226. Rizk, M.A. and Elghobashi, S.E. (1985). The Motion of a Spherical Particle Suspended in a Turbulent Flow near a Plane Wall. Phys. Fluids, Vol. 20, pp. 806–817.
227. Rouhiainen, P.O. and Stachiewiz, J.W. (1970). On the Deposition of Small Particles from Turbulent Streams. J. Heat Transfer, Vol. 92, pp. 169–177.
228. Raabe, O.G., Yeh, H.C., Schum G.M., and Phalen, R.F. (1976). Tracheobronchial Geometry: Human, Dog, Rat and Hamster, LF53, Lovelace Foundation Report, New Mexico.
229. Skalak, R., Tozeren, A., Zarda, P.R., and Chien, S. (1973). Strain Energy Function of Red Blood Membranes. Biophys. J., Vol. 13, pp. 245–264.
230. Skalak, R., Ozkaya, N., and Skalak, T.C. (1989). Biofluid Mechanics. Ann. Rev. Fluid Mech., Vol. 21, pp. 167–204.
231. Squire, J.M., Chew, M., Nneji, G., Neal, C., Barry, J., and Michel, C. (2001). Quasi-Periodic Substructure in the Microvessel Endothelial Glycocalyx: A Possible Explanation for Molecular Filtering? J. Struct. Biol., Vol. 136, pp. 239–255.
232. Takeshita, S., Inoue, N., Ueyama, T., Kawashima, S., and Yokoyama, M. (2000). Shear Stress Enhances Glutathione Peroxidase Expression in Endothelial Cells. Biochem. Biophys. Res. Commun., Vol. 273, pp. 66–71.
233. Tanganelli, P., et al. (1993). Distributions of Lipid and Raised Lesions in Aortas of Young People of Different Geographic Origins (WHO-ISFC PBDAY Study). World Health Organization-International Society and Federation of Cardiology. Pathobiological Determinants of Atherosclerosis in Youth. Arterioscler. Thromb., Vol. 13, pp. 1700–1710.
234. Tarbell, J.M. (2003). Mass Transport in Arteries and the Localization of Atherosclerosis. Ann. Rev. Biomed. Eng., Vol. 5, pp. 79–118.
235. Saffman, P.G. (1965). The Lift on a Small Sphere in a Slow Shear Flow. J. Fluid Mech., Vol. 22, pp. 385–400.
236. Saffman, P.G. (1968). Corrigendum to the Lift on a Small Sphere in a Slow Shear Flow. J. Fluid Mech., Vol. 31, p. 264.
237. Schamberger, M. R., Peters, J. E., and Leong, K. H. (1990). Collection of Prolate Spheroidal Aerosol Particles by Charged Spherical Collectors. J. Aerosol Sci., Vol. 21, pp. 539–554.
238. Schiller, C.F., Gebhart, J., Heder, J., Rudolf, G., and Stahlhofen,W. (1986). Factors influencing total deposition of ultrafine aerosol particles in the human respiratory tract. Journal of Aerosol Science, Vol. 17, pp. 328-332.
239. Schreck, S., Sullivan, K.J., Ho, C.M., and Chang, H.K. (1993). Correlations Between Flow Resistance and Geometry in a Model of the Human Nose. J. Appl. Physiol., Vol. 75, pp. 1767–1775.
240. Sehmel, G.A. (1973). Particle Eddy Diffusities and Deposition Velocities for Isothermal Flow and Smooth Surfaces. J. Aerosol Sci., Vol. 4, pp. 125–138.
241. Seinfeld, J.H. (1986). Atmospheric Chemistry and Physics of Air Pollution, Wiley, New York.
242. Shapiro, M. and Goldenberg, M. (1993). Deposition of Glass Fiber Particles from Turbulent Air Flow in a Pipe. J. Aerosol Sci., Vol. 24, pp. 65–87.
243. Shams, M., Ahmadi, G., and Rahimzadeh, H. (2000). A Sublayer Model for Deposition of Nano- and Micro-Particles in Turbulent Flows. Chem. Eng. Sci., Vol. 55, pp. 6097–6107.
244. Scherer, P.W., Keyhani, K., Mozell, M.M. (1994). Nasal Dosimetry Modeling for Humans. Inhal. Toxicol., Vol. 6, pp. 85–97.
245. Snyder, W.H. and Britter, R.E. (1987). Atmos. Environ., Vol. 21, pp. 735–751.
246. Snyder, W.H. and Lumley, J.L. (1971). Some Measurements of Particle Velocity Autocorrelation Functions in a Turbulent Flow. J. Fluid Mech., Vol. 48, pp. 41–71.
247. Soltani, M. and Ahmadi, G. (1995). Direct Numerical Simulation of Particle Entrainment in Turbulent Channel Flow. Phys. Fluid A, Vol. 7, pp. 647–657.
248. Soltani, M., Ahmadi, G., Ounis, H., and McLaughlin, J.B. (1998). Direct Numerical Simulation of Charged Particle Deposition in a Turbulent Channel Flow. Int. J. Multiphase Flow, Vol. 24, pp. 77–92.
249. Soltani, M. and Ahmadi, G. (2000). Direct Numerical Simulation of Curly Fibers in Turbulent Channel Flow. Aerosol Sci. Technol., Vol. 33, pp. 392–418.

250. Spurny, K.R. (1986). Physical and Chemical Characterization of Individual Airborne Particles, Wiley, New York.
251. Squires, K.D. and Eaton, J.K. (1991). On the preferential concentration of solid particles in turbulent. Phys. Fluid A, Vol. 3, pp. 1169–1178.
252. Stapleton, K.W., Guentsch, E., Hoskinson, M.K., and Finlay, W.H. (2000). On the Suitability of k–ϖ Turbulence Modeling for Aerosol Deposition in the Mouth and Throat: A Comparison with Experiment. J. Aerosol Sci., Vol. 31, pp. 739–749.
253. Strong J.C. and Swift D.L. (1987). Deposition of Ultrafine Particles in a Human Nasal Cast. Proceedings of the First Conference of Aerosol Society, Loughborogh University of Technology, Loughborogh, pp. 109–112.
254. Subramaniam, R.P., Richardson, R.B., Morgan, K.T., Kimbell, J.S., and Guilmette, R.A. (1998). Computational Fluid Dynamics Simulations of Inspiratory Airflow in the Human Nose and Nasopharynx. Inhal. Toxicol., Vol. 10, pp. 91–120.
255. Swift, D.L. and Proctor, D.F. (1977). Access of Air to the Respiratory Tract. In Respiratory Defense Mechanisms, Part 1, Edited by Brian, J.D., Proctor, D.F., and Reid, L.M., Dekker, New York, pp. 63–93.
256. Swift, D.L., Montassier, N., Hopke, P.K., Karpen-Hayes, K., Cheng, Y.S., Su, Y.F., Yeh, H.C., and Strong, J.C. (1992a). Inspiratory Deposition of Ultrafine Particles in Human Nasal Replicate Casts. J. Aerosol Sci., Vol. 23, pp. 65–72.
257. Swift, D.L., Montassier, N., Hopke, P.H., Karpen-Hayes K., Cheng, Y.S., Su, Y.F., Yeh, H.C., and Strong, J.C. (1992b). Inspiratory Deposition of Ultrafine Particles in Human Nasal Replicate Cast. J. Aerosol Sci., Vol. 23, pp. 65–72.
258. Swift, D.L. and Strong, J.C. (1996). Nasal Deposition of Ultrafine ^{218}Po Aerosols in Human Subjects. J. Aerosol Sci., Vol. 27, pp. 1125–1132.
259. Tavoularis, S. and Corrsin, S. (1985). Effects of Shear on the Turbulent Diffusivity Tensor. Int. J. Heat Transfer, Vol. 28, pp. 265–274.
260. Tchen, C.M. (1947). Mean Value and Correlation Problems Connected with the Motion of Small Particles Suspended in a Turbulent Field. Ph.D. Thesis, University of Delft, Martinus Nijhoff, Hague.
261. Thomson, D.J. (1987). Criteria for the Selection of Stochastic Models of Particle Trajectories in Turbulent Flows. J. Fluid Mech., Vol. 180, pp. 529–556.
262. Tian, L., Ahmadi, G., Mazaheri, A., Hopke, P.K., and Cheng, S.-Y. (2005). Particle Deposition in 3-D Asymmetric Human Lung Bifurcations. 79th ACS Colloid and Surface Science Symposium, Clarkson University, Potsdam, NY.
263. Tian, L. and Ahmadi, G. (2007). Particle Deposition in Turbulent Duct Flows - Comparisons of Different Model Predictions, J. Aerosol Science, Vol. 38, pp. 377–397.
264. Tsusa, A., Butler, J.P., and Fredberg, J.J. (1994a). Effects of alveolated duct structure on aerosol deposition in the pulmonary acinus. Part I: Diffusion in the absence of gravity. Journal of Applied Physiology, Vol. 76, pp. 2497–2509.
265. Tsusa, A., Butler, J.P., and Fredberg, J.J. (1994b). Effects of alveolated duct structure on aerosol deposition in the pulmonary acinus. Part II: Gravitational sedimentation and inertial impaction in the absence of diffusion. Journal of Applied Physiology, Vol. 76, pp. 2510–2516.
266. Trinh, E.H. (1998). Acoustic Streaming in Microgravity: Flow Stability and Mass and Heat Transfer Enhancement. Proceeding of the Fourth Microgravity Fluid Physics and Transport Phenomena Conference, NASA Lewis, Cleveland, OH, pp. 117–118.
267. Twomey, S. (1976). Atmospheric Aerosols, Elsevier, Amsterdam.
268. Upchurch, G.J., Ford, J.W., Weiss, S.J., Knipp, B.S., Peterson, D.A., Thompson, R.W., Eagleton, M.J., Broady, A.J., Proctor, M.C., and Stanley, J.C. (2001). Nitric Oxide Inhibition Increases Metalloproteinase-9 Expression by Rat Aortic Smooth Muscle Cells in Vitro. J. Vasc. Surg., Vol. 34, pp. 76–83.
269. Vincent, J.H. (1995). Aerosol Science for Industrial Hygienists, Pergamon, Oxford, UK.
270. Vink, H. and Dulling, B.R. (1996). Identification of Distinct Luminal Domains for Macromolecules, Erythrocytes, and Leukocytes within Mammalian Capillaries. Circ. Res., Vol. 79, pp. 581–589.

271. Wang, L.-P. and Stock, D.E. (1992). Stochastic Trajectory Models for Turbulent Diffusion: Monte-Carlo Process versus Markov Chains. Atmos. Environ., Vol. 26, pp. 1599–1607.
272. Wang, L.-P. and Stock, D.E. (1993). Dispersion of Heavy Particles by Turbulent Motion. J. Atmos. Sci., Vol. 50, pp. 1897–1913.
273. Wang, Q., Squires, K.D., Chen, M., and McLaughlin, J.B. (1994). On the Role of the Lift Force in Turbulence Simulations of Particle Deposition. Int. J. Multiphase Flow, Vol. 23, pp. 749–763.
274. Wang, Z., Hopke, P., Baron, P., Ahmadi, G., Cheng, Y., Deye, G., and Su, W.-C. (2005). Fiber Classification and the Influence of Average Air Humidity. Aerosol Sci. Technol., Vol. 39, pp. 1056–1063.
275. Weibel, E.R. (1963). Morphometry of the Human Lung, Academic Press, New York.
276. Weinbaum, S., Zhang, X., Han, Y., Vink, H., and Cowin, S.C. (2003). Mechanotransduction and flow across the endothelial glycocalyx. PNAS, Vol. 100, pp. 7988–7995.
277. Womersley, J.R. (1975). An Elastic Tube Theory of Pulse Transmission and Oscillatory Flow in Mammalian Arteries, Wright Air Development Center, Wright-Patterson Air Force Base, OH.
278. Wood, N.B. (1981). A Simple Method for Calculation of Turbulent Deposition to Smooth and Rough Surfaces. J. Aerosol Sci., Vol. 12, pp. 275–290.
279. Wood, N.B. (1981). The Mass Transfer of Particles and Acid Vapour to Cooled Surfaces. J. Inst. Energy, Vol. 76, pp. 76–93.
280. Yaglom, A.M. and Kader, B.A. (1974). Heat and Mass Transfer Between a Rough Wall and Turbulent Fluid Flow at High Reynolds and Peclet Numbers. J. Fluid Mech., Vol. 62, pp. 601–623.
281. Yu, G., Zhang, Z., and Lessmann, R. (1998). Fluid Flow and Particle Diffusion in the Human Upper Respiratory System. Aerosol Sci. Technol., Vol. 2, pp. 146–158.
282. Zahmatkesh, I., Abouali, O., and Ahmadi, G. (2006). Numerical Simulation of Turbulent Airflow and Particle Deposition in Human Upper Oral Airway. FEDSM2006-98309, ASME Second Joint U.S.–European Fluids Engineering Summer Meeting, Miami, FL.
283. Zamankhan, P., Ahmadi, G., Wang, Z., Hopke, P.K., Su, W.-C., Cheng, Y.-S., and Leonard, D., (2006). Airflow and Deposition of Nano-Particles in Human Nasal Cavity. Aerosol Sci. Technol., Vol. 40, pp. 463–476.
284. Zhang, H. and Ahmadi, G. (2000). Aerosol Particle Transport and Deposition in Vertical and Horizontal Turbulent Duct Flows. J. Fluid Mech., Vol. 406, pp. 55–80.
285. Zhang, H., Ahmadi, G., Fan, F.-G., and McLaughlin, J.B. (2001a). Ellipsoidal Particles Transport and Deposition in Turbulent Channel Flows. Int. J. Multiphase Flows, Vol. 27, pp. 971–1009.
286. Zhang, Z. and Kleinstreuer, C. (2004). Airflow Structures and Nano-Particle Deposition in a Human Upper Airway Model. J. Comput. Phys., Vol. 198, pp. 178–210.
287. Zhang, Z., Kleinstreuer, C., and Kim, C.S. (2001b). Flow Structure and Particle Transport in a Triple Bifurcation Airway Model. J. Fluid Eng. Trans. ASME, Vol. 123, pp. 320–330.
288. Zhang, Z., Kleinstreuer, C., and Kim, C.S. (2002). Micro-Particle Transport and Deposition in a Human Oral Airway Model. J. Aerosol Sci., Vol. 33, pp. 1635–1652.
289. Zhang, Z., Kleinstreuer, C., Donohue, J.F., and Kim, C.S. (2005). Comparison of Micro- and Nano-Size Particle Depositions in a Human Upper Airway Model. J Aerosol Sci., Vol. 36, pp. 211–233.

XPS Analysis of Biosystems and Biomaterials

Michel J. Genet, Christine C. Dupont-Gillain, and Paul G. Rouxhet

Introduction

Chemical analysis is usually understood as the identification and quantification of constituents in terms of elements, chemical functions, and molecules. However, the proper characterization of many systems of practical concern, e.g., materials and biosystems, requires information at other levels. A ceramic is indeed not a mere mixture of SiO_2, Al_2O_3, K_2O, etc; a polymer material is not a random accumulation of macromolecules; a living cell is not a mere mixture of carbohydrates, proteins, and lipids. Understanding the properties of materials and biosystems requires the knowledge of the spatial distribution of their constituents. For materials of moderate complexity, this involves the nature, the amount, and the distribution of solid phases. For mesomorphic solids, this deals with the association of constituents at different scales above the molecular size: in semi-crystalline polymers, the key scales of organization are macromolecules ($\sim$1 nm), lamellae ($\sim$10 nm thickness), which are linked by tie molecules, and spherulites ($\sim$10 μm);[1] in starch, the macromolecules ($\sim$1 nm) form lamellae ($\sim$10 nm) separated by "amorphous" layers, which form blocklets ($\sim$100 nm) assembled in shells ($\approx$500 nm) to constitute a granule ($\sim$10 μm).[2] In the case of living cells, a crucial level of organization is the self-assembly of phospholipids in layers, producing membranes ($\sim$5 nm thick) that create a barrier with respect to the environment and define different compartments in the cell.[3]

Considering the properties of materials and biosystems thus involves (a) thinking in terms of colloids, interfaces, and space distribution, (b) keeping in mind that interfaces are zones where interaction forces are not balanced, which is responsible for an excess free energy.[4] Porous and dispersed solids (adsorbents, catalysts) represent a case in which different levels of characterization are clear-cut:

1. Bulk chemical composition in terms of elements and solid phases
2. Texture, which designates the space distribution of the solid (particle size, surface area, pore size distribution)

3. Chemical composition at the surface, which may be different from that of the bulk
4. Physico-chemical properties (zeta potential, hydrophobicity) controlling interfacial phenomena such as adsorption, adhesion, flocculation.

The situation is more complex for many materials and even more for biosystems.

For a long time, the complexity of materials and biosystems was mainly scrutinized by the combination of direct observations by optical and electron microscopy, with the separation of constituents. This approach led to the discovery of the main cellular organites.[5] The electron microprobe (1951) opened the way to chemical analysis with high spatial resolution and chemical imaging.[6] In this case, the chemical information is resolved in a plane perpendicular to the direction of observation i.e. essentially along the sample surface. X-ray photoelectron spectroscopy (XPS) (1967) opened another door by allowing the chemical composition to be determined in a thin layer at the surface of a solid, as opposed to the bulk.[7] In other words, the chemical information is resolved in the direction perpendicular to the surface. Atomic force microscopy (1986) and other near-field microscopies considerably broadened the possibilities to match spatial distribution, chemical composition, and molecular interactions by imaging a surface according to its physico-chemical properties.[8] Numerous other methods are now available to provide chemical information with spatial selectivity either laterally or according to depth: Infrared and Raman microscopy, Auger Spectroscopy, Time of Flight Secondary Ion Mass Spectroscopy, Confocal Laser Scanning Microscopy, etc.

XPS is the first technique to consider regarding surface chemical analysis, owing to the attractive balance between qualitative information, quantification and surface selectivity. It has indeed become a technique of major importance in the field of material science, owing to the importance of interfacial phenomena such as adsorption and adhesion. This includes the domain of biomaterials, the performances of which rely strongly upon the interactions between their surface and cells or biological fluids. It can be applied to systems of biological nature (microbial cells, food products) but this requires particular precautions as the samples are exposed to high vacuum during the analysis.

The aim of this contribution is to present a review of the use of XPS for the chemical analysis of biosurfaces. These are understood here as surfaces of living systems, materials covered by bioconstituents, and materials designed to be brought in contact with living cells and biological fluids. The review is designed to provide an account of recent developments and also to give an introduction to beginners in the field and to help ordinary users in developing a good practice.

An overview of XPS is presented in the following section with a brief examination of instrument components and recent developments. The basic methodology used for any XPS analysis is recalled and updated in *Basic Methodology and information*. In the section *Data Interpretation and Evaluation*, specific methodological aspects are discussed and practical reference data are given with a focus on biosystems and biomaterials, emphasizing the critical mind needed. Thus the above-mentioned sections are organized according to methods and levels of information. A thorough review of the application of XPS analysis to microorganisms and related systems is presented in *Surface Analysis of Microorganisms and Related Systems*; the leading thread is the information obtained, its relevance, and its application. The section *XPS Study of Systems Related to Biomedical Applications* provides an illustration of the use of XPS in the field of biomedical applications; this survey follows a progression according to the complexity of the investigated systems. Finally, a conclusion points out the potentialities, the limitations and the perspectives offered by XPS in order to achieve a better knowledge of the space distribution of constituents in systems related to the living world.

Overview of XPS

Background

The first experimental XPS spectrometer was developed by Siegbahn and his team at the University of Uppsala (Sweden) during the second half of the fifties and the beginning of the sixties.[9,10] He was the first to measure core levels chemical shifts in 1957 and to use the ESCA acronym (Electron Spectroscopy for Chemical Analysis), which is still used now, but in a less extent, concomitantly with XPS. Figure 1 shows the progression of the publications involving the technique and illustrates the respective use of the two acronyms. The first commercial instruments were constructed in 1969–1970. Kai Siegbahn obtained the Nobel price in physics in 1981 "for the development of high resolution electron spectroscopy" as did his father, Manne Siegbahn, in 1924, for his "discoveries and research in the field of X-ray

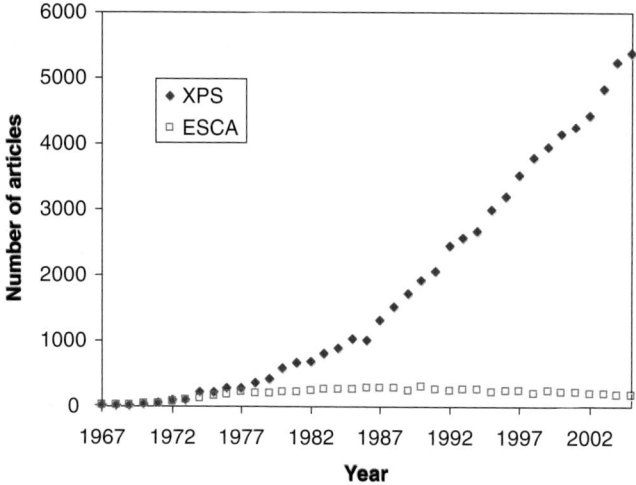

Fig. 1 Number of articles published per year from 1967 to 2005. Search through SciFinder Scholar software in CAPLUS and MEDLINE data sources, with "ESCA" and "XPS" keywords

spectroscopy." The place of XPS in a more detailed history of surface science can be found elsewhere.[11–13]

This spectroscopy is now extensively applied (Fig. 1) to materials such as minerals,[14] catalysts,[15] metals,[16] and polymers,[17] and also to biomaterials,[7,18,19] pharmaceutical products,[20,21] microbial cells, and biofilms.[22–25] A series of references are providing a detailed presentation of principles and applications of the technique.[26–31]

Principle and Specificity of XPS

A Method of Elemental Analysis

In the photoelectric effect, an atom A irradiated by photons of a given energy $h\nu$ gives rise to an ion in an excited state A^{+*} and a free electron e, which is expelled with a certain amount of kinetic energy (E_k).

$$A + h\nu \rightarrow A^{+*} + e \qquad (1)$$

The principle of energy conservation requires that the kinetic energy E_k of the emitted electron is related to its binding energy (E_b) by the equation

$$E_k = h\nu - E_b \qquad (2)$$

The binding energy E_b is the energy necessary to bring the electron from its original level to a state of rest outside the atom without relaxation of the ion A^{+*} (frozen orbital Koopmans approximation).[32]

Photoelectron spectroscopy is based on recording the kinetic energy spectrum of the electrons (number of electrons *vs* kinetic energy) emitted under irradiation of a material by photons with a given energy $h\nu$. The most widely spread form is XPS, in which the excitation is insured by X-rays, which are able to eject electrons from core shells; in Ultraviolet Photoelectron Spectroscopy (UPS), only electrons from outer orbitals can be ejected. Figure 2 presents a scheme of an XPS spectrometer with the X-ray source, the sample, the kinetic energy analyzer, the detector, and the recorded spectrum. The whole system is under ultra-high vacuum (UHV).

Actually, the kinetic energy measured by the spectrometer E'_k is different from that of the electron leaving the atom E_k. If a photoemitted electron is collected without undergoing inelastic collision, its kinetic energy is given by

$$\begin{aligned} E'_k &= E_k - \Phi_{sp} - E_c \\ &= h\nu - E_b - \Phi_{sp} - E_c \end{aligned} \qquad (3)$$

The meaning of the different terms is illustrated by Fig. 3: the charging term E_c represents the work spent to escape the attraction exerted by the charged sample surface, and the spectrometer work function Φ_{sp} represents the work further required

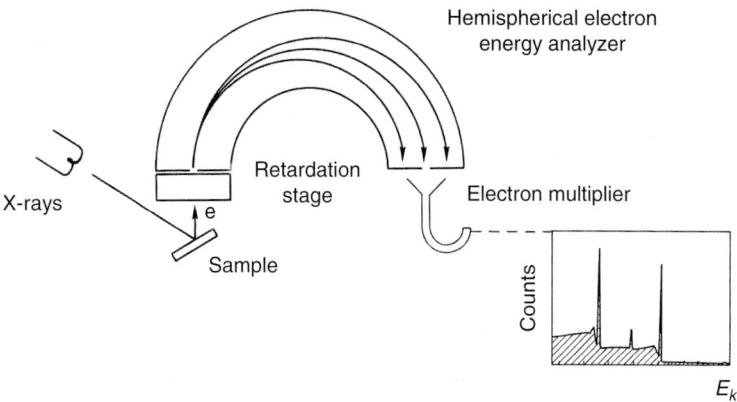

Fig. 2 Schematic representation of an XPS spectrometer. Reprinted from Ref. 22

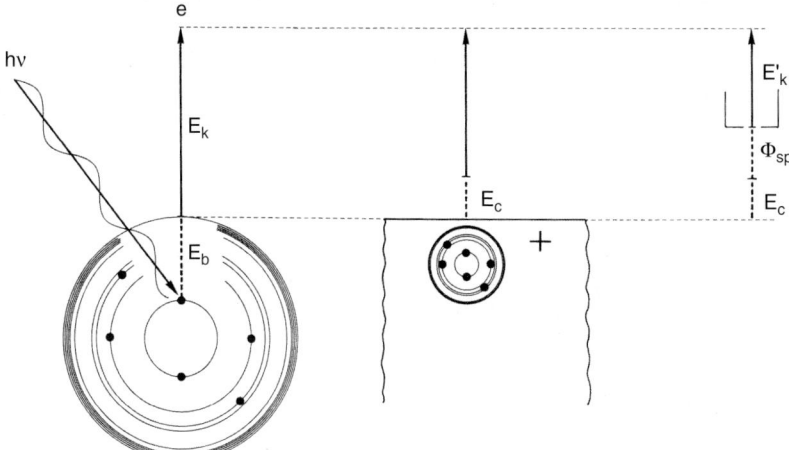

Fig. 3 Illustration of the electron photoejection process from 1s level of carbon and the defined energy losses that determine the relationship between the measured kinetic energy E'_k and the energy of the photon $h\nu$; the continuous arrows represent the photon energy and the electron kinetic energy at each step. (*Left*) Ejection from the atom; loss of E_b, the binding energy. (*Centre*) Escape from the sample; loss of E_c, due to charging effect. (*Right*) Travel from the sample to the kinetic energy analyzer; loss of Φ_{sp}, the instrument work function. Adapted from Ref. 22

to bring the electron from the zero attraction by the sample to the entrance of the energy analyzer. For a conducting sample, $E_c = 0$. As long as $h\nu$, Φ_{sp}, and E_c are fixed or calibrated, the position of an XPS peak on the kinetic energy scale is determined by E_b, which is specific of a given energy level of a given element. An XPS peak is thus characteristic of a given element and the XPS spectrum provides an elemental analysis.

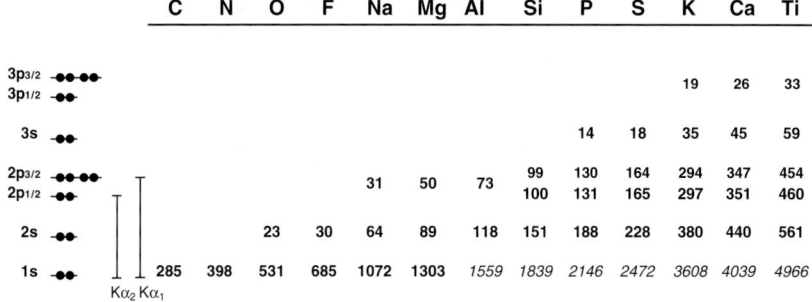

Fig. 4 Diagram of the electron energy levels with their maximum occupancy; binding energy of the XPS peaks for elements that are of interest in the study of biosystems and biomaterials (*bold*) and can be analyzed with the commercial spectrometers; energy of other 1s levels (*italic*); and sketch of the transitions responsible for the Kα_1 and Kα_2 X-ray lines[33,34]

The notation used in XPS for the electron energy levels and their maximum occupancy is recalled in Fig. 4. The approximate binding energy of the XPS peaks is also given for elements of common occurrence in biological systems and in materials of related interest, and for Mg and Al, which are the targets commonly used in the X-ray sources of commercial spectrometers. Typical XPS survey spectra are shown in Fig. 5. Note that hydrogen would give a H 1s peak at a binding energy of 14 eV; however, its sensitivity is 3 orders of magnitude lower than most peaks, so that hydrogen is not detected by XPS.[35]

As the energy level positions depend on the chemical environment of the atom, XPS peak position is sensitive to the chemical state of the element. This will be detailed in sections dedicated to *Chemical Functions*.

A Method of Surface Analysis

Let us onsider an electron photoemitted at depth z in a flat and smooth solid in a direction making the angle θ with the normal to the surface of the solid. The probability Q that the photoelectron leaves the solid in this direction without energy loss is given by

$$Q = \exp(-z/\lambda \cos \theta), \quad (4)$$

where $z/\cos \theta$ is the distance d traveled by the photoelectron in the solid and λ is the inelastic mean free path (IMFP), i.e., the average distance between inelastic collisions:

$$\lambda = -\frac{\mathrm{d}d}{\mathrm{d}Q}. \quad (5)$$

The IMFP will be discussed further in dedicated sections. It depends on the kinetic energy of the photoemitted electron and on the matrix, essentially its density. The photoelectrons that undergo inelastic collisions but still reach the detector contribute

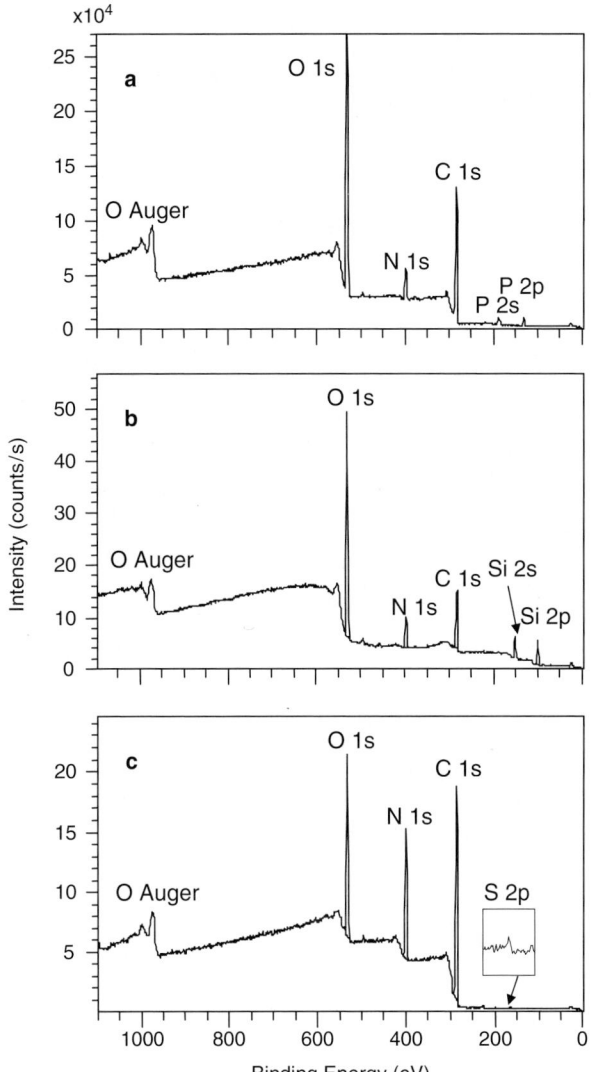

Fig. 5 Wide scan XPS spectra: (**a**) *Bacillus subtilis*; (**b**) glass with adsorbed human serum albumin; and (**c**) polyacrylonitrile with adsorbed fibrinogen. (Kratos Axis Ultra, monochromatized Al$_{K\alpha}$ source, pass energy = 160 eV)

to the background on the low kinetic energy side of the peak. This explains that every peak is accompanied by a rise of the background.

If the analyzed solid is flat and smooth and has a homogeneous chemical composition, the contribution dI_A of a layer of thickness dz, located at depth z, to the peak intensity of element A (Fig. 6 top), is given by

$$dI_A = P_A C_A \, dz \, \exp(-z/\lambda_A \cos\theta), \tag{6}$$

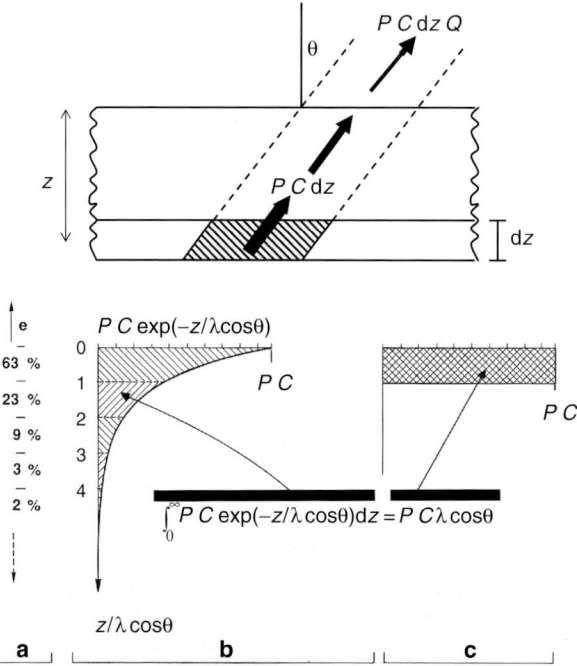

Fig. 6 *Top*: illustration of the contribution of a layer at depth z to the XPS signal collected along a direction making an angle θ with respect to the normal to the surface, taking account of the attenuation of photoelectrons along their trajectory. *Bottom*: illustration of the depth explored by XPS (case of homogeneous solid). (**a**) Contribution to the intensity of successively deeper layers of thickness equal to $\lambda \cos\theta$; (**b**) variation of the contribution to the intensity as a function of depth z; and (**c**) representation of a model situation in which the intensity would originate only from a layer of thickness equal to $\lambda \cos\theta$, in which the photoelectrons would not be subject to inelastic collisions (shaded areas are equivalent in **b** and **c**)

where C_A is the concentration of element A and P_A is a constant for a given element in fixed recording conditions. The exponential decrease of the contribution as a function of depth is illustrated by Fig. 6 (bottom): 63% of the signal originates from a layer of thickness equal to $\lambda \cos\theta$, 86% from a layer of thickness $2\lambda \cos\theta$, and 95% from a layer of thickness $3\lambda \cos\theta$. The IMFP is commonly in the range of 1.5–3 nm. Thus XPS provides an analysis of a very thin layer at the surface of the sample, usually considered to have a thickness up to $3\lambda \cos\theta$ (i.e., 1–10 nm).

As a matter of fact, if the solid has a homogeneous chemical composition over the depth that contributes to the information, the peak intensity is given by

$$I_A = P_A C_A \int_0^\infty \exp(-z/\lambda_A \cos\theta) dz = P_A C_A \lambda_A \cos\theta. \tag{7}$$

As illustrated by Fig. 6c, everything happens as if the peak was due to a layer of thickness $\lambda_A \cos\theta$, inside which the photoelectron intensity would not be attenuated by inelastic scattering, the layers situated deeper giving no contribution to the signal. This explains the frequently used but inaccurate expression that "XPS probes a surface layer with a thickness equal to $\lambda \cos \theta$."

Instrumentation

Vacuum

An XPS spectrometer is constituted of at least two chambers, one to introduce samples (c in Fig. 7) and a second one to proceed to the analysis (d in Fig. 7); the latter is always kept under ultra high vacuum. Chambers are made of stainless steel with a series of valves and windows. High vacuum is maintained, thanks to

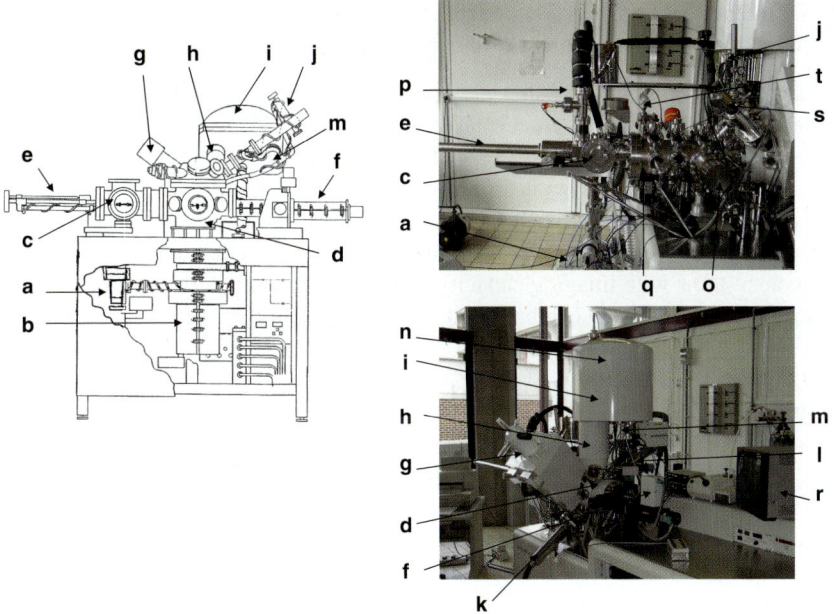

Fig. 7 X-ray photoelectron spectrometer. *Left*: schematic view of a SSX 100/206 (Surface Science Instruments). *Right*: photographs of a Kratos Axis Ultra (Kratos Analytical) with the introduction and intermediate chambers (*top*) and analysis chamber (*bottom*). **a**, Turbomolecular pump; **b**, cryogenic pump; **c**, introduction chamber; **d**, sample analysis chamber (SAC); **e**, transfer probe; **f**, automatized X, Y, Z manipulator; **g**, X-ray monochromator; **h**, electrostatic lens; **i**, hemispherical analyzer (HSA); **j**, ion gun; **k**, aluminum anode (with monochromator); **l**, aluminum–magnesium twin anode; **m**, detector. *Left*: channel plate. *Right*: 8 channeltrons (Spectroscopy mode), phosphor screen behind a channel plate with a video camera (Imaging mode); **n**, spherical mirror analyzer (SMA); **o**, parking facility in the sample transfer chamber; **p**, sample cooling device for the introduction chamber; **q**, sample transfer chamber; **r**, monitor interconnected with the video camera viewing samples in the SAC; **s**, video camera in the SAC; **t**, high temperature gas cell (catalyst pretreatment)

knife-edge flanges, which are compressed against copper gaskets. Electron trajectories are screened from earth magnetic field by using a metal with high magnetic permeability for internal walls of the analysis chamber. Additional chambers can also be added to allow sample pretreatment or parking (q, t in Fig. 7).

Vacuum is required for several reasons:

1. The photoelectrons leaving the sample should not undergo inelastic collisions with gas molecules. This imposes a residual pressure lower than about 10^{-4} Pa.[26]
2. Residual gases may contaminate the sample surface. Although this factor is often mentioned, it is of limited relevance for many applications as the samples have been exposed to air before and are already contaminated by chemisorbed species and adsorbed organic compounds. It is of greater concern when the ambition is to study atomically clean surfaces, for instance after cleaning by etching in the spectrometer, as these must be protected from subsequent reoxidation, chemisorption of reactive gases, or adsorption of organic compounds.
3. Different spectrometer components (X-ray source, gauges, detectors) require high vacuum to insure proper functioning and to avoid early deterioration.

In practice, it is aimed to reach a pressure of the order of 10^{-7} Pa and a pressure only slightly above 10^{-6} Pa is acceptable.

Pumping systems used on recent XPS spectrometers are typically (1) turbomolecular pumps coupled with a rotary or membrane pump for the introduction chamber (a in Fig. 7) and (2) a cryogenic pump (b in Fig. 7) or an ionic pump with Ti sublimator for the analysis chamber; the second one being preferred to cryogenic pump on systems with imaging capacity to avoid vibrations. A detailed description of the vacuum technology can be found elsewhere.[36]

Sample Handling

Self-supporting specimens (metals, polymers, glasses, etc.) are attached on the sample holder with a double-sided tape, clips, screws, or conductive glue used in scanning electron microscopy (SEM). On recent XPS systems, the largest analyzed area is around 1 mm^2; the sample size is often about 5×5 mm^2. Powder specimens (catalysts, minerals, freeze-dried microbial cells, etc.) can be deposited in small stainless steel troughs or on an indium foil, or alternatively on a piece of double-sided tape, taking care to insure complete screening of the tape by the powder. A sample holder may accommodate several specimens.

Pumping down the introduction chamber takes usually between half a hour and a few hours depending on the type, composition, and number of samples. High surface area solids and solids releasing water or other volatile compounds may need longer pumping time; they may require to reduce the number of samples introduced together and to reduce sample size. Samples are transferred to the analysis chamber with a transfer probe (e in Fig. 7). On recent systems, automatic analysis of series of samples is feasible, thanks to motorized X, Y, Z displacement (f in Fig. 7).

Some XPS spectrometers are equipped with specific accessories such as ion gun (j in Fig. 7), sample fracture stage, sample heating and cooling (p in Fig. 7), high pressure and high temperature gas cell (t in Fig. 7), plasma treatment chamber, etc. The ion gun uses ions of an inert gas (commonly Ar^+) to clean surfaces or etch the sample for depth profiling. This device is commonly used for metal plates and almost never used on powders or on organic solids, owing to problems of degradation and differential sputtering. A cooling accessory may be particularly useful for biosystems and will be described later (*Surface Reorganization and Low Temperature Analysis*).

X-Ray Sources

In common X-ray sources, a metal target is bombarded with high energy electrons produced by a tungsten filament or an electron gun. The emitted polychromatic X-ray beam is composed of a series of lines and a continuous background called Bremsstrahlung.

The most popular targets (anodes) are made of aluminum or magnesium. For Al the $K_{\alpha1,2}$ doublet (sketched in Fig. 4) is at 1,486.71 and 1,486.29 eV, giving a line centered at 1,486.57 eV with a width of about 0.85 eV. For Mg, the $K_{\alpha1,2}$ doublet is at 1,253.69 and 1,253.44 eV, giving a line centered at 1,253.60 eV with a width of about 0.70 eV.[29,37] Both lines are narrow and give intense photoelectron spectra compared to other less common targets that are used in specific cases: $Zr_{L\alpha}$ (151.4 eV), $Si_{K\alpha}$ (1,739.6 eV), $Ag_{L\alpha}$ (2,984 eV), $Ti_{K\alpha}$ (4,510.9 eV), $Cu_{K\alpha}$ (8,047.8 eV).[38] At present, most XPS analyses are performed using a monochromatized X-ray beam ($Al_{K\alpha}$) of about 1–2 mm diameter (or less on spectrometers with a microfocused X-ray beam) focused onto the sample. Therefore, the polychromatic beam of an Al source (k in Fig. 7) is directed to a bent quartz crystal (g in Fig. 7), where it is diffracted, allowing the Bremsstrahlung and unwanted lines to be eliminated and the energy width of the $Al_{K\alpha}$ line to be reduced to about 0.26 eV. The XPS peak intensity obtained with a monochromatic source is about 10% relative to a conventional source, but the signal to noise ratio is improved owing to the removal of the Bremsstrahlung radiation.

The use of two different sources (Mg and Al) (l in Fig. 7), possibly combined in a dual target, helps avoiding the overlap between an Auger peak (see the section on *Auger Peaks*) and an XPS peak.[39] The use of a nonmonochromatic Al source allows the Bremsstrahlung to induce Auger peaks[40] and to determine the Auger parameter.[40,41]

Tunable synchrotron sources are increasingly used. They can produce[42] a variable photon energy with a high intensity allowing XPS spectra to be recorded with an energy resolution of a few meV and with a resolution of less than one degree on the electron collection angle θ; thereby a high lateral resolution can be achieved in photoemission microscopy (100 nm or better compared to 3–5 µm on the most recent conventional imaging XPS spectrometers). This potentiality offers promising perspectives in spectromicroscopy to achieve surface chemical imaging of biomaterials and biosystems, for which the chemical variations at the scale of 1 μm and below may indeed be of great practical importance.[43–47] When high spatial resolution and high-energy resolution are achieved simultaneously, radiation damages may be a severe limitation, unless the data are recorded quickly enough.[42]

Lens System and Energy Analyzer

Electrons ejected from the sample are collected by an electrostatic lens (h in Fig. 7), and they are sorted out according to their kinetic energy in an energy analyzer (i in Fig. 7), as schematized in Fig. 2.

The electrostatic *hemispherical analyzer* (HSA), also called concentric hemispherical analyzer (CHA), is now fitting all commercial XPS spectrometers (i in Fig. 7). It consists of two concentric hemispheres of radius R_1 and R_2 respectively; a section comprising the entrance slit and the center is shown in Fig. 8. Electrical potentials V_1 and V_2 are applied, so that the electrons are attracted by hemisphere 1 and repelled by hemisphere 2.[26,29]

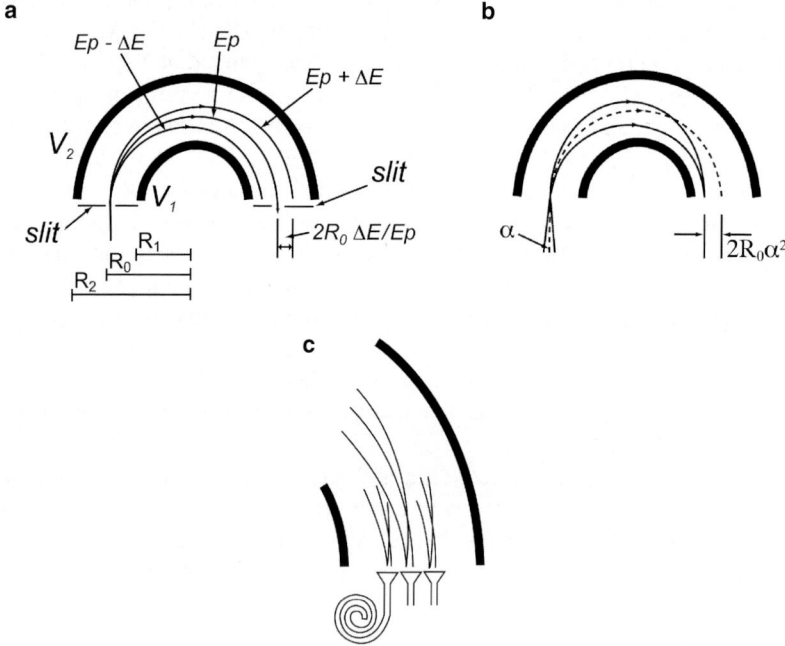

Fig. 8 Section through a hemispherical analyzer (HSA), along any plane including the slits and the center: (**a**) Trajectories followed by electrons of different kinetic energies entering tangentially to the median sphere of radius R_0; (**b**) influence of the angle ($\pm\alpha$) of electron injection on the position in the exit plane; and (**c**) series of three channeltrons collecting electrons at the analyzer exit. Inspired from Refs. 29 and 30

Electrons entering the analyzer tangentially to the median surface of radius R_0 follow a trajectory of radius R_0, provided their kinetic energy is equal to E_p, where E_p is called the pass energy defined as

$$E_p = e(V_1 - V_2)\left(\frac{R_1 R_2}{R_2^2 - R_1^2}\right), \qquad (8)$$

where e is the charge of the electron and $\left(\frac{R_1 R_2}{R_2^2 - R_1^2}\right)$ is called the spectrometer constant.

Electrons with a lower and higher kinetic energy than E_p will follow a path with a radius smaller and larger than R_0, respectively, and will reach the exit plane closer to or further away from the center by the distance $\Delta R = 2R_0 \Delta E / E_p$ (Fig. 8a). This is the essence of the dispersive property of the HSA,[29] the dispersion being $\frac{\Delta R}{\Delta E} = \frac{2R_0}{E_p}$.

Considering a width W for the entrance and exit slits, the absolute resolution of the HSA is then

$$\Delta E = (E_p / 2R_0) W. \qquad (9)$$

This equation shows that injection of photoelectrons as emitted by the sample would have the important disadvantage that scanning the XPS spectrum by sweeping $V_1 - V_2$ would give a variation of pass energy and thus of absolute resolution over the spectrum. Moreover, recording peaks of high kinetic energy with the desired resolution would require very high ratios of analyzer dimension over the slit width.

In practice, at the stage of the electrostatic lens, electrons are accelerated or retarded to a defined energy equal to a desired E_p. The difference of potential between the hemispheres is maintained constant and the spectrum is scanned by sweeping the voltage of the stage. This *constant analyzer energy* (CAE) mode is used in XPS. Equation (9) shows that the resolution ΔE is improved when the pass energy decreases. For instance, for a widespread instrument, the range of kinetic energy dispersed on the exit plane (where the exit slit is located) is about 18, 12, 6, and 3 eV for pass energies of 150, 100, 50, and 25 eV, respectively.

In addition, electrons of energy E_p injected at an angle of α radians to the tangential direction will reach the exit plane nearer to the center of the analyzer, at a distance $2R_0 \alpha^2$ from the tangential ray as illustrated by Fig. 8b. The angular spread of the electron beam will thus further deteriorate the resolution by increasing ΔE. The role of the electrostatic lens is to convey as many photoelectrons as possible from the X-ray illuminated spot into the analyzer, with a direction as tangential as possible to the median surface of the HSA.

In recent spectrometers, the specimen is immersed within a magnetic field created by an additional magnetic lens. The current through the magnetic lens is varied according to the electron kinetic energy being analyzed. The main advantage of this system is to collect electrons emitted in a wide range of directions, thereby increasing the sensitivity.

Electron Detectors

Two types of electron multipliers (m in Fig. 7) are used: channeltrons and channel plates. *Channeltrons* are spiral-shaped glass tubes through which a high difference of potential (order 2–4 kV) is applied, allowing acceleration and multiplication of incoming electrons in the tube (cascade effect). The typical electron gain is 10^6–10^8. If a series of channeltrons is arranged radially across the exit plane (Fig. 8c), each channeltron will collect electrons of a given energy range and the system allows simultaneous counting, which increases sensitivity.

An alternative is to use *channel plates* that consist of disks with an array (hundred to thousands) of small holes (a few tens of micrometers in diameter), each hole playing the same role as a combination of a single slit and a channeltron.

Recently, a device, which was first used with synchrotron radiation,[48] has been applied to a laboratory instrument to identify the spot of a channel plate hitted by an photoelectron and thus the energy of the latter. Secondary electrons from the channel plate are impinging a copper wire where they induce an electrical pulse that propagates to the two ends of the wire (Delay Line Detector, DLD). The signal is detected and positioned along the length of the wire by measuring, with an electronic clock, the difference in the arrival time of twin pulses at the ends of the wire. Ultrahigh-speed multichannel detectors for use in synchrotron are in development.[49]

Analyzed Area

The analyzed area is determined in two ways depending on the spectrometer geometry:

1. The area from which photoelectrons are collected (as small as 10–15 μm on recent spectrometers) is defined by the size of an aperture introduced in the electrostatic lens and the magnification of the photoelectron image by the magnetic lens. In this case, the analyzed area is smaller than the bombarded area. It is important to be aware that, if the sample is sensitive to X-ray degradation, the degraded area will be larger than the analyzed area.
2. The monochromatized X-ray beam is focused onto the sample as a small spot from where photoelectrons are collected. In this case, the analyzed area may correspond to the X-ray irradiated area. On certain spectrometers the focused X-ray beam is used to scan a given area on the sample.

XPS Imaging

One can take advantage of reducing the analyzed area to produce XPS images showing the distribution of a given element or even a given chemical state of the element at the sample surface. Image acquisition can be performed (1) by serial acquisition, either moving the sample stage along X and Y or scanning the sample with the focused X-ray beam (space resolution down to about 10 μm) or scanning the imaged area via the electrostatic lens without moving the sample, and (2) by parallel acquisition, collecting simultaneously data from different spots of a given sample area (space resolution about 3 μm). In one instrument, a phosphor screen (scintillator) is placed behind the channel plate at the exit slit of a spherical mirror analyzer and light pulses are recorded with a video camera (m in Fig. 7 right). In more recent developments, the phosphor screen is replaced by a set of two delay line detectors perpendicular to each other.

Angle-Resolved XPS (ARXPS)

In the section *A Method of Surface Analysis* and Fig. 6, it has been shown that the explored depth was depending on λ and on the collection angle θ. With flat and smooth samples, the collection angle can be varied by tilting the sample. Collection at grazing angle provides information on a thinner layer near the surface. One commercial instrument is capable of performing a simultaneous collection of angle-resolved data without tilting the specimen. ARXPS is a nondestructive method, allowing concentration gradients near the surface to be revealed, as discussed in the section *Semi-Quantitative Use of Angle Resolved Analysis*.

Sample Charging and Charge Stabilization

On *conducting samples*, the charge developed at the sample surface by the photoejection process is immediately compensated by electrons coming through the grounded sample (Fig. 9a). In this case, $E_c = 0$ in (3).

On *insulating samples*, as are many biosystems and related materials, photoelectron emission creates a positive charge at the surface. The charging term E_c may vary according to the position on the sample and as a function of time, which may lead to peak broadening, peak distortion, or peak shift during the analysis. Charge stabilization is achieved in different ways, depending on the manufacturer and on the X-ray source used.

With a nonmonochromatic source, secondary electrons generated by the Bremsstrahlung on the X-ray tube window and metal surfaces in the vicinity of the sample will stabilize the positive charge developed on the sample surface and will maintain E_c constant and positive. The recorded peaks will be shifted by a few eV to a lower kinetic energy as illustrated by Fig. 9b. In this case, the charging voltage may differ according to the surface composition. This is illustrated by Fig. 10 in which results obtained on human dental enamel treated by mouth rinse solutions and/or saliva are presented.[50] The apparent carbon/calcium concentration ratio was measured by XPS and the thickness t of the organic layer built up at the surface by adsorption of saliva and mouth rinse constituents was deduced as described in the section *Data Interpretation Through Simulation*. The decrease of E_c as t increased may be attributed

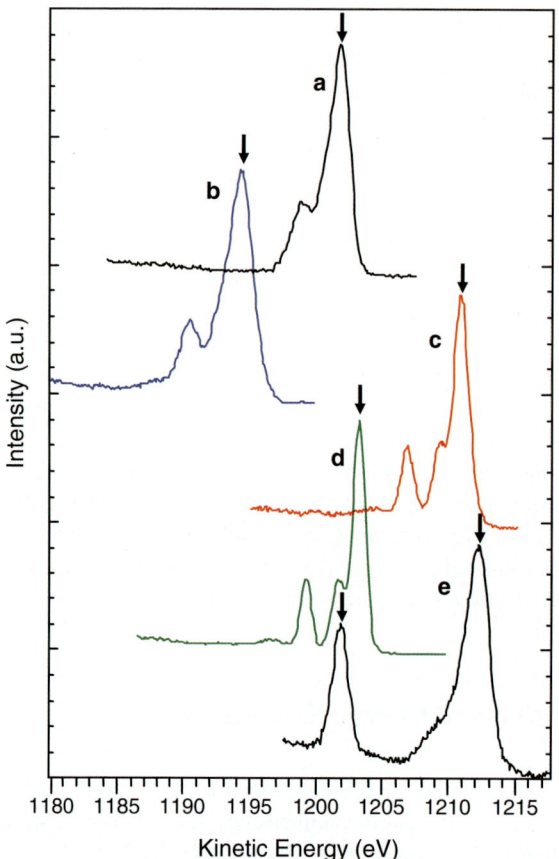

Fig. 9 Illustration of the charging effect on C 1s peaks recorded on different samples with different spectrometers. Arrows are pointing the $\underline{C}$—(C, H) component of the C 1s peaks, the binding energy of which is at 284.8 eV. (**a**) Organic contamination on stainless steel (conductor), Kratos, $E_c = 0$ eV; (**b**) poly(ethylene terephthalate) (PET, insulator), VG, $E_c = +7.2$ eV; (**c**) PET, SSI, flood gun 6 eV, $E_c = -9.0$ eV; (**d**) PET, Kratos with charge neutralizer (-2.8 eV) and magnetic lens, $E_c = -1.4$ eV; and (**e**) *Azolla* fern (size $\sim$250 μm, insulator) on a gold coated electron microscope grid (conductor), SSI, flood gun 8 eV: occurrence of differential charging effect, $E_c = 0$ for the organic contamination on the grid (peak on the left) and $E_c = -10.4$ eV for the *Azolla* fern (peak on the right). Kratos: Kratos Axis Ultra, monochromatized Al$_{K\alpha}$; VG: VG ESCA 3 MkII, non-monochromatized Al$_{K\alpha}$; SSI: SSX 100/206, monochromatized Al$_{K\alpha}$ with flood gun and grid above the sample

to a decrease of the flux of photoelectrons, due to the lower cross-section of carbon compared to calcium.[51]

With a monochromatized X-ray beam, the charging effect is very serious. In this case a specific accessory called flood gun or charge neutralizer is used to bombard the sample with a flux of low energy electrons. The electrical charge at the surface is overcompensated ($E_c < 0$) but the charging term is maintained constant. The recorded peaks are shifted to a higher E_k as illustrated by Fig. 9c and d for two

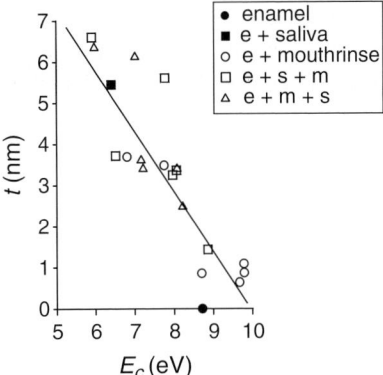

Fig. 10 Correlation between the thickness t of adsorbed layers, estimated from C/Ca ratios, and the charging voltage E_C. XPS analysis of human dental enamel samples (e) treated in different ways [*filled circle*, untreated; *closed square* + saliva (s); *open circle*, + mouth rinse (m); *open square*, + saliva + mouth rinse; *open triangle*, + mouth rinse + saliva].VG ESCA 3 MkII, non-monochromatized Al$_{K\alpha}$. Redrawn from Ref. 50

commercial spectrometers using different charge neutralizer devices, which allow reliable charge stabilization on insulators. It is recommended to adjust the flood gun energy to obtain XPS peaks as narrow as possible and to pay attention to possible peak distortion. The flood gun can be used as such or in combination with a metallic fine mesh grid placed a few millimeters above the sample (spectrometers from Surface Science Instruments).[51]

The charging shift measured with NaCl samples (SSX-100/206 spectrometer with a monochromatic X-ray beam)[52] was shown to be influenced by three different factors: (1) the total photoelectron flux, E_c being about 0.6 eV less negative when the total photoemitted electron count increased by a factor of 3 owing to an increase of the X-ray source potential; (2) the flood gun energy, an increase of which made E_c becoming more negative by about the same increment without affecting the total photoemitted electron count; (3) the presence of an organic overlayer, the thickness and regularity of which affected both the total photoelectron flux from the sample and the flux of electrons from the overlayer to the substrate. Upon irradiation by X-rays, the overlayer acts as an additional electron source, which makes the substrate more negative and reduces inhomogeneities of the local potential affecting the full-width at half maximum (FWHM) of peaks. Overlayer charging may be slightly different from substrate charging (vertical differential charging). The charging shift of poly(dimethyl siloxane) (PDMS) overlayers was the same as that of NaCl substrate, while the charging shift of polystyrene (PS) overlayers was 0.5–1.0 eV more negative compared to the substrate.

When sample surfaces present both insulating and conductive phases (for instance a thin insulating polymer layer covering part of a conductive wafer), a lateral differential charging can occur, generating peak broadening or leading to two well-separated peaks due to the same element on the two phases. This may be used to make the spectral information more selective as illustrated by Fig. 9e, which shows the C 1s peak recorded for a cuticle of *Azolla* fern (size about 250μm) placed on a transmission electron microscope grid. This allowed the C 1s and O 1s peaks of the microscopic sample to be recorded without interference of the support, although the sample size was smaller than the analyzed zone.[53] Most often, lateral differential charging is a complication and is prevented by mounting the specimen on an insulating holder such as a ceramic.

Advanced charge stabilization devices are fitted on the most recent spectrometers. For instruments in which the sample is immersed in a magnetic field, low energy photoelectrons from the sample are reflected by a plate positioned a few centimeters above the sample. They are guided by the magnetic field lines to return back to the sample near their point of origin. As fewer electrons are returning to the sample than those leaving, additional electrons are injected into the magnetic field from a filament situated above the sample close to the electrode plates. In this configuration, the sample surface is only slightly overcharged (1–2 eV) as illustrated by Fig. 9d. Another advanced charge compensating device is combining electrons generated by a flood gun and low energy positive ions (<50eV), which are used to reduce negative charging near the area irradiated by the X-ray beam.

A survey of the evolution of the commercial spectrometers over 35 years, market considerations, and perspectives was published recently.[54]

Basic Methodology and Information

Overview of an XPS Spectrum

Energy Scale Referencing

The zero of the kinetic energy scale is set by recording the spectrum of a clean standard gold specimen and fixing the binding energy of the Au $4f_{7/2}$ peak at 83.96 eV. As pointed out in section *Sample Charging and Charge Stabilization*, no charging is built up on conducting sample surfaces; thus $E_c = 0$ in (3). This allows Φ_{sp} to be determined. The linearity of the spectrometer energy scale must also be checked and adjusted by recording specific well-separated peaks on gold or copper. The recommended reference binding energies are reported in an ISO standard procedure.[55]

On conducting samples, the peak position is measured directly on the spectrometer energy scale properly calibrated and the peak kinetic energy is immediately converted into peak binding energy.

On insulators, the charging shift E_c must be accounted for by assigning a given binding energy to a line used as reference. Au $4f_{7/2}$ peak of gold deposited as a thin layer on the sample was used previously but was discredited mainly because of differential charging and masking effects. For samples containing organic constituents, including the organic overlayer contaminating the surface of oxides and metals, the common practice is to take as reference the component of the C 1s peak due to carbon only bound to carbon and hydrogen [C̲—(C, H)]. Different reference binding energy values for $\underline{C}$—(C,H) are reported in the literature; the most common databases are using 284.8[33,56] or 285.0 eV.[57] When the choice of a reference line is not straightforward or when the $\underline{C}$—(C,H) component is poorly defined, the determination of the peak binding energy requires different trials, with a critical examination of the consistency found within a spectrum or in a series of spectra.

Core Photoelectron Peaks and Background

Electrons photoejected from an atom and escaping the sample without energy loss give rise to discrete peaks (no-loss peaks), the kinetic energy of which follows (3) (Fig. 5). All peaks with a binding energy lower than the photon energy $h\nu$ of the X-ray source can be observed. The non-s levels may give resolved doublets depending on the element. All elements of the periodic table can be detected except hydrogen and helium for which the sensitivity is too low.

The background shows a rise at the low kinetic energy side, i.e., high binding energy side, of each peak (Figs. 5 and 11). This is due to electrons that have the same origin as those collected under the peak but have suffered inelastic collisions along the travel out of the sample. The typical energy loss in a single scattering event is 15–30 eV depending on the solid.[58,59] On spectra recorded with a nonmonochromatized X-ray beam, electrons excited by the Bremsstrahlung also contribute to the background.

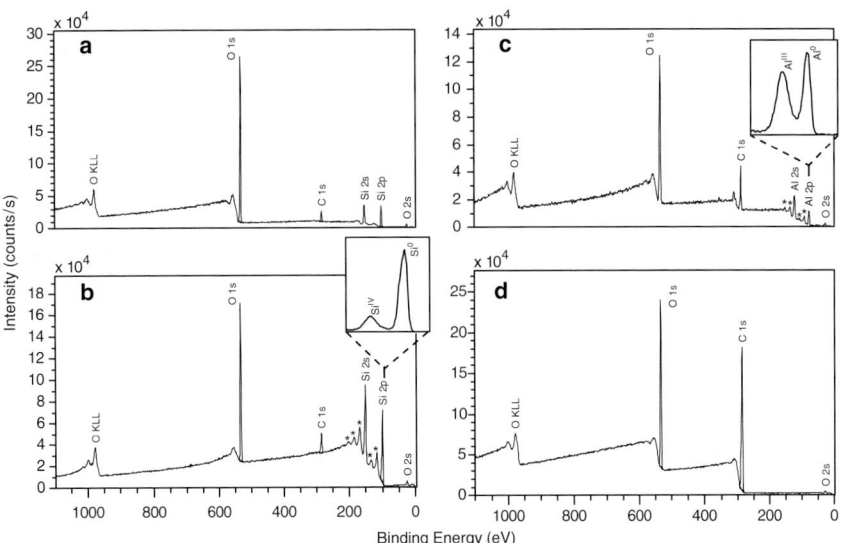

Fig. 11 Wide scan spectra recorded on (**a**) amorphous silica slab; (**b**) silicon wafer (apparent $Si^0/Si^{IV} = 3.71$; (**c**) scraped aluminum slab (apparent $Al^0/Al^{III} = 0.76$); and (**d**) poly(methyl methacrylate). Plasmon peaks are indicated by *. Kratos Axis Ultra, monochromatized $Al_{K\alpha}$, pass energy 160 eV

The background rise depends on the distribution of the concerned element within the outermost depth range. For instance, the background intensity next to the Au 4f doublet increases relative to the peak height as the thickness of the gold film deposited on graphite increases; this is due to the fact that, in average, Au 4f photoelectrons have to traverse greater distances to reach the surface and that a larger proportion undergo inelastic collisions, as explained in section *A Method of Surface Analysis*.[58,60] This phenomenon is illustrated in Fig. 11. The background rise near the O 1s and O Auger peaks, relative to the intensity of the peaks, is larger for bulk silica (a) compared to a silicon wafer (b), on which a thin layer of silicon oxide covers silicon. Similarly, the background rise observed next to the C 1s peak of a polymer (Fig. 11d) or a microbial cell (Fig. 5a) is relatively higher compared to the C 1s peak of the thin layer of organic contamination present on a silica or on a silicon wafer (Fig. 11a, b) or on glass with adsorbed protein (Fig. 5b). The influence of the distribution of an element near the surface (layer thickness, depth, lateral heterogeneity) on the background profile can be simulated on the basis of an accurate description of electron transport in the solid.[58,59]

Valence Band

The valence band is the binding energy region between 0 and 20 eV, consisting of a series of closely spaced peaks. The corresponding electrons are involved in delocalized or bonding orbitals (molecular orbitals). The valence band thus depends on the chemical structure of the analyzed compounds and its interpretation requires molecular orbital simulation.[61]

Figure 12a presents the valence band spectra of linear low-density polyethylene (LLDPE) with two peaks attributed to C 2s contributions and a broad C 2p contribution.[62] In polypropylene (PP), a third intense peak appears near 14.5 eV (Fig. 12b).

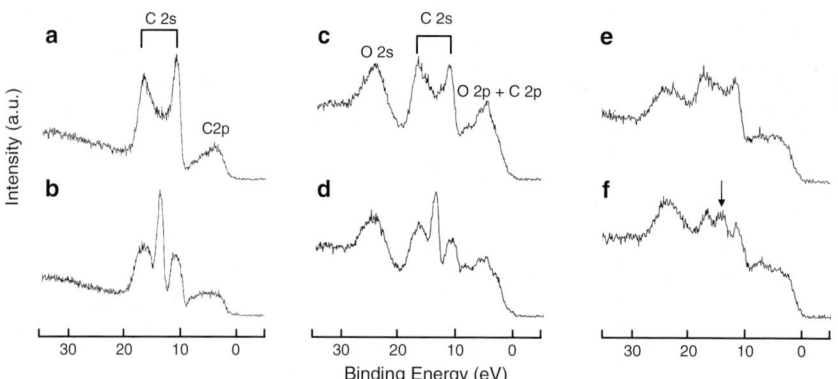

Fig. 12 Valence band spectra recorded on (**a**) LLDPE; (**b**) PP; (**c**) oxygen-plasma treated LLDPE; (**d**) oxygen-plasma treated PP; and nitrogen-plasma treated PE at (**e**) 20° and (**f**) 70° photoelectron collection angle θ. The arrow points to the propylene-like feature. Adapted from Ref. 62 with permission, copyright (1991) John Wiley and Sons Limited

Treating these polymers by an oxygen plasma leads to the appearance of a broad O 2s band and of an O 2p contribution (Fig. 12c, d). The spectra of LLDPE treated by nitrogen plasma, recorded at $\theta = 20°$ and $70°$ photoelectron collection angles differ according to the intensity near 14.5 eV (Fig. 12e, f). This is much higher at $70°$ (arrow on Fig. 12f) compared to $20°$, indicating the appearance of PP-like feature within the C 2s valence band structure and providing evidence for cross-linking within the uppermost surface layers. Surface oxidation is frequently used to increase the polarity of polymers. The associated crosslinking may be an important factor to slow down the hydrophobicity recovery that takes place upon subsequent aging, owing to the tendency of the surface layer to reorganize in order to decrease the surface energy.[63]

The spectrum in the valence band region is influenced by the source of radiation used: UV radiation (UPS), X-ray conventional source, or X-ray synchrotron radiation.[42] UV radiation provides higher photoelectron cross-sections in the valence band region than the X-ray radiation. The lower energy of UV photons leads to lower electron kinetic energies and lower IMFPs, thus providing more surface-sensitive spectra. As a consequence, the surface of interest may easily be masked by contaminants, which explains the limited use of UPS. A tunable energy synchrotron X-ray radiation can help in separating contributions of different electronic states in the valence band.

Auger Peaks

The photoionization process described in the section *A Method of Elemental Analysis* leaves an ion in an excited state, which may relax in two ways illustrated by Fig. 13a for oxygen: (a) electron transition to fill the position left unoccupied at the lower level, the energy conservation being insured by the emission of a photon; this is responsible for X-ray fluorescence, and will not be discussed here; and (b) Auger electron emission, which is described using the symbols of energy levels (K, L, M, etc.) used in X-ray spectroscopy.

In the Auger effect, two electrons are involved, belonging to subshells L_1 and $L_{2,3}$ (i.e. 2s and 2p, respectively) in the case of oxygen sketched in Fig. 13b. One fills the hole at K level (1s) and the other one is ejected, leaving a doubly ionized atom. The ejected electron is called the Auger electron and named according the energy levels involved KLL'.

$$E_{k,KLL'} = E_{b,K} - E_{b,L} - E_{b,L'} - \Phi_{sp} - E_c + W, \qquad (10)$$

where E_k and E_b are kinetic energy and binding energy respectively, and W accounts for the displacement of energy levels upon the formation of a doubly ionized atom. In XPS spectra, both photoelectrons and Auger electrons are detected as illustrated in Fig. 14, the KL_1L_1, $KL_1L_{2,3}$, and $KL_{2,3}L_{2,3}$ Auger lines of oxygen appearing at a kinetic energy of about 475, 490, and 511 eV, respectively. Changing the nature of the anode keeps the Auger peaks at the same kinetic energy E_k and the XPS peaks at the same binding energy E_b. For atoms with a small atomic number, relaxation

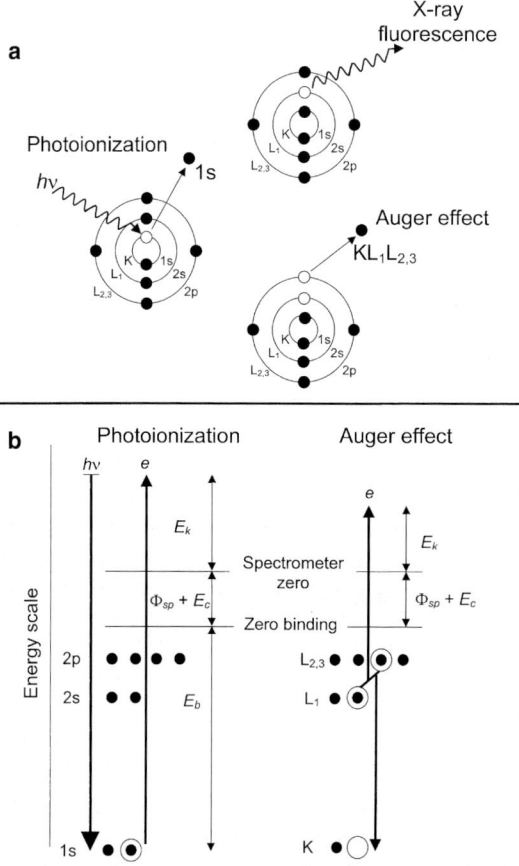

Fig. 13 (a) Schematic illustration, for oxygen, of photoionization and of relaxation by X-ray emission (X-ray fluorescence) and by Auger emission; and (b) sketch of energy diagrams with an illustration of photoionization and Auger effect ($KL_1L_{2,3}$ line)

by X-ray fluorescence following creation of a core hole in the K shell is negligible, compared to Auger electron emission. As the atomic number increases, the probability of the former increases and the probability of the latter decreases.[26]

In XPS, the main concern for Auger peaks is that they sometimes overlap with photoelectron peaks (e.g., interference of Ti KL_1L_1 with the Na 1s peak at 1,072 eV recorded with an Al anode). Note that the Auger electron kinetic energy is independent of the excitation, in contrast with the kinetic energy of a photoelectron (Fig. 14). Thus the overlap between Auger peaks and the photoelectron peaks can be solved by changing the energy of the X-ray source.

The Auger parameter is defined as the kinetic energy difference between an Auger peak and the photoelectron peak associated to the initial vacancy responsible for the Auger effect[41]

$$\alpha = E_{k,\text{Auger}} - E_{k,\text{XPS}} \tag{11}$$

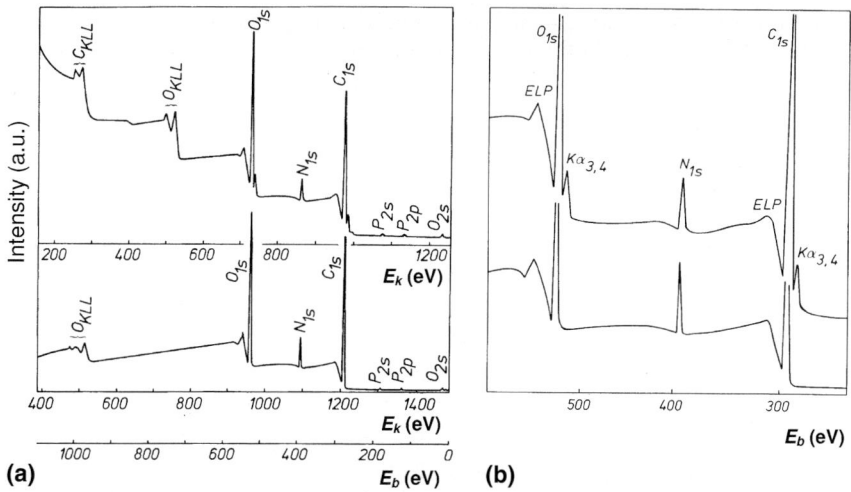

Fig. 14 (a) Wide scan XPS spectra of *Streptococcus thermophilus* recorded with a spectrometer equipped with a Mg anode without a monochromator (*upper curve*) and an Al anode with a monochromator (*lower curve*); attribution of the XPS (1s, 2s, 2p) and Auger (KLL) peaks of the elements; the binding energy scale E_b is the same for both spectra. **(b)** Enlarged spectra with detail of energy loss peaks (ELP) and $K\alpha_{3,4}$ satellites

and the modified Auger parameter as

$$\alpha' = \alpha + h\nu. \tag{12}$$

For the Auger peak KLL′, Eqs (3), (10), and (11) lead to

$$\alpha = (E_{b,K} - E_{b,L}) + (E_{b,K} - E_{b,L'}) - h\nu + W. \tag{13}$$

The Auger parameters of an element are more sensitive to its chemical state (e.g., coordination number in the case of Al[40]) than the photoelectron binding energy. Moreover, their values are independent of the way of defining the zero of the energy scale.

Secondary Lines

Source Satellites

In XPS spectra recorded with non-monochromatized X-ray, each peak is accompanied by a series of satellites on the low binding energy side of the peak, corresponding to electrons photoemitted by X-ray lines other than the $K\alpha_{1,2}$ line. The $K\alpha_{3,4}$ satellite shown in Fig. 14 is generated by the $K\alpha_{3,4}$ line of the magnesium X-ray source, the energy of which is about 10 eV lower than the $K\alpha_{1,2}$ line and is due to relaxation of doubly ionized Mg atoms. The intensity of the Al $K\alpha_{3,4}$ satellite is about 6% the intensity of the main peak. Similarly, a K_β satellite may be observed at 48.5 eV from the main XPS peak, with an intensity 200 times smaller than the latter.[64]

Multiplet Splitting

Multiplet splitting is due to spin coupling between the remaining electron left in a core level after photoionization and the spin of unpaired electrons in valence levels. This is responsible for the multiplication or broadening of XPS peaks of transition elements.[65]

Shake-Up Satellites

For certain compounds, the photoejection of an electron from a core level may be accompanied by excitation of valence electrons. The kinetic energy of the photoelectron is reduced accordingly and a so-called shake-up satellite appears at a higher binding energy with respect to the parent peak. For aromatic organic materials, a shake-up structure with intensities of up to 5–10% of the main peak, due to π-π^* transition, is visible.[66] This is illustrated by the C 1s and O 1s peaks of a commercial sample of poly(ethylene terephthalate) (PET) presented in Fig. 15.[57]

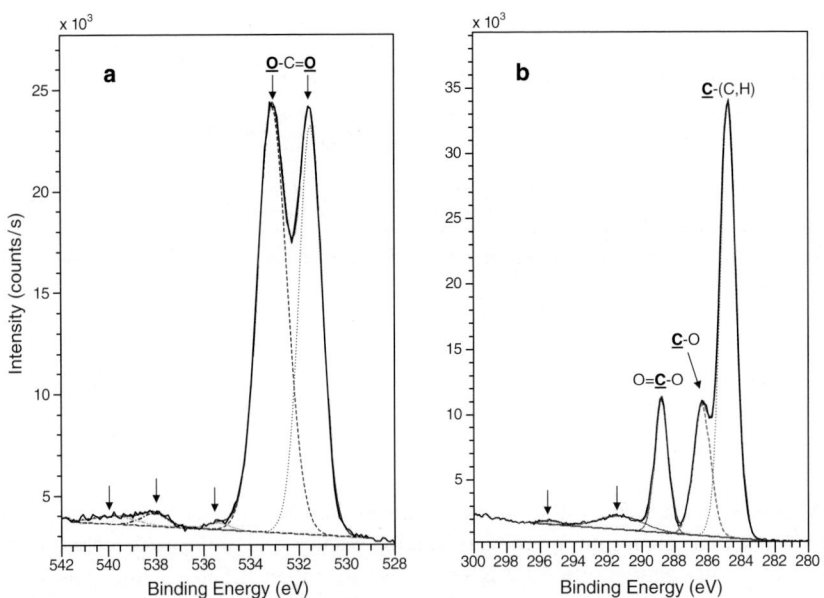

Fig. 15 O 1s and C 1s peaks of poly(ethylene terephthalate); Kratos Axis Ultra, monochromatized Al$_{K\alpha}$, pass energy 40 eV. Arrows point to shake-up satellites

The presence and characteristics of the shake-up peak structure are dependent on the chemical structure in which the element is involved.[67] A shake-up satellite is observed for the C 1s peak of polystyrene (PS); surface oxidation by plasma treatment reduces the intensity of the C 1s shake up,[68] while the incorporated oxygen does not show any shake-up satellite.[69] A strong shake-up satellite is observed at about 8.7 eV from the main Cu $2p_{3/2}$ and Cu $2p_{1/2}$ peaks on CuO but not on Cu_2O.[70]

Other Energy Loss Structures

Additional features are frequently observed near the background rise at the high binding energy side of photoelectron peaks. Figs. 11b and c show large plasmon structures next to the Si and Al peaks. Plasmons are quantized excitations of the conduction electrons and are thus typical of metals and doped semiconductors.[65] Note that silicon dioxide (Fig. 11a) does not show the plasmon structure observed on the silicon wafer. The background increase on the high E_b side of peaks of insulators often shows a structure, referred to as energy loss peak (ELP), which recalls the plasmon structure. This appears near the O 1s peak of silica and PMMA in Fig. 11a and d and near C 1s peaks of different systems in Figs. 5 and 11. Note that a broad maximum near the background rise may result from random inelastic scattering, depending on the distribution of the element as a function of depth.[58,71]

For graphite[72] as well as for metals,[66] a tail is observed on the high binding energy side of the main peak instead of a discrete satellite, and attributed to the interaction of the positive core hole with the conduction electrons.

An asymmetry, called *vibrational fine structure* and attributed to C—H vibrational excitation, is observable on C 1s spectra recorded on polymers containing saturated hydrocarbon components as polyethylene.[73]

Bases of Quantification

Basic Equations

The aim of this section is

1. To show that the quantification is complex but can be handled in a simple way
2. To point out the approximations involved in the simple common practice
3. To give the background for more sophisticated approaches (modeling rough samples, variation of sample orientation, etc.)

Figure 16a represents a specimen irradiated by X-rays and emitting photoelectrons. Figure 16b gives a three-dimensional sketch showing a particular direction X of incident X-rays and a particular direction G of photoelectron collection, making respectively angles δ and θ with the normal (z axis) to the sample-holder plane. Figure 16c illustrates, in three dimensions, a portion of an analyzed sample with

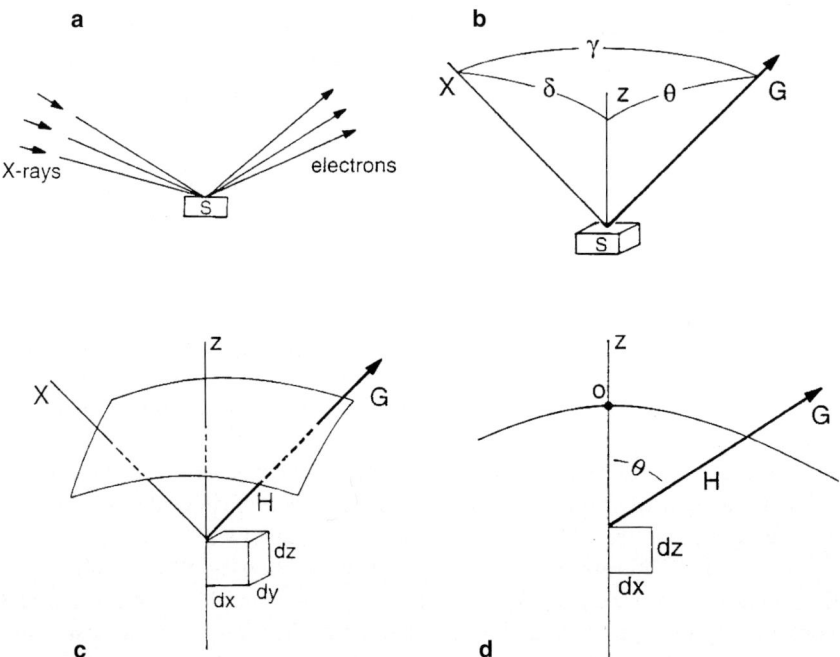

Fig. 16 Representation of a sample irradiated by X-rays and emitting photoelectrons: (**a**) Overall view; (**b**) three-dimensional representation of a particular X-ray direction X and a particular direction G of photoelectron collection; (**c**) contribution of an infinitesimal volume of a sample with a curved surface, showing the way H traveled by photoelectrons; and (**d**) section of c along the zG plane, such that the sample surface at $z = 0$ is parallel to the sample holder plane

a curved surface, the interaction between X-rays (incident direction X) and an infinitesimal volume ($dx\,dy\,dz$) leading to emission of photoelectrons along G. Figure 16d shows a section in 16c along the zG plane in the particular situation where the sample surface at $z = 0$ is parallel to the sample-holder plane.

The contribution of the infinitesimal volume $dx\,dy\,dz$ to photoelectrons counted in peak a of element A can be written as

dI_{Aa} = photoelectrons emitted in the adequate direction $(I) \times$ probability that they get out of the sample $(II) \times$ probability that they reach the detector and are counted by the latter (III)

$= [I] \times [II] \times [III]$

$$= \left[J' \sigma_{Aa} L'_{Aa} C_A dx\,dy\,dz\right] \left[\exp\left(-\int_H dH/\lambda_{Aa}\right)\right] [T'_{Aa} D_{Aa}], \qquad (14)$$

where J' is the X-ray flux along direction X; σ_{Aa} is the photoemission cross-section, i.e., the probability of ejection of an electron from energy level a of element A when a photon hits A; L'_{Aa}, called angular asymmetry factor and accounting for the anisotropy of electron emission, depends on the electron orbital involved and on the angle γ between the photoemission direction (G) and the incident X-ray direction (X); H represents (see Fig. 16c and d) the path followed in the sample by photoelectrons emitted along direction G, and collected in the same direction; λ_{Aa} is the IMFP of photoelectrons Aa in the material; T'_{Aa} is the transmission function of the spectrometer for the photoelectron direction G; and D_{Aa} is the efficiency of the detector.

The factor $\exp\left(-\int_H dH/\lambda_{Aa}\right)$ comes from the fact that the attenuation of photoelectrons along a path H composed of different sections $dH_1, dH_2, \ldots dH_i$ is given by

$$\exp(-dH_1/\lambda_1)\exp(-dH_2/\lambda_2)\ldots\exp(-dH_i/\lambda_i).$$

The total intensity of peak Aa is the integral of (14) over

1. the whole sample, i.e., according to x, y, z;
2. all the directions of the incident X-ray, i.e., according to X; and
3. all the directions of photoelectron collection, i.e., according to G.

The range of x and y determines the lateral space resolution of the spectrometer. It is currently in the order of 1 mm; it can be reduced down to about 15 µm in certain spectrometers in spectroscopy mode and to less than 5 µm in imaging mode. The distribution of X and G directions vary according to the spectrometer design. For the SSX 100/206 spectrometer (Surface Science Instruments, USA), the G directions are within a cone characterized by an angle of 30° (acceptance angle for photoelectron collection)[74] and an axis making an angle $\theta = 55°$ with the vertical, which is the normal to the sample-holder plane in ordinary recording conditions. Thus,

$$I_{Aa} = \int_x \int_y \int_z \int_X \int_G dI_{Aa}. \tag{15}$$

In the common practice, this equation is used in a considerably simplified form.

1. Consider that the sample is smooth (no roughness), that its surface is flat and that there is no variation of composition along x and y. In this case one may separate z from other variables and integrate separately along z, on the one hand, and along x, y, X, G, on the other hand.

$$I_{Aa} = \int_z \left\{\int_x \int_y \int_X \int_G dI_{Aa}\right\} \tag{16}$$

2. Consider that the variations of J', L'_{Aa}, and T'_{Aa} according to x, y, X and G can be accounted for by an average value J, L_{Aa}, T_{Aa}.

$$\int_x \int_y \int_X \int_G J'L'_{Aa} dx\, dy T'_{Aa} \approx JL_{Aa}T_{Aa}Sf(\theta), \tag{17}$$

where S represents the irradiated area. The function $f(\theta)$ accounts for the difference between the irradiated area and the area from which photoelectrons can be collected. It depends on the spectrometer design and on recording conditions (sample tilting, X-ray spot size, aperture of the electrostatic lens, resolution). For a given spectrum, the variation of the analyzed area according to the photoelectron energy and thus the variation between peaks is included in T, so that $f(\theta)$ is considered to be the same for all elements.

3. Consider that the variation of sample properties as a function of depth is weak enough so that the variation of λ_{Aa} is negligible along H. Then

$$\exp\left(-\int_H dH/\lambda_{Aa}\right) = \exp(-z/\lambda_{Aa}\cos\theta), \quad (18)$$

where the origin of the z axis coincides with the surface and z is the depth at which the infinitesimal volume $dx\,dy\,dz$ is located.

The following relationship is then obtained

$$I_{Aa} = k'JSf(\theta)\sigma_{Aa}L_{Aa}T_{Aa}D_{Aa}R_{Aa}\int_0^\infty C_A \exp(-z/\lambda_{Aa}\cos\theta)\,dz, \quad (19)$$

where k' is a constant. The factor R_{Aa} accounts for approximations made in 1). Certain parameters $[k', J, S, f(\theta)]$ cannot be determined practically but are the same for all peaks. So analytical information is always obtained by using the ratio of the intensities of two peaks, I_{Aa}/I_{Bb}.

Photoionization Cross-Section

The most widely used cross-sections (σ) are those of Scofield,[35] who calculated absolute values for all subshells of all elements and for the most widely used X-ray radiations. Other values, however, can also be found. In a review paper on quantitative XPS parameters, Seah[75] concluded that the experimentalists should continue to use Scofield's data for sake of consistency.

Angular Asymmetry Factor

The angular asymmetry factor, accounting for anisotropy of the photoelectron emission, depends on the angle γ between the X-ray beam and the direction of photoelectron collection (Fig. 16b).

$$L_{Aa} = 1 + 0.5\beta_{Aa}(1.5\sin^2\gamma - 1), \quad (20)$$

where β depends on the photon energy, the atomic number, and the subshell of the atom. The widely used values are those proposed by Reilman et al.[76] These are given only for atomic numbers with an increment of 5, which requires performing interpolations to obtain data for intermediate atomic numbers. In the NIST data base,[77] the values already proposed in 1979 by Band et al.[78] for all atomic numbers are used.

Note that $\beta = 2$ for subshells s. So L cancels when using intensity ratios between C 1s, N 1s, and O 1s peaks, which are the major peaks of interest for biosystems. Moreover for spectrometers with γ close to 54.7° (the "magic angle"), L is equal or close to 1. For the SSX 100/206 and the Kratos Axis Ultra spectrometers, $\gamma = 71°$ and $60 \pm 5°$, respectively. The effect of a difference between L_{Aa} values of two elements will further tend to be attenuated by elastic scattering (cf. the section on *Electron Mean Free Path*).

Inelastic Electron Mean Free Path

The IMFP (λ) is the average distance that an electron with a given energy travels between successive inelastic collisions.[79,80] Curves of λ vs. the photoelectron kinetic energy have been compiled by several authors;[81–83] an overview is presented in Fig. 17. The scatter of the data is in part due to the influence of the matrix; for a given kinetic energy, the IMFP decreases as the matrix density increases.

From such compilations, numerous authors have proposed to compute ratios of λ by considering that, in a given matrix and for $E_k > 100\,\text{eV}$, λ is proportional to E_k^m with $0.50 < m < 0.82$. A value close to $m = 0.7$ has been proposed by different authors: 0.73,[84] 0.723,[85] 0.65,[86] 0.75,[87] and 0.66.[88] The ratio $\lambda_{Aa}/\lambda_{Bb}$ may thus be evaluated by $(E_{k\text{-}Aa}/E_{k\text{-}Bb})^m$, where m is about 0.7. According to a recent

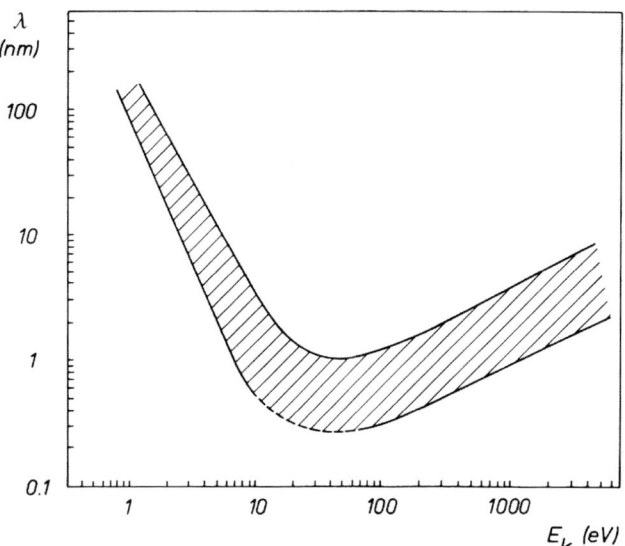

Fig. 17 Variation of the electron inelastic mean free path (IMFP), λ, as a function of kinetic energy for elemental solids. Adapted from Ref. 83 with permission, copyright (1979) John Wiley and Sons Limited

publication,[89] a value of $m = 0.79$ may be recommended for polymers and other organic materials. When calculating molar concentration ratios of two elements, the uncertainty on m is not crucial when the kinetic energies of the two corresponding XPS peaks are not very different, which is the case for C 1s, N 1s, and O 1s.

Analyzer Transmission Function

The intensity/energy response function (IERF) of a spectrometer is the product of the area from which photoelectrons are collected, of the transmission function (T), and of the detector efficiency (D). With a microfocused X-ray beam, such as in SSX 100/206 spectrometer, the area of collection is defined by the X-ray spot size and is smaller than the acceptance area of the analyzer. The relative intensity of the peaks is sensitive to the position of the sample on the vertical axis.[90] When a broad X-ray source is used, the IERF includes also the variation of the area of collection as the latter may depend on the energy of the photoelectrons detected (cf. the section on *Basic Equations*).

The electron transmission function is determined by the lenses, which influence the distribution of acceptance angles, and by the kinetic energy of the electrons in the HSA. In a given spectrum it may thus change according to the kinetic energy of the photoelectrons; moreover, it is influenced by the selected pass energy.[91] It can be given by the following general expression[26]

$$T = k'' E_p f(E_p/E_k), \qquad (21)$$

where k'' is a constant for a spectrometer. The dependence on the E_p/E_k ratio is often considered to be of the form

$$T = k'' E_p (E_p/E_k)^n, \qquad (22)$$

with $0 < n < 1$.[92] The value of n is characteristic of the instrument design and is given by the manufacturer or in the literature. In general, a single n value is assumed to apply across the entire range of E_p and E_k.[93,94]

As long as the pass energy is the same for recording the various peaks, the ratio of transmission functions is given by

$$T_{Aa}/T_{Bb} = (E_{k-Bb}/E_{k-Aa})^n. \qquad (23)$$

For the SSX 100/206 spectrometer, T can be assumed to be independent of E_k for $E_p = 150\,\text{eV}$, while complex expressions of n should be considered for the other pass energies.[95] Once again the uncertainty on n is not crucial when the kinetic energies of the two peaks are not too different from each other.

The IERF function of a given spectrometer can be determined by recording wide scan spectra on clean Au, Ag, and Cu standard specimens (as those provided by the NPL [SCAA 90 National Physical Laboratory, Teddington, Middlesex, UK]), which present photoelectron and Auger lines throughout the useful kinetic energy range, and by comparing them with those obtained on a Metrology Spectrometer[96] specially constructed by the National Physical Laboratory (Teddington, UK). A special software package is also available, allowing the analyst to perform the calibration.[97,98]

On the Kratos Axis Ultra and on some other recent spectrometers, the IERF has been established by the manufacturer for each machine with respect to their own reference machine on which the experimental sensitivity factors (cf. the section on *Empirical Sensitivity Factors*) were evaluated. The function is included in each file containing XPS data and depends on kinetic energy, pass energy, and lens combination used. This method seems to be less reliable than the NPL method when quantitative results are compared between different spectrometers.

Quantification Considering the Analyzed Zone as Homogeneous

General Expression

This approach comes to neglecting the variation of composition as a function of depth in the "explored zone" and to determine the "composition of a surface layer as seen by XPS."

In (19), J is not known a priori but is constant for a given record; $f(\theta)$ is also considered to be independent of the electron kinetic energy and is thus the same for each peak. Consequently,

$$I_{Aa} = k\sigma_{Aa} L_{Aa} T_{Aa} D_{Aa} R_{Aa} \lambda_{Aa} C_A, \tag{24}$$

where $k = k' JS f(\theta) \cos\theta$ is the same for any peak and C_A is the concentration of element A "as seen by XPS." It turns out that only intensity ratios can be treated, leading to concentration ratios

$$\frac{I_{Aa}}{I_{Bb}} = \frac{\sigma_{Aa} L_{Aa} T_{Aa} D_{Aa} R_{Aa} \lambda_{Aa} C_A}{\sigma_{Bb} L_{Bb} T_{Bb} D_{Bb} R_{Bb} \lambda_{Bb} C_B}. \tag{25}$$

In common practice, it is assumed that the influence of roughness is the same on both intensities, in other words that $R_{Aa}/R_{Bb} = 1$. When the kinetic energy analyzer is used in the constant-analyzer energy mode (CAE) (cf. the section on *Lens System and Energy Analyzer*), which is required for quantitative analysis, the residual kinetic energy of the electron reaching the detector is the same whatever its initial kinetic energy; therefore, $D_{Aa}/D_{Bb} = 1$. The above equation may then be rewritten

$$\frac{I_{Aa}}{I_{Bb}} = \frac{\sigma_{Aa} L_{Aa} T_{Aa} \lambda_{Aa} C_A}{\sigma_{Bb} L_{Bb} T_{Bb} \lambda_{Bb} C_B} \tag{26}$$

$$= \frac{i_{Aa} C_A}{i_{Bb} C_B}, \tag{27}$$

where i is a sensitivity factor. The ratio of molar concentrations C_A/C_B may be deduced from the intensity ratio I_{Aa}/I_{Bb} either by evaluating the ratio of σ, L, T, λ in (26) or by experimental calibration of the sensitivity factors in (27).

First Principles Approach

The relation between intensity ratios and concentration ratios may be directly evaluated from σ, L, T, and λ, which is often called the first-principles approach. Ratios of λ are computed as explained in the section *Inelastic Electron Mean Free Path*, and T ratios can be evaluated using (23) or with respect to a reference spectrometer as stated in the section *Analyzer Transmission Function*. The instruments softwares include an evaluation of T for data treatment; this should be considered with a critical mind.

According to the section *Inelastic Electron Mean Free Path* and (23) and (26), the ratio of sensitivity factors is essentially

$$\frac{i_{Aa}}{i_{Bb}} = \frac{\sigma_{Aa}}{\sigma_{Bb}} \left(\frac{E_{k,Aa}}{E_{k,Bb}}\right)^{m-n}. \tag{28}$$

A recent computation of m for 36 inorganic compounds provided a value of 0.719.[99] The spectrometer used in that paper was a VG Microtech characterized by $n = 0.762$; thus, $m - n = -0.043$.

Empirical Sensitivity Factors

The sensitivity factors presented in (27) may be determined by using standards that offer guarantees regarding surface composition. A list of sensitivity factors, related to the sensitivity factor for F 1s peak, has been published by Wagner.[100,101] It should be emphasized that they were determined with an instrument characterized by a transmission function T for which $n = 1$ in (22); these sensitivity factors must be corrected accordingly when another instrument is used.[102]

One can determine the sensitivity factor of a given spectrometer by running oneself adequate standards. It should be kept in mind that ratios of intensities may be influenced by sample roughness, as pointed out above. More crucial, they may also be influenced by the presence of adsorbed contaminants.

Influence of Adsorbed Contaminants

The surface of high surface-energy solids, typically metals and oxides is always contaminated by adsorbed organic compounds.[103] In ultrahigh vacuum, surface contamination takes place quickly,[104] presumably because organic contaminants represent a high proportion in the residual gas. Three main sources of contamination in UHV are mentioned in the literature. The first source is the pumping system that can release hydrocarbons and constituents of lubricant oils;[105] in the Kratos Axis Ultra spectrometer, the only oil pump is the forepump on line with the turbo pump of transfer chamber (Fig. 7q), which reduces the effect of that source of contamination.[26, 106] Secondly, residual gases may contribute to the contamination of a surface.[107–109] Finally, contamination may arise from the desorption of molecules from the walls of the vacuum chamber and their readsorption on the sample.[103] These compounds may originate from the atmosphere or from samples analyzed before, and thus possibly depend on the history of spectrometer utilization.

From an analytical point of view, the surface contamination adds foreign elements (C, O) to the composition of the surface of interest. More troublesome, it may alter the apparent concentration ratio of elements constituting the solid surface itself as the signal of interest is screened to an extent that depends on the electron kinetic energy. Furthermore, the contamination overlayer may differ, particularly in thickness, between the standards used for sensitivity factor determination and the unknown samples. The influence of the organic overlayer on the relative error in the surface composition of a model compound and its dependence on the line separation is illustrated in a recent paper.[99]

Obtaining exact concentration ratios C_X/C_Y for a surface requires particular approaches if X and Y photoelectrons possess strongly different kinetic energies and thus different IMFP in the contamination overlayer. One way is to assume that the contamination overlayer is the same for the analyzed samples and for the standards used to determine the sensitivity factors. Another way is to follow the first-principles method and to account for attenuation by the overlayer, considering a reduced thickness determined from the ratio of the intensities of the C 1s peak and of the Auger peak of carbon ($KL_{23}L_{23}$).[110] An approximative method is based on deducing the overlayer thickness from the mole fraction of carbon.[99] Its application to 11 oxides reduced the average error from ±8.2 to ±5.6%. However, the relative error was not strongly reduced for some elements and was appreciably increased for other elements, even for the same compound. Another way is to deduce a corrected intensity ratio I_X/I_Y computed as the slope of the plot of I_X/I_{C1s} as a function of I_Y/I_{C1s} for standards and samples on which the contamination overlayer thickness is deliberately varied.[111]

For hydrophobic polymers, which are non polar and are characterized by a low surface energy, the surface contamination does not seem to be a problem. The situation is less clear in the case of organic surfaces that expose polar groups such as OH- or COOH-terminated self-assembled monolayers. It is indeed difficult to decide whether unexpected results should be attributed to preparation failure or to contamination subsequent to preparation (unpublished results).

Illustration

Considering a priori the probed zone as having a homogeneous chemical composition while the aim is to analyse a surface is somehow contradictory. However this is

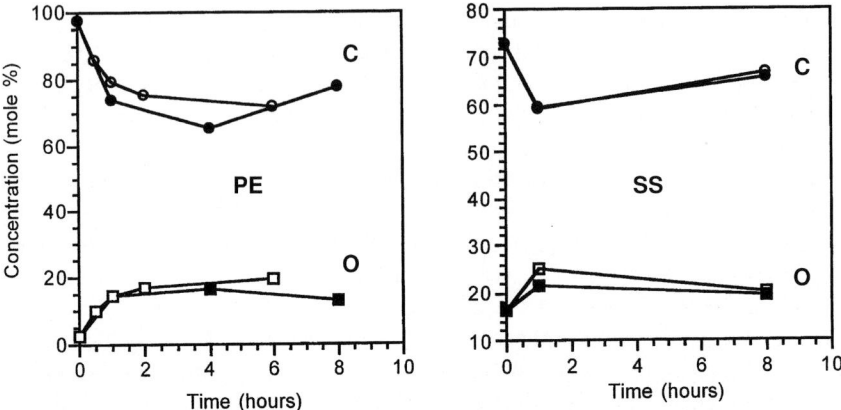

Fig. 18 Evolution of elemental composition (mole fraction %) of polyethylene (PE) and stainless steel (SS) coupons exposed to cooling water in bottom position (closed symbols) and vertical position (open symbols)

the most simple and widely used approach. For instance, a comparison between concentration ratios (P/N, N protonated/N total) determined by XPS and expected from the overall composition revealed a surface enrichment of dipalmitoyl phosphatidylcholine (DPPC) for powders made of DPPC, albumin, and a sugar or polyalcohol, spray-dried under certain conditions.[21] The relative decrease of the shake up contribution in the C 1s peak of polystyrene resulting from substrate screening was taken as a semi quantitative indicator of the presence of adsorbed protein.[112]

As another example, Fig. 18 shows the evolution of apparent surface composition of polyethylene (PE) and stainless steel (SS) coupons immersed in river water circulating in the heat exchanger of a power plant (to be published). This shows that, in a few minutes, the surface is modified by adsorption of constituents that contain an appreciable concentration of oxygen. Other elements appearing with small concentrations are Si, N, and Ca. Similar results were obtained on poly(vinyl chloride) (PVC) and on another type of SS. The composition is the same for coupons exposed on the bottom and on the lateral side of the pipe, indicating that the surface modification is not due to sedimentation. This surface conditioning occurred before adhesion of bacteria and other particles, such as diatom skeletons, which lead to the development of a biofilm. Note that the initial carbon concentration on SS is very high due to organic contamination and that the adsorbates are characterized by a higher oxygen concentration, compared to the initial organic contaminants. In this case, information is extracted from the XPS data considering the analyzed zone as homogeneous, despite the fact that the sample is a complex system made of superposed layers: adsorbed compounds/organic contamination/oxides/metal.

Data Interpretation and Evaluation

Data Treatment

Peak Smoothing

Smoothing reduces the noise and clarifies the information provided by the recorded spectrum. However any smoothing procedure may introduce distortion in the spectrum and should be used with caution. It is prohibited when peak decomposition is performed. The least-squares central point smoothing technique proposed by Savitsky and Golay[113] is found on most recent data treatment systems.

Background Subtraction

As pointed out before (*Core Photoelectron Peaks and Background*), the background contains contributions of the photoelectrons ejected by the continuous X-ray radiation and contributions of the electrons that have encountered inelastic collisions.

A very simple and widely used method for background subtraction is to draw a straight line between the two sides of a peak (Fig. 19). The choice of the binding energies at which the straight line meets the recorded spectrum is sometimes difficult, particularly when the peak-to-noise ratio is low.

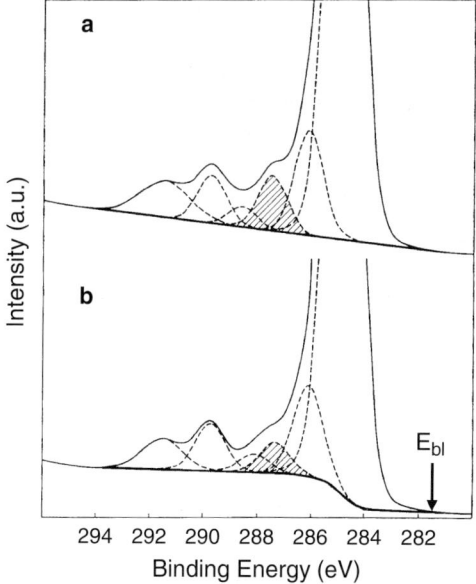

Fig. 19 Detailed record of C 1s peak of a tissue culture dish; influence of the mode of background subtraction (*thick line*) on the intensity of the various components (*dashed line*) resulting from a tentative decomposition. (**a**) Linear baseline; and (**b**) increase of the baseline proportional to the area of the peak situated at lower binding energy (Shirley). Note the crucial influence of the chosen baseline on the intensity of the shaded component

When the background level is very different from one side to the other, linear background subtraction may be unsatisfactory and a more sophisticated approach is desirable. In that case (Shirley procedure) (Fig. 19), the baseline at any binding energy, E_b, is drawn in such a way that the increase with respect to its level on the low binding energy side of the peak, E_{bl}, is related to the area of the peak between E_b and E_{bl}.[60,114,115] The choice of the binding energy at which the baseline meets the recorded spectrum is not as critical as that for linear background subtraction. The universal cross-section computation from Tougaard[116,117] gives more control over the background shape than the Shirley procedure. Values for the three parameters used in the universal cross-section algorithm have been published for different classes of materials including polymers.[59]

Most of the data treatment systems have at least both linear and Shirley background subtraction facilities. Background subtraction becomes more and more prone to error as the energy range over which the baseline is drawn becomes larger.

Peak Decomposition

Decomposition of a peak after background subtraction involves the addition of several components in different proportions and obtaining a best fit by iteration. This should not be confused with deconvolution,[118,119] which aims at correcting for instrumental peak broadening (X-ray and analyzer broadening, cf. the section on *Peak Width*). A nonlinear least-square curve fitting is used. Each component is often described by a mixed Gaussian/Lorentzian sum or product function.[120] The procedure consists in minimizing a chi square or root mean square function. Note that this function depends on the spectral intensity; therefore, its absolute value does not allow a comparison to be made regarding the quality of the fits performed on different spectra.[121]

Each individual component is defined by at least three parameters: height or area, FWHM, and position (E_b). The Gaussian/Lorentzian proportion is generally kept equal for all peaks. It depends on the relative importance of different instrumental contributions detailed in the section *Peak Width*. Constraints can be introduced in the iteration process, such as peak energy differences, peak intensity ratios, equality of FWHM, imposed FWHM, and E_b. Different types of asymmetric line shapes can also be used in certain softwares.[120] As in other forms of spectroscopy, peak decomposition should be used with much caution due to the high number of variables involved. *The best fitting given by a computer does not necessarily mean that the solution is unique and/or that it has a physical meaning* as illustrated in the section *Peak Components of Biochemical Compounds*. Complementary information is helpful, and common sense is required when appreciating the variation of the best-fitting parameters from one spectrum to another one.

Details on errors in curve-fitting approaches can be found in references.[122,123] The following example may be taken as a guideline.[123] If two components are characterized by a product of Lorentzian and Gaussian lines with a L/G ratio of 0.5, by equal intensities and by a separation that is 2/3 of the FWHM, a single maximum is observed but the intensity of a component can be determined with a good precision.

When the separation is equal to the FWHM, the presence of two components is clear. The noise has also a dramatic influence on the precision of decomposition. A useful tool to evaluate the reliability of peak decomposition may be found in Ref. 124

Figure 19 presents a tentative decomposition of the C 1s peak of a tissue culture dish with an ordinate expansion, allowing an investigation of the contribution of carbon bound to oxygen or nitrogen. It shows that the mode of background subtraction may appreciably affect the decomposition. A linear baseline tends to overestimate the intensity of small peaks located on the high binding energy side of an intense peak.

Chemical Shifts and Chemical Functions

Introduction

As already stated, the position of a no-loss peak is characteristic of a given element. However the atoms bound to an element may give rise to measurable binding energy differences, i.e., "chemical shifts." Initial state and final state effects are involved in the chemical shift. In the *initial state effect*, the electron density on the atom before photoemission is a key factor influencing the magnitude of the chemical shift. A decrease of electron density provokes an increase in binding energy. Different *final state effects* can occur after photoelectron emission: core hole screening, relaxation of electron orbitals, and polarization of surrounding ions. An overview of the theory and interpretation of core-level binding energy shifts observed in XPS can be found elsewhere.[125] Correlations are found between binding energy and electrical charge computed by different methods, including calculations based on Pauling electronegativity and on quantum mechanics. A review was published recently with a historical survey. The chemical shifts produced on the C 1s peak by different groups and the deduced group electronegativities are also given.[126]

For most elements, the binding energy increases with oxidation, as illustrated below for Si (*Data Interpretation Through Simulation* section) and Al (*Semi-Quantitative Use of Angle Resolved Analysis* section). There are some exceptions for which the final state effect is so large that this chemical shift is in the direction opposed to that expected for oxidation; for instance, the Ce $3d_{3/2}$ peak of Ce^{IV} in CeO_2 and of Ce^0 appears at 881.8 and 883.9 eV respectively. If the shift is small, it may be remembered that the modified Auger parameter is more sensitive to the chemical state and is independent of the way of defining the zero of the binding energy scale (cf. the section on *Auger Peaks*).

Reference Data for Polymers

Figure 15 presents the O 1s and C 1s peaks of PET, illustrating the presence of components that are typical of chemical functions. Three components are clearly visible in the C 1s peak beside the shake up contributions. They are due to carbon only bound to carbon and hydrogen [$\underline{C}$—(C,H)], to carbon bound to oxygen of the ester [$\underline{C}$—O—(C=O)] and to carbon belonging to the ester function [($\underline{C}$=O)—O]. The contributions of the two oxygen atoms involved in the ester are also well separated.

The binding energy characteristic of elements in different chemical environments is given in Table 1 for representative materials. A compilation of chemical shifts for C 1s, O 1s, N 1s, F 1s as well as some other elements present in polymers (Si 2p, Cl 2p, S 2p, P 2p) has been published[57] and a CD-ROM is now available.[129] The binding energy is influenced by the formal oxidation stage. Each bond of carbon with oxygen shifts the C 1s line upward by 1–1.5 eV, as illustrated by the larger lines interval used for CHO polymers in Table 1. However, the key factor is the net charge on the atom rather than the formal oxidation state.[9,126] For instance, the binding energy of the C 1s peak of —$\underline{C}F_3$ is higher than that of —($\underline{C}$=O)—OR, although the formal oxidation stage is +3 in both cases; the higher electronegativity of fluorine compared to oxygen, makes the electron density on the carbon atom of —CF_3 lower as compared with the ester.

Table 1 illustrates also the influence of aromatic rings and of second nearest neighbors. A secondary C 1s chemical shift relative to hydrocarbon is 0.4 eV for carbon bound to a ketone ($\underline{C}$—(C=O)—$\underline{C}$; PVMK) and 0.7 eV for carbon bound to an ester ($\underline{C}$—(C=O)—O—C; PMMA). The influence of the second nearest neighbor X on the C 1s peak of $\underline{C}$—O—X moieties (COH, COCH$_3$, COC$_2$H$_5$, etc.) is weak except in the case where X stands for NO$_2$, which produces an additional increase of 0.9 eV.[121]

The chemical shift on oxygen is less pronounced than on carbon owing to the more limited range of possible chemical bonds. For nitrogen, protonation of amine leads to a positive shift of about 1 eV; the binding energy is much higher for oxidized forms.[28,130–133]

Peak Components of Biochemical Compounds

Table 2 presents the components of C 1s, N 1s, and O1s peaks and their assignment, based on the spectra recorded for different biochemical compounds: polyalcohol, sugar derivatives, organic acids and salts, homopeptides,[134] and triglyceride (unpublished). Peak shape, peak decomposition, and peak component assignments are illustrated by Fig. 20, which shows spectra recorded on model compounds representative of polysaccharides, polypeptides and lipids: maltodextrin [H—(C$_6$H$_{10}$O$_5$)$_n$—OH],[135] poly(L-serine) [H—(NH—CH(CH$_2$OH)—CO)$_n$—OH],[136] collagen, and DPPC [CH$_2$(O—CO—C$_{15}$H$_{31}$)—CH(O—CO—C$_{15}$H$_{31}$)—CH$_2$(O—POO$^-$—O—CH$_2$—CH$_2$—N$^+$(CH$_3$)$_3$)]. It may be noted that the species $\underline{C}$=O (peptide link) and O—$\underline{C}$—O (acetal of polysaccharides) both contribute to the component near 287.8–288.0 eV, while the species $\underline{O}$=C and C—$\underline{O}$—C are responsible for components at 531.3 and 532.6 eV, respectively.

Figure 21 shows the S 2p doublet, near 164 eV, of human serum albumin adsorbed on PS. The polypeptide residue obtained after treatment with Renalin™ (reagent containing hydrogen peroxide and peroxyacetic acid used for cleaning

Table 1 *Top*: Binding energy of common elements in chemical functions of solids, referred to the <u>C</u>—(C,H) component of the C 1s peak of aliphatic hydrocarbon or adventitious contamination set at 284.8 eV, Ref.[57] unless otherwise specified. *Bottom*: Repeat unit of the polymers listed

Element and function	Position (eV)	Compound of reference
Carbon	C 1s	
CHO polymers		
<u>C</u>—(Si—O)	284.2	PDMS
<u>C</u>$_{aro}$—(C,H)	284.3–284.6	PC, PEEK, PET, PS
<u>C</u>$_{ali}$—(C,H)	284.8	PS, PE, PC, PVA, PMMA, PVMK, organic contamination
<u>C</u>—(C=O)—<u>C</u>	285.2	PVMK
<u>C</u>—(C=O)—O—C	285.5	PMMA
O—(C=O)—O—<u>C</u>$_{aro}$	286.0	PC
Φ—(C=O)—O—<u>C</u>	286.0	PET
Φ—O—<u>C</u>$_{aro}$	286.1	PEEK
<u>C</u>—O	286.3	PVA, PEG
<u>C</u>—(C=O)—O—<u>C</u>	286.6	PMMA
Φ—(<u>C</u>=O)—Φ	286.9	PEEK
C—(<u>C</u>=O)—C	287.8	PVMK
Φ—(<u>C</u>=O)—O—C	288.5	PET
C—(<u>C</u>=O)—O—C	288.8	PMMA
O—(<u>C</u>=O)—O	290.2	PC
Shake up	~291.4	PS, PEEK, PC, PET[127]
Nitrogen-containing polymers		
<u>C</u>—C—(C=O)—N—C	284.8	PA-6 (Nylon^{TM})
C—<u>C</u>—(C=O)—N—C	285.1	PA-6 (Nylon^{TM})
C—C—(C=O)—N—<u>C</u>	285.8	PA-6 (Nylon^{TM})
C—C—(<u>C</u>=O)—N—C	287.8	PA-6 (Nylon^{TM})
Chlorine-containing polymers		
H—<u>C</u>—C—Cl	285.7	PVC
H—C—<u>C</u>—Cl	286.8	PVC
Fluorine-containing polymers		
<u>C</u>—C—O—(C=O)—CF$_3$	284.8	PVTFA
<u>C</u>—CF$_2$	286.2	PVdF
C—<u>C</u>—O—(C=O)—CF$_3$	286.5	PVTFA
C—C—O—(<u>C</u>=O)—CF$_3$	289.3	PVTFA
C—<u>CF</u>$_2$	290.7	PVdF
(<u>CF</u>$_2$)$_n$	292.3	PTFE (Teflon^{TM})
C—C—O—(C=O)—<u>C</u>F$_3$	292.5	PVTFA
Oxygen	O 1s	
Inorganic compounds		
M—<u>OH</u>	530.8–532.8	Inorganic hydroxide[56]
<u>O</u>—Si—<u>O</u>	533.1	Silica (quartz)[128]
Polymers		
Φ—(C=<u>O</u>)—Φ	531.1	PEEK
C—C—(C=<u>O</u>)—N—C	531.2	PA-6 (Nylon^{TM})
Φ—(C=<u>O</u>)—O—C	531.4	PET
Si—<u>O</u>—Si	531.8	PDMS
C—(C=<u>O</u>)—O—C	532.0	PMMA

Table 1 (continued)

Element and function	Position (eV)	Compound of reference
O—(C=**O**)—O	532.1	PC
C—(C=**O**)—C	532.2	PVMK
O—(C=**O**)—CF$_3$	532.4	PVTFA
C—**O**H, C—**O**—C	532.5–532.6	PVA, PEG
Φ—(C=O)—**O**—C	533.0	PET
Φ—**O**—Φ	533.1	PEEK
C—(C=O)—**O**—C	533.6	PMMA
C—C—**O**—(C=O)—CF$_3$	533.6	PVTFA
O—(C=O)—**O**	533.8	PC
Nitrogen	N 1s	
C—C—(C=O)—**N**—C	399.6	PA-6 (Nylon^{TM})
Fluorine	F 1s	
C—C**F**$_2$	688.0	PVdF
C—C—(O—(C=O)—C**F**$_3$)	688.0	PVTFA
(C**F**$_2$)$_n$	689.5	PTFE (Teflon^{TM})
Chlorine	Cl 2p$_{3/2}$	
C**Cl**—C	200.4	PVC
Cl$^-$	197.7–199.4	Salts[56]
Silicon	Si 2p	
C—(**Si**—O)	101.6	PDMS
O—**Si**—O	103.8	Silica (quartz)[128]

PA-6 (Nylon™): Polyamide 6 −(CH$_2$−(CH$_2$)$_3$−CH$_2$−C(=O)−NH)−

PC: Poly(bisphenol A carbonate)

PDMS: Poly(dimethylsiloxane)

PE: Polyethylene −(CH$_2$−CH$_2$)−

PEEK: Poly(aryl ether ether ketone)

PEG: Poly(ethylene glycol) −(CH$_2$−CH$_2$−O)−

PET: Poly(ethylene terephthalate)

PMMA: Poly(methyl methacrylate)

PS: Polystyrene

PTFE: (Teflon™): Poly(tetrafluoroethylene) −(CF$_2$−CF$_2$)−

PVA: Poly(vinyl alcohol)

PVC: Poly(vinyl chloride)

PVdF: Poly(vinylidene fluoride) −(CF$_2$−CH$_2$)−

PVMK: Poly(vinyl methyl ketone)

PVTFA: Poly(vinyl trifluoroacetate)

blood dialysis systems) shows a S 2p peak shifted to a higher binding energy ($\sim$168.5 eV), revealing an oxidation of all the sulfur.

The chemical information generated by decomposition of a given peak is free from uncertainties concerning the sensitivity factors. In absence of any reason to proceed otherwise, the following two-step procedure may be recommended for peak decomposition, once the shape of the components (combinaison of Gauss and

Table 2 Binding energy of elements in chemical functions of biochemical compounds, Ref. 134 unless otherwise specified

Element and function	Position (eV)	Compound of reference
Carbon		
C—(C,H)	284.8	Hydrocarbon, adventitious contamination
C—N, (C=O)—N—C	286.1	Amine; amide, peptidic link
C—O	286.3	Alcohol
(C=O)—O—C	286.8	Ester[a]
C=O, O—C—O	287.8	Aldehyde, (hemi)acetal
(C=O)—N—C, O=C—O⁻	288.0	Amide, peptidic link; carboxylate
(C=O)—O—C	289.0	Ester[a]
(C=O)—OH	289.0	Carboxylic acid
Oxygen		
O=C—O⁻	531.1	Carboxylate
(C=O)—N	531.3	Amide, peptidic link
(C=O)—OH	531.8	Carboxylic acid
(C=O)—O—C	531.9	Ester[a]
C—OH, C—O—C—O—C	532.6	Alcohol, (hemi)acetal
(C=O)—O—C	533.4	Ester[a]
(C=O)—OH	533.4	Carboxylic acid
Nitrogen		
C—NH₂	399.3	Amine
(C=O)—NH	399.8	Amide, peptidic link
C—NH₃⁺	401.3	Protonated amine

[a]Determined on glycerol tristearate (unpublished)

Lorentz functions depending on the experimental conditions; cf. the section on *Peak Decomposition*) and the number of components have been tentatively selected.

1. A first curve fitting is performed without constraint regarding binding energy, area and FWHM of the components.
2. A second fitting is performed by imposing certain constraints. Constraints may concern the FWHM of components of a given peak. Depending on the results of the first fit and on information available on the sample or on related samples, the choice may be to impose the same FWHM to all or some components of a peak. This FWHM may be let free during curve fitting, or may be fixed according to a certain criterion (value obtained for the best resolved component in the first fit, value given by the average of the FWHM obtained for all or some components in the first fit, etc.). Step 2 has sometimes to be repeated, with adjusted FWHM or binding energy values, in order to improve fitting. Other constraints may be imposed such as distance or area ratio between certain components. Sophisticated softwares also allow constraints to be imposed between components of two different peaks.

This two-step decomposition procedure offers a reasonable balance between avoiding bias in curve fitting, insuring the reproducibility required for a comparison of samples and providing a consistent basis for quantification. Whatever the procedure, it should be clearly stated when quantitative data are extracted from peak decomposition.

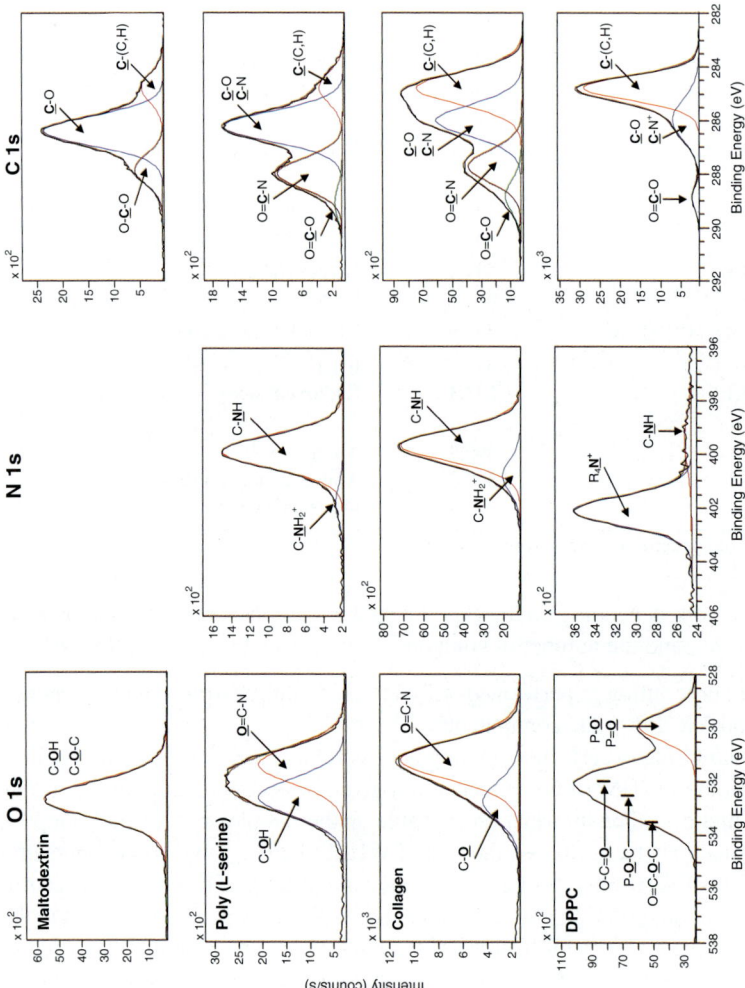

Fig. 20 O 1s, N 1s, and C 1s peaks of maltodextrin[135] and poly(L-serine)[136] [SSX 100/206 spectrometer, monochromatized Alkα, pass energy 50 eV, Gaussian/Lorentzian ratio 85/15], and type I collagen and dipalmitoyl phosphatidylcholine (DPPC) [Kratos Axis Ultra spectrometer, monochromatized Alkα, pass energy 40 eV, Gaussian/Lorentzian ratio 70/30]

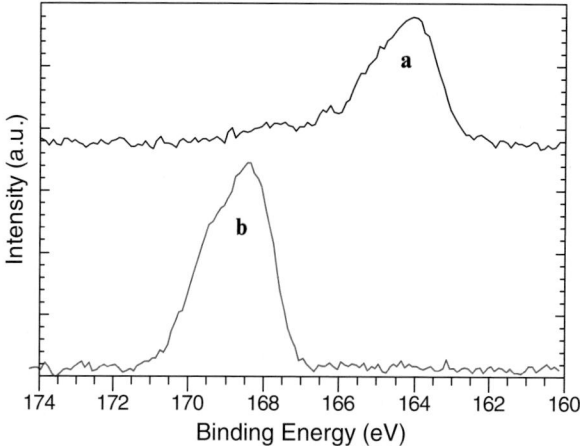

Fig. 21 S 2p peak recorded on human serum albumin (HSA) adsorbed on polystyrene (PS) (**a**) before and (**b**) after treatment with an oxidizing reagent (Renalin^{TM})

The following examples illustrate the need to be critical in performing peak decomposition, even regarding the number of components per peak. The C 1s peak of maltodextrin and poly(L-serine) in Fig. 20 shows clearly a component, which is due to $\underline{C}$—(C,H) of surface contaminants and has been set at 284.8 eV. If the structural formula of poly(L-serine) was not known, the presence of two components in the O 1s peak could only be suggested from comparison with the FWHM's obtained for O 1s components of other related compounds in identical recording conditions.

As another example consider DPPC in Fig. 20. The O 1s component near 530 eV represents about one fourth of the whole O 1s peak intensity. Consideration of bond polarity, suggests its attribution to the oxygen making a double bond with phosphorous (P=$\underline{O}$) and to the oxygen of the deprotonated acid (P—$\underline{O}^-$). Data available for esters indicate that the other part of the O 1s peak (531–534 eV) must contain the contributions of the two types of oxygen of the ester functions as indicated in Fig. 20. It must also contain the contributions of the oxygen of ester phosphate (P—$\underline{O}$—C), presumably closer to oxygen doubly bound to carbon. It is obvious that there are many ways to decompose this broad spectral feature into three components, unless constraints are imposed. The broad component of the C 1s peak near 286 eV could also tentatively be separated into distinct contributions for carbon singly bound to carbon and oxygen bound to N^+. Accurate decomposition and assignment would require the analysis of a set of closely related compounds.

Figure 22 presents the peaks recorded on glycerol mono-, di- and tristearate [CH$_2$R—CHR—CH$_2$R, R being O—(CO)—(CH$_2$)$_{16}$—CH$_3$], figuring as models for oils and fats. It illustrates further the potentialities and limitations of peak decomposition, and the advantage of knowing the stoechiometry and examining sets of related samples to impose relevant constraints. (1) The C 1s and O 1s peaks of tristearate were first decomposed by constraining O—$\underline{C}$=O and $\underline{C}$—O components of C 1s peak to be of equal area, and the two components of the O 1s

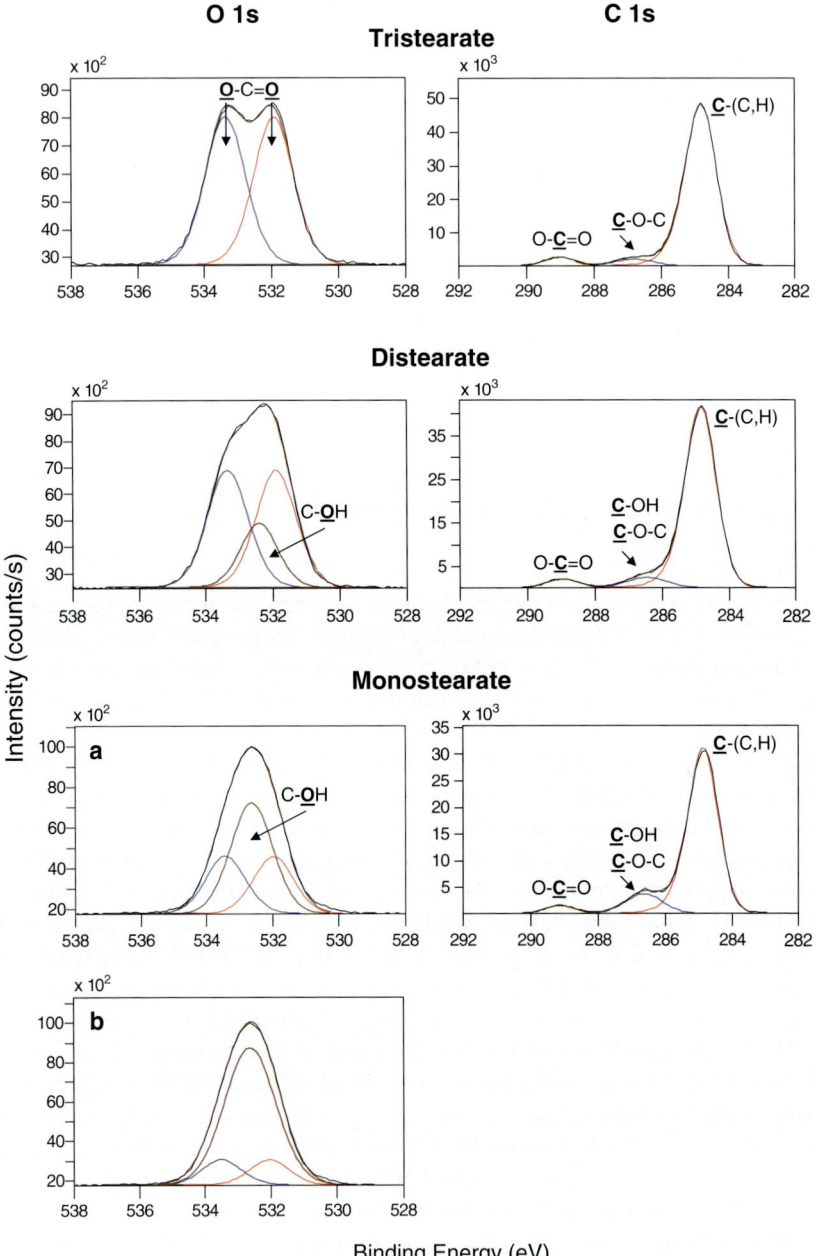

Fig. 22 O 1s and C 1s peaks of glycerol tri-, di-, and monostearate. Kratos Axis Ultra spectrometer, monochromatized Al$_{K\alpha}$, pass energy 40 eV; decomposition as explained in the text

peak to be of equal area. A small asymmetry was added on the high binding energy side of the $\underline{C}$—(C,H) component, accounting for the secondary chemical shift of $\underline{C}$—(C=O)—O—C as mentioned in the section *Reference Data for Polymers*. (2) The C 1s and O 1s peaks of distearate were then decomposed with the same asymmetry of the $\underline{C}$—(C,H) component as on tristearate. The FWHM of $\underline{C}$—O component was constrained between 1.3 and 1.4 eV to be close to the value obtained (1.39 eV) from the decomposition of the C 1s peak of tristearate. The distance between $\underline{O}$—C=O and O—C=$\underline{O}$ components was fixed at 1.45 eV and their FWHM was fixed at the value of 1.41 eV, as obtained for tristearate. The area of the C—$\underline{O}$H contribution was fixed equal to half the area of the $\underline{O}$—C=O component, according to stoichiometry. (3) Finally the C 1s and O 1s peaks of monostearate were decomposed with the same constraints as for distearate except the area of the C—$\underline{O}$—H contribution which was fixed equal to twice the area of the O—C=$\underline{O}$ component as presented in spectrum (a). In contrast, best-fitting decomposition of the O 1s peak without fixing a constraint for the C—$\underline{O}$—H area (b) gives proportions between the O 1s peak components that are very far from those expected. It is obvious that peak decomposition of an unknown sample is not reliable if the number of components is large compared to the number of minima or inflexion points in the peak shape. This is frequently forgotten by XPS users.

For the three stearates presented in Fig. 22, the contribution of the components due to carbon bound to oxygen, $C_{ox}/C = [(\underline{C}—O) + (O—\underline{C}=O)]/C$, was not constrained and was found to be about 10% lower than the expected value. This may due to several causes: limited precision of the measurements, constraint of the same asymmetry for the $\underline{C}$—(C,H) component of the three compounds, and preferred orientation of the hydrocarbon chains near the surface in order to decrease the surface energy.

Tentative Consideration of the Heterogeneity of the Analyzed Zone

Semi-Quantitative Use of Angle Resolved Analysis (ARXPS)

Figure 6 shows that the angle of photoelectron collection θ from a flat and smooth sample influences the explored depth, which depends on $\lambda \cos \theta$. As θ increases, the relative contribution of the uppermost layers rises. Thus exploration of the heterogeneity according to depth may be performed by tilting the sample along an axis perpendicular to the direction of photoelectron collection. When doing so, it must be made sure that the rotation axis lies in the plane of the surface, which requires an adequate design of the sample holder. Otherwise the zone explored by XPS will vary considerably as a function of tilting.

Table 3 presents illustrative data, obtained at collection angles θ of 0° and 60° on films of PP of interest in food packaging, used to protect food from oxygen and humidity: polymer film treated by corona or plasma discharge to increase surface polarity, aluminum coating, metal side and polymer side obtained after a peeling test. The results give the apparent mole fraction "as seen by XPS," considering that the "analyzed zone" is homogeneous. Note that the latter is twice thinner

Table 3 Apparent composition of surfaces relevant to aluminum-coated biooriented polypropylene films, obtained at two collection angles θ and computed by considering that the analyzed zone is homogeneous: film submitted to an oxidizing treatment, aluminum-coated film, surfaces obtained after the peeling test. Roughness rms measured by AFM on $2 \times 2\,\mu m^2$ images

Film treatment	Surface Analyzed	$\theta = 0°$ Mole fraction (%)						$\theta = 60°$ Mole fraction (%)						R_{rms} (nm)
		Al^0	Al^{III}	O	N	C	C_{ox}/C^a	Al^0	Al^{III}	O	N	C	C_{ox}/C^a	
Corona	Polymer before metalization	–	–	9.7	–	90.3	0.11	–	–	11.3	–	88.7	0.12	–
	Aluminum coating	18.9	20.3	38.5	–	22.4	0.20	8.3	17.0	41.1	–	33.6	0.17	–
	After rupture													
	metal side	8.2	11.3	22.8	–	57.7	0.20	2.0	6.0	16.4	–	75.5	0.15	6.6
	polymer side	bdl	bdl	1.0	–	99.0	0.01	bdl	bdl	1.1	–	98.9	0.04	2.6
Plasma	Polymer before metalization	–	–	4.3	5.6	90.2	0.07	–	–	5.3	6.5	88.2	0.08	–
	Aluminum coating	15.8	21.3	36.0	–	26.9	0.14	5.8	16.0	35.2	–	43.0	0.12	–
	After rupture													
	metal side	10.2	13.8	22.7	1.6	51.8	0.09	2.3	7.7	16.0	1.6	72.4	0.07	3.8
	polymer side	bdl	bdl	1.5	1.2	97.4	0.02	bdl	bdl	1.3	1.1	97.6	0.01	5.6

bdl below detection limit

[a] Proportion of the C 1s peak due to carbon bound to oxygen or nitrogen

at 60° ($\cos 60° = 0.5$) compared to 0°. The oxygen and nitrogen concentrations measured on the treated films are only slightly higher at 60° than at 0°, indicating that the polymer modification resulting from the surface treatments takes place over a depth larger than the IMFP. For the aluminum coating, the comparison of apparent carbon concentrations between 60° and 0° indicates the presence of a thin overlayer of organic compounds. The increase of the Al^{III}/Al^0 ratio with the collection angle reflects the fact that aluminum metal is coated by a thin layer of aluminum oxide. The comparison between the metal side and the polymer side obtained after the peeling test indicates that the rupture does not take place in the metal layer (no Al on the polymer side). The polymer layer left on the metal side after rupture is very thin, as revealed by the observation of an appreciable apparent aluminum concentration and confirmed by the higher apparent surface concentration of carbon measured at 60° compared to 0°. The difference between the metal side and the polymer side regarding the relative concentrations of carbon bound to oxygen or nitrogen (C_{ox}/C) (cf. the section on *Peak Components of Biochemical Compounds*) further shows that the rupture essentially takes place between the oxidized and the intact polymer. On the metal side, aluminum is also covered by a thin layer of oxide. It is not known whether this was produced during the metal coating process or after the peeling test.

These qualitative deductions, which are based on considering a smooth surface, lead to a contradiction as the polymer surface-treated before metalization appears to be oxidized over an appreciable depth whilst the oxidized polymer layer left on the metal side after the rupture does not screen completely aluminum. Actually an AFM examination has shown that the surfaces obtained after rupture are not smooth. They present a nodular morphology with a rms roughness (measured on $2 \times 2\ \mu m^2$) in the range of 3–7 nm, i.e., of the order of the IMFP. Further analysis of the results would require modeling, using the kind of approach described below.

The variation of absolute intensities as a function of θ depends on the spectrometer design. For many instruments, the space resolution is controlled by acting on the electron collection system, and thus $f(\theta)$ is not a constant and is not known. Nevertheless, the significance of apparent concentration ratios should be examined in the light of variations of absolute intensities. One should be particularly careful at extreme θ values as shadowing effects as well as intersection between the sample surface and the cone of acceptance angles for photoelectron collection may be responsible for weak intensities, the ratio of which is neither accurate nor precise.

Information Provided by the Background

The shape of the background in the range of kinetic energy up to 100 eV below a peak depends on the distribution of the element as a function of depth, as explained and illustrated in the section *Core Photoelectron Peaks and Background*. Background simulation and comparison with the recorded spectrum provides information on the concentration–depth profile up to a depth about 5 times the IMFP.[58,59] Background simulation relies on considering the electron transport in solids. It is less frequently used than simulation of peak intensity ratio, which is more readily accessible and more informative for the chemist.

Data Interpretation Through Simulation

This approach is of great interest to analyze adsorbed phases and composite materials. Information on the heterogeneity of the zone analyzed by XPS can be generated by using the following strategy:

1. A hypothetical model is considered, figuring the distribution of different constituents in space
2. Peak intensity ratios are computed on basis of the considered model and adequate data: $(I_A/I_B)^{com}$
3. Computed intensity ratios are compared with experimental ratios: $(I_A/I_B)^{exp}$
4. The process is repeated for different samples constituting a series within which certain experimental parameters vary (e.g., amount of adsorbed compound, sample hydrophobicity, sample handling after adsorption, etc.)
5. Alternative models are considered and the sensitivity of the computed intensity ratios to model type and parameters is explored
6. The different results are analyzed with a critical mind: *a model that fits is not necessarily a model that represents the reality*

Instead of comparing intensity ratios, a convenient practice may be to compute apparent molar concentration ratios $(A/B)^{com}$, which would be the ratios characteristic of homogeneous solids giving the same intensity ratios (cf. the section *General Expression*)

$$\left(\frac{A}{B}\right)^{com} = \left(\frac{I_A}{I_B}\right)^{com} \frac{i_B}{i_A} \tag{29}$$

$$= \left(\frac{I_A}{I_B}\right)^{com} \frac{\sigma_B T_B \lambda_B}{\sigma_A T_A \lambda_A} = \left(\frac{I_A}{I_B}\right)^{com} \frac{\sigma_B T_B E_{k,B}^{0.7}}{\sigma_A T_A E_{k,A}^{0.7}}. \tag{30}$$

Obviously, the way of considering the transmission function must be consistent with that used to treat experimental data. Here the influence of the photoemission anisotropy parameter (L) is dropped, which is strictly correct when s peaks are used or γ is close to the "magic angle" (cf. the section on *Angular Asymmetry Factor*).

Simple models, illustrated by Fig. 23, are considered here. They correspond to different formulations of the intensity and are relevant to practical situations.

Continuous adlayer

Silicon wafers are frequently used in protein adsorption studies owing to their smoothness, which makes them suitable for atomic force microscope investigations. Consider the silicon substrate (su) covered by a continuous layer (ad) of oxide with a constant thickness t.

Figure 24a presents the variation of the ratio of the peak intensity of pure silicon (Si^0 2p at 99.1 eV) to the peak intensity of oxidized silicon (Si^{IV} 2p at 103.4 eV). The spectra were recorded with a SSX 100/206 spectrometer. In this instrument the X-ray beam is microfocused (spot size 300 µm in this case) and the zone irradiated by the X-ray beam is always smaller than the zone from which photoelectrons can

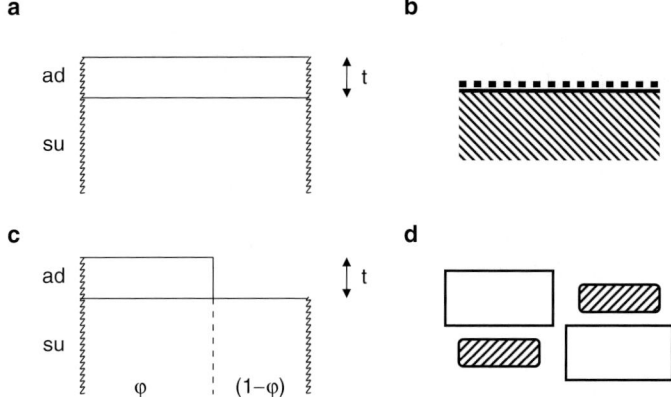

Fig. 23 Cross sections of models used to account for the heterogeneity of the zone analyzed by XPS: (**a**) Continuous adlayer of constant thickness t; (**b**) distribution of an element at the atomic scale; (**c**) adlayer of thickness t and fractional coverage φ; and (**d**) random mixture of two types of particles

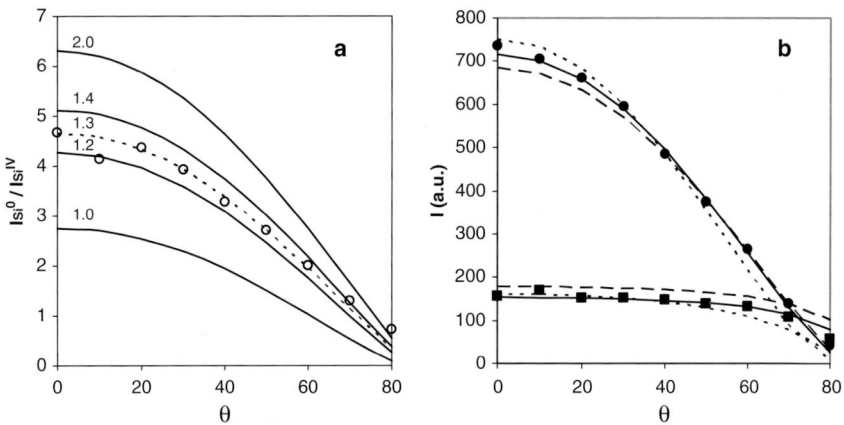

Fig. 24 XPS results obtained on a silicon wafer at different angles, θ, of photoelectron collection. (**a**) Ratio of the Si 2p peak intensities Si^0/Si^{IV}: experimental results (*open circle*), curves computed using Eqs (31) and (32) with a thickness of oxide, t, equal to 1.0, 1.2, 1.3 (*dotted line*), 1.4 and 2.0 nm; $\lambda_{Si}^{ad} = 3.9$ nm; and $\lambda_{Si}^{su} = 3.2$ nm. (**b**) Absolute intensity of each peak (a.u.): experimental results (*filled circle*, Si^0; *filled square*, Si^{IV}); curves computed using Eqs (31) and (32) with $t = 1.3$ nm, the same λ as in a) and $k^* = 3.8$ (*continuous line*); with $t = 1.3$ nm and the same λ as in a), considering the screening by a 1 nm thick organic contamination overlayer and $k^* = 5.0$ (*dotted line*); and with $t = 1.3$ nm, $\lambda_{Si}^{ad} = 4.7$ nm, $\lambda_{Si}^{su} = 2.5$ nm and $k^* = 4.3$ (*dashed line*)

be collected, as schematized in Fig. 25. The X-ray beam lies in a plane defined by the vertical and the tilting axis. When the sample is horizontal, the collection angle θ is 55°; the incident angle δ is also 55°, the angle γ between the X-ray beam, and the collection direction being 71°. As the collection angle varies, the irradiated area varies according to $1/\cos(55° - \theta)$ but the X-ray flux varies inversely. Consequently $S f(\theta)$ in (17) is a constant.

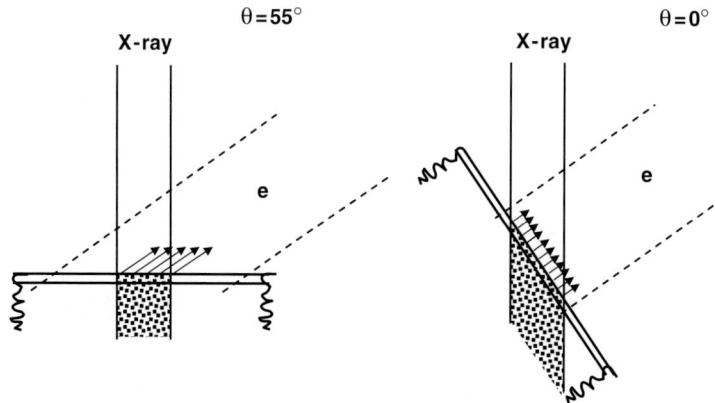

Fig. 25 Schematic representation of a slab with a thin coating analyzed in the SSX-100/206 spectrometer in the horizontal position ($\theta = 55°$) and at zero collection angle ($\theta = 0°$). The tilting axis is perpendicular to the sheet. *X-ray* represents the projection of the X-ray beam on the plane perpendicular to the tilting axis and containing the direction of photoelectron collection. The dotted area represents the irradiated zone; the thin interrupted lines represent the area from which electrons can be collected; the arrows represent the emission of photoelectrons

According to (4) and (19), the intensities may be computed as follows:[7,137]

$$I_{Si^{IV}}^{com} = k^* C_{Si}^{ad} \lambda_{Si}^{ad} \cos\theta \left\{ 1 - \exp\left(-t/\lambda_{Si}^{ad} \cos\theta\right) \right\}, \tag{31}$$

$$I_{Si^0}^{com} = k^* C_{Si}^{su} \lambda_{Si}^{su} \cos\theta \left\{ \exp\left(-t/\lambda_{Si}^{ad} \cos\theta\right) \right\}, \tag{32}$$

where k^* is a constant and the exponents designate the phase to be considered for λ and C.

The concentrations C_{Si}^{su} and C_{Si}^{ad} are 82.96 mmole cm^{-3} and 36.55 mmole cm^{-3}, respectively, taking a density of 2.196 g cm^{-3} for silicon oxide.[138] Values of λ_{Si}^{ad} and λ_{Si}^{su} of 3.9 and 3.2 nm, respectively, were calculated according to Tanuma et al.[139,140] A convenient way to evaluate the thickness of the oxide layer is to fit the experimental intensity ratio by the computed ratio, adjusting the thickness t or the reduced thickness t/λ_{Si}^{ad}. Figure 24a shows that an excellent fit is provided by a thickness of 1.3 nm.

To evaluate the significance of angle-resolved measurements it may be recommended to examine the evolution of individual peak intensities. These are presented in Fig. 24b for the silicon wafer. The same figure presents the intensity computed according to (31) and (32), considering a thickness of 1.3 nm and adjusting the k^* value. An excellent fit is obtained for the intensity variation of each peak and not only for their intensity ratio. To examine the sensitivity of the results to λ, Fig. 24b shows the intensity variations computed using a value of λ_{Si}^{ad} 20% higher and λ_{Si}^{su} 20% lower. In (31) and (32), the influence of the organic contamination present at the surface of the wafer was neglected. The latter can be taken in account by multiplying the right hand side member of the two equations by $\exp\left(-t^*/\lambda_{Si}^{cont} \cos\theta\right)$ where the contamination (cont) is modeled by $C_{15}H_{28}O_4$ with a density of 0.9 g cm^{-3}, leading to $\lambda_{Si}^{cont} = 4.4$ nm. Figure 24b shows the variation of the computed intensities, if the initial values of λ_{Si} are taken and if a thickness $t^* = 1$ nm is considered for the contamination overlayer. The three curves presented in Fig. 24b concerning Si^0 show that, in simulations, the consequences of errors (in this case a difference of λ) or simplifications (in this case neglecting the organic contamination) may be non apparent because their effects compensate each other. If an X-ray spot size of $1,000$ μm is used, the agreement between computed and measured intensity variations is lower; appreciable deviations are found at low θ values. For many instruments the variation of absolute intensities cannot be computed because the geometry of the system (irradiated zone, zone contributing to photoelectron collection) is too complex. However their plots is useful to detect anomalous behaviors at the extremities of the θ range.

Equations similar to (31) and (32) are used to evaluate the thickness of organic contamination on inorganic solids and to correct the apparent surface composition as explained in the section *Influence of Adsorbed Contaminants*. The range of apparent concentrations of $C/Si = 0.50$–0.65 for glass leads to a range of 0.9–1.1 nm for the organic overlayer; the range of $C/Cr = 2.4$–4.7 for SS leads to a thickness of 2.6–3.9 nm; the range of $C/Au = 0.9$–2.8 for gold leads to a thickness of 1.6–3.9 nm. The values computed for the overlayer thickness are only rough indications, owing not only to the experimental variability but also to the uncertainties on the parameters involved in the model. To evaluate the significance of the difference of overlayer thickness between glass and the two other materials, the computations were performed using λ_{Si}^{su} 20% higher, and λ_{Cr}^{su} and λ_{Au}^{su} 20% lower than the values given above. This lead to ranges of overlayer thickness of 1.1–1.3 nm, 2.3–3.5 nm, and 1.4–3.4 on glass, SS and gold, respectively. It thus appears that the adventitious contamination overlayer is thinner on glass compared to more dense materials, which is attributed to weaker van der Waals forces and a lower Hamaker constant for glass.[104] A value of 1.1 nm was reported for the thickness of the contamination overlayer on NaCl.[111] Values in the range of 2–3 nm were measured on SS as received or cleaned with solvents, using XPS analysis at different collection angles.[141] They were decreased to the range of 0.5–1.0 nm by polishing or etching the surfaces; correlatively, the surface energy of the material, as evaluated by contact angle measurements, was found to increase.

Oxidation of PS is commonly used to prepare tissue culture plates[112,142] or to control the distribution or behaviour of mammalian cells (*Surface-Modified Materials*). Eqs (31) and (32) refer to the boxcar model, in which the profile considered for the concentration of the element or chemical function of interest vs. depth is a rectangle. More sophisticated models have been considered to simulate ARXPS data obtained on surface-modified polymers: triangular depth profile, trapezoid depth profile, profile involving several linear segments, or a continuous function.[143,144] In all cases, the significance of models and of fitting parameters must be examined, considering not only the sensitivity of the computed results to their variations[145] but also the possible errors on the experimental data and on the fixed parameters introduced.

Distribution at the Atomic Scale

If the thickness is much smaller than $\lambda \cos\theta$, Eqs (31) and (32) reduce to

$$I_{\text{Si}^{\text{IV}}}^{\text{com}} = k^* C_{\text{Si}}^{\text{ad}} t \tag{33}$$

and

$$I_{\text{Si}^0}^{\text{com}} = k^* C_{\text{Si}}^{\text{su}} \lambda_{\text{Si}}^{\text{su}} \cos\theta. \tag{34}$$

If the substrate contains element Y and the adlayer is constituted by adsorbed ions X or molecules lying flat at the surface and containing element X (Fig. 23b), (19) leads to

$$I_X^{\text{com}} = k\sigma_X T_X \Gamma_X, \tag{35}$$

where Γ_X is the surface concentration of X and

$$I_Y^{\text{com}} = k\sigma_Y T_Y C_Y^{\text{su}} \lambda_Y^{\text{su}} \cos\theta. \tag{36}$$

In a study of the surface modification of poly(bisphenol A carbonate) by sulfochromic acid, application of this approach to the I_S/I_C intensity ratio led to a density of 4–10 sulphur atom cm^{-2}, which was found unrealistic and lead to the conclusion that the polymer was attacked and that sulfate was incorporated over a certain depth. This pretreatment made the zeta potential of the material more negative. It increased the uptake of ferric hydroxide colloids in subsequent adsorption tests, which in turn made the zeta potential more positive. As a consequence the promotion of yeast cell adhesion by the ferric hydroxide treatment was less affected by drying.[146]

Discontinuous Adlayer

Consider a substratum (su) and an adlayer (ad) of thickness t covering the fraction φ of substratum surface (Fig. 23c); the adlayer thickness is considered to be the same on all the covered areas. Expressions below are worked out in the case where proteins would be adsorbed on a fluorinated polymer.

The peak intensity computed for an element that is present only in the adsorbed layer (ad), for instance N, is given by

$$I_N^{\text{com}} = k\sigma_N T_N \varphi \lambda_N^{\text{ad}} C_N^{\text{ad}} \left[1 - \exp\left(-t/\lambda_N^{\text{ad}} \cos\theta\right)\right]. \tag{37}$$

The peak intensity for an element that is present only in the substratum, F in the example considered here, is given by

$$I_F^{\text{com}} = k\sigma_F T_F \lambda_F^{\text{su}} C_F^{\text{su}} \left\{(1-\varphi) + \varphi \exp\left(-t/\lambda_F^{\text{ad}} \cos\theta\right)\right\}. \tag{38}$$

The peak intensity for an element that is present in both the substratum and the adsorbed layer, for instance C, is given by

$$I_C^{\text{com}} = k\sigma_C T_C \left\{ \varphi \lambda_C^{\text{ad}} C_C^{\text{ad}} \left[1 - \exp\left(-t/\lambda_C^{\text{ad}} \cos\theta\right)\right] \right. \\ \left. + \varphi \lambda_C^{\text{su}} C_C^{\text{su}} \left[\exp\left(-t/\lambda_C^{\text{ad}} \cos\theta\right)\right] + (1-\varphi) \lambda_C^{\text{su}} C_C^{\text{su}} \right\}. \tag{39}$$

The comparison between the experimental and computed values of N/C intensity ratios provides a set of $\varphi - t$ pairs that are compatible with the measurements.

On the other hand the amount adsorbed per unit area is given by

$$Q = \rho \varphi t, \qquad (40)$$

where ρ is the density of the adsorbed layer. This provides another set of φ–t pairs.

This approach is illustrated by Fig. 26, which concerns the analysis of polymers and surface oxidized polymers after collagen adsorption.[147] The full lines and dashed lines present the sets of φ–t pairs compatible with the N/C ratios measured by XPS and with the adsorbed amount measured by radiolabeling, respectively. For each polymer, the intersection between the two curves indicates the pair of values which satisfies both measurements. To assess the sensitivity of the results to the data

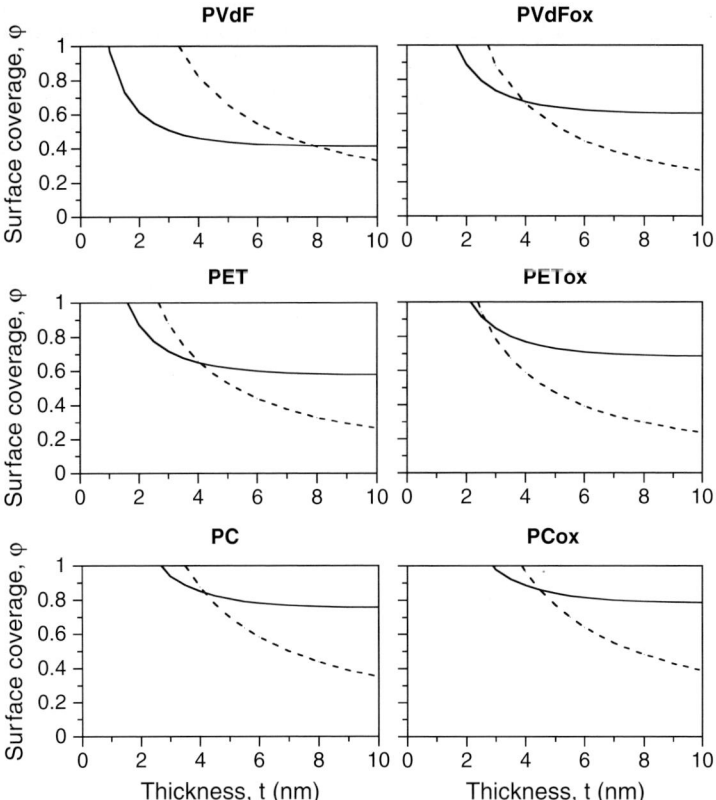

Fig. 26 Plot of the surface coverage, φ, as a function of the thickness, t, of the layer of adsorbed collagen, showing the (φ, t) pairs that are compatible with N/C ratios measured by XPS (*solid lines*) and with amounts of adsorbed collagen measured by radiolabeling (*dashed lines*). Adsorption on poly(vinylidene difluoride) (PVdF), poly(ethylene terephthalate) (PET), poly(bisphenol A carbonate) (PC), and the same modified by oxygen plasma discharge[147]

precision, ranges of satisfying φ–t pairs were computed, considering an error of ±10% on both Q and N/C concentration ratio (not shown here). Values of φ–t pairs obtained for PVdF and PVdFox indicated two different surface morphologies: on PVdF, collagen formed a discontinuous layer covering only 40% of the surface, while on PVdFox, the surface coverage was about 70% and the adsorbed layer was about twice thinner. This difference in the film organization was directly supported by AFM images that revealed a discontinuous layer with holes on PVdF and a continuous layer on PVdFox. The trend observed for PVdF and PVdFox was also found for PET and PETox and again confirmed by AFM. In contrast, φ–t values computed for PC and PCox indicated the presence of a layer, about 4 nm thick, covering about 85% of the surface on both substrata, in agreement with the observation of a continuous layer on both substrata by AFM.

The previous example demonstrates the complementarity of AFM, XPS, and radiolabeling to probe the spatial organization of adsorbed proteins. The combination of XPS and radiolabeling provides a quantitative description of the layer of adsorbed collagen based on a simple geometric model. AFM allows this organization to be confirmed by direct observation of the continuous or discontinuous character of the adsorbed layer and, in some cases, of the layer thickness. Figure 27 presents the evolution, according to the collagen concentration, of φ–t pairs, providing a fit between N/C ratio and the amount adsorbed on PET. The incidence of a variation of 20% in the experimental data is also presented. The figure indicates a surface coverage in the range of 0.7–1.0 and a thickness increasing from 2 nm to about 6 nm according to the initial concentration.[148]

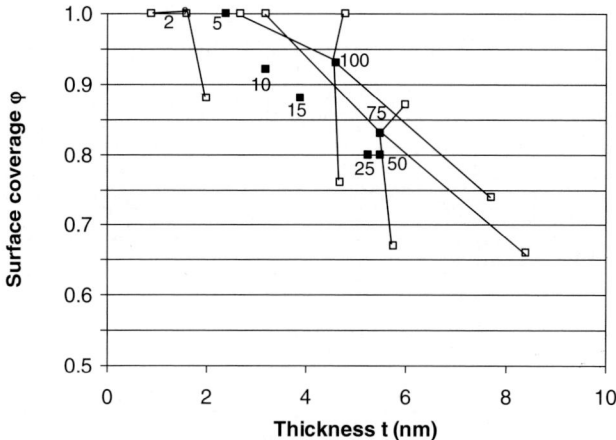

Fig. 27 Surface coverage-thickness (φ – t) pairs providing a fit with both the N/C ratio measured by XPS and the amount of collagen adsorbed on poly(ethylene terephthalate), using the indicated values (2, 5, 10, 15, 25, 50, 75, and 100 μg mL^{-1}) of the initial collagen concentration. *Closed symbols*: values obtained from experimental N/C ratio and adsorbed amount; *open symbols*: values obtained from data differing from the experimental data by 20% (both lower, both higher, one higher and the other one lower) for concentrations of 2, 75, and 100 μg mL^{-1} [148]

The combination of XPS with another technique (AFM and, or radiolabeling) to characterize adsorbed phases is further illustrated for adsorption of proteins on polymers[112, 149, 150] and on self-assembled monolayers,[151] and for mixed Langmuir-Blodgett films of DPPC and surfactin deposited on mica.[152]

Mixtures of Particles

Consider a mixture of particles I containing elements A and C and particles II containing elements B and C (Fig. 23d). This may be representative of biofilms, crusts, ceramics, and catalysts, in which the particle size is appreciably larger than the IMFP. Let the particles expose respective surface areas S^I and S^{II}. If the effect of surface roughness is neglected and the mixture is random, the particles contribute to the spectrum proportionally to the respective areas exposed.

$$I_A^{com} = k\sigma_A T_A \lambda_A^I C_A^I S^I \tag{41}$$

$$I_B^{com} = k\sigma_B T_B \lambda_B^{II} C_B^{II} S^{II} \tag{42}$$

$$I_C^{com} = k\sigma_C T_C \{\lambda_C^I C_C^I S^I + \lambda_C^{II} C_C^{II} S_C^{II}\}. \tag{43}$$

The surface areas exposed by one type of particles may be related to their amount in the mixture and to the external area developed by a unit amount of particles. This approach is thus useful to evaluate the respective distribution of the particles in the material: random distribution of primary particles, autoaggregation of one type of particles, and coating of particles of one type by particles of another type.

Rough Surfaces and Porous Solids

The influence of roughness on XPS data will not be developed here. It involves more complex simulations accounting for the screening of photoelectrons emitted from valleys by neighbouring hills. Details can be found in the literature.[111, 153, 154]

The particular case of porous solids will also not be developed here. It may be kept in mind that, if the thickness of the walls becomes smaller than the IMFP, XPS tends to probe the bulk as well as the surface. If the solid is considered as an accumulation of sheets, the sheet thickness is given by $2/\rho S$ where ρ is the solid density and S the surface area. According to this model, the wall thickness of a silica adsorbent gets smaller than the IMFP of Si 2p photoelectrons in silica, i.e. 3.6 nm, if the surface area is higher than $250 \, m^2 g^{-1}$. When this occurs, the XPS analysis tends to loose its selectivity in terms of surface with respect to bulk. However, it is still selective regarding the analysis of the external surface of the grains, in contrast with the inside of the grains. Further details can be found elsewhere.[153, 155]

Electron Mean Free Path

Absolute values of electron IMFPs are needed to simulate XPS intensities for considering surface heterogeneity (*Data Interpretation Through Simulation*).[139] Tanuma et al. have proposed a formula (TPP-2) to calculate IMFP's on a series of compounds for which experimentally determined optical dielectric functions were known from the literature.[156] This formula was subsequently improved (TTP-2M), particularly to be better adapted to organic compounds. It is believed to be satisfactory for estimating IMFP's of elements, inorganic compounds, and organic compounds.[157] The IMFP can be calculated according to TPP-2M, as well as with algorithms proposed by other authors, from a dedicated NIST data base.[140] The computation is based on the photoelectron kinetic energy, the density, the number of valence electrons per atom or molecule, the molar mass and the energy band gap for non conductors. The main trend is that it increases as the electron kinetic energy increases and as the density decreases. The band gap for many compounds can be found in a number of references. Examples are given in Table 4; reference[161] provides band gaps for numerous elements and inorganic compounds.

Cumpson recently proposed another equation (QSPR) applicable to polymers and other organic materials, and provided IMFP values computed accordingly.[89] IMFP at a kinetic energy of 1,000 eV is computed on the basis on the structural formula alone (number of aromatic six-membered rings and atoms other than hydrogen in the molecule or polymer repeating unit, molecular indices). The IMFP is

Table 4 Energy band gap (in eV) used for computing the inelastic mean free paths

	Band gap (eV)
Elements and oxides	
Si	1.12,[158] 1.16–1.17[161]
SiO_2	8,[160,161] 9[159]
Al	0[158,162]
Al_2O_3	7,[159] 9.5[161]
Ti	0[158]
TiO_2	
Rutile	3,[159,161,163,164] 3.1,[160] 3.75[161]
Anatase	3.2,[160,163] 3.3[160]
C	
Graphite	0.1[160]
Diamond	5.4,[162] 5.5,[161] 7.0[160]
Polymers and organic compounds	
Polyethylene (PE)	7,[157] 7.5[165]
Polystyrene (PS)	4.5[157,165]
Poly(methyl methacrylate) (PMMA)	5[157,165]
Bovine plasma albumine	5.5[157,165]
Deoxyribonucleic acid (DNA)	4[157,165]
26-*n*-paraffin	6[157,165]

extended to other kinetic energies by considering that λ is proportional to $E_k^{0.79}$. The values computed for different polymers and biochemical compounds, in the range of 2.4–3.5 nm, are correlated with the values obtained from the TPP-2M equation.

In reality, electrons suffer inelastic collisions, as already stated, but also elastic collisions by which their trajectory is altered without any energy change. The *attenuation length*, AL, is defined as the "quantity l in the expression $\Delta d/l$ for the fraction of a parallel beam of electrons removed in passing through a thin layer Δd of a substance in the limit as Δd approaches zero." This accounts for both elastic and inelastic scattering and is of obvious concern when an electron beam is sent through a thin layer of a solid. Detailed discussions on AL and its application in different conditions can be found in references.[80,166]

The closeness between IMFP and AL is illustrated by computations performed with the NIST database.[77] Furthermore, the difference between using IMFP and AL for computing intensities is reduced by the fact that photoelectrons are emitted in all directions in the solid. It should vanish when the angular asymmetry factor L is equal to 1, i.e. with spectrometers in which γ is close to 54.7°.

Sample Preparation and Integrity

The aspects that concern microorganisms will be treated in the section *Surface Analysis of Microorganisms and Related Systems*.

Materials with Adsorbed Layers

The need to bring the sample under high vacuum imposes severe constraints regarding the XPS analysis of biological samples. In the study of adsorbed biomolecules, the sample is frequently dried by a flow of nitrogen. It must be kept in mind that an adsorbed layer may be altered by dewetting or evaporation of the liquid film. The sensitivity of an adsorbed protein film to the conditions of drying is illustrated by Fig. 28.[167] It shows AFM images of PS recorded after contact for increasing time with a $7\,\mu g\,mL^{-1}$ of collagen in phosphate buffer saline, rinsing, and either fast drying (performed with a nitrogen flow) or slow drying (performed under 95% relative humidity). The images obtained by AFM in the contact mode show that the film morphology obtained is profoundly altered after slow drying when the adsorption time is below 2 h. This is also revealed by a lower nitrogen/carbon molar ratio measured by XPS and a higher water contact angle. Such alteration is attributed to dewetting, the liquid film being ruptured and adsorbed collagen being displaced by the water meniscus. It did not occur at higher adsorption time or after fast drying; it was also prevented by treating the PS surface by oxygen plasma. It may be noted that samples examined after an adsorption time of 2 h and slow drying showed a poor reproducibility, with all the characteristics (XPS, contact angle, AFM) being

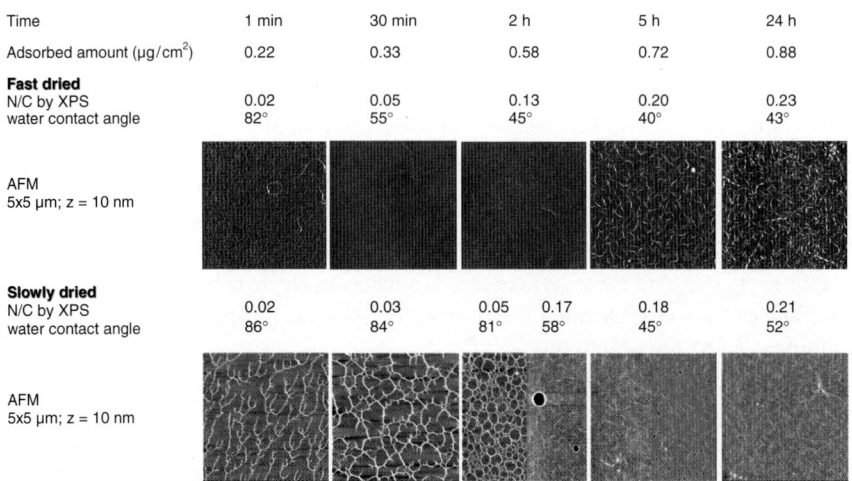

Fig. 28 Influence of the conditions of drying (fast with a nitrogen flow; slow under 95% relative humidity) on the properties of a film of type I collagen adsorbed on polystyrene for increasing times from a $7\,\mathrm{mg\,L^{-1}}$ solution in phosphate buffer saline: adsorbed amount, N/C ratio determined by XPS, water contact angle, and AFM image in the contact mode. Adapted from Ref. 167

sometimes those of a continuous layer, sometimes those of a patterned layer. This example will be further discussed in the section *Adsorption from Single Protein Solutions* via system modeling and data simulation.

Surface Reorganization and Low-Temperature Analysis

Surface reorganization may take place upon dehydration as a result of the tendency of systems to minimize the interfacial free energy. A striking example is presented in Fig. 29, which shows the C 1s peak recorded on a silicone rubber substrate grafted with poly(2-hydroxyethyl methacrylate) (PHEMA).[168, 169] If the material is analyzed without particular precaution the recorded C 1s peak is typical of silicone rubber. If the material is freeze-dried and if the analysis is performed at low temperature, it is characteristic of PHEMA; after raising the temperature to 25 °C, a C 1s peak typical of silicone rubber is again recorded. At low temperature, the interface exposed to water was frozen and the PHEMA grafts were maintained exposed at the surface after freeze-drying. Bringing the sample at room temperature allowed a reorganization to take place, bringing silicone at the surface and burying the more polar grafts to decrease the surface energy.

An analogous phenomenon is the hydrophobicity recovery that occurs when a surface-oxidized polymer[68, 170] is stored in air. For instance, the O/C concentration ratio was 0.01, 0.26, 0.25, 0.18 and the water contact angle was 96, 36, 50, 65°, respectively, for PS, and for plasma-oxidized PS analyzed within 24 h, analyzed

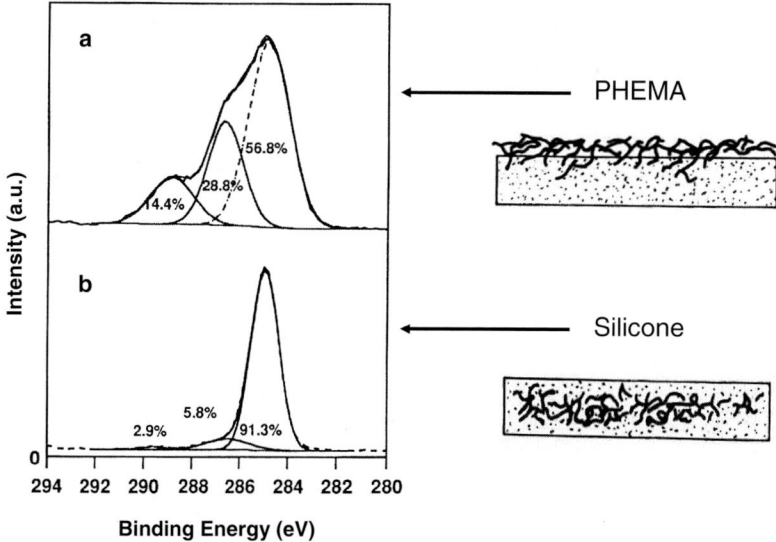

Fig. 29 C 1s peak of silicone grafted with PHEMA, recorded (**a**) at low temperature after freeze drying and (**b**) at room temperature. Adapted from Ref. 168 with permission, copyright (1978) John Wiley & Sons, Inc.; and from Ref. 169, with permission, copyright (1994) Elsevier Limited

after storage for about 10 days in a dessicator at room temperature and analyzed after storage for about 10 days at 95% relative humidity and room temperature.[171] It may thus be useful to freeze a sample in the hydrated state and to analyze it by XPS at different stages such as covered by ice, at low temperature during and after ice sublimation, and at room temperature.

XPS analysis at low temperature, in conditions allowing the surface to be frozen as exposed to water has not been much utilized since 1978,[168] owing to the difficulty to run the experiment and to the length of spectrometer immobilization. The requirements are (1) to bring the frozen sample to the introduction chamber and then to the analysis chamber while maintaining the low temperature, and (2) to control the rate of ice sublimation, which is strongly dependent on sample size, thermal conductivity, and rate of evacuation. However, besides allowing a surface reconstruction to be monitored directly as a function of environment, time or temperature, the method offers the possibility to perform an XPS analysis of the interface between water and any other phase (solid, gel, non miscible liquid of low volatility). This is illustrated by Fig. 30, obtained by spreading a layer of triolein containing DPPC and depositing a drop of water. After freezing and sublimation of ice in the introduction chamber, the XPS spectrum was recorded on different spots of the sample maintained at low temperature. Phosphorous and nitrogen were only found in the crater marking the location of the water drop, with a N/P ratio close to 1, demonstrating that the DPPC accumulated at the oil–water interface.[172]

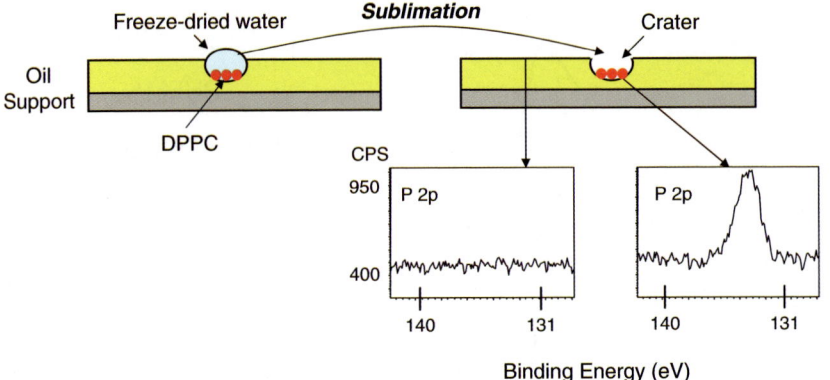

Fig. 30 Schematic illustration of the sample used to analyse the oil–water interface, before (*left*) and after (*right*) ice sublimation. P 2p peak recorded outside (*left*) and inside (*right*) the crater

Sample Alteration During Analysis

Although XPS is often considered as a non destructive technique, biological specimens and polymers may be subject to degradation in the analysis conditions. Sample damage is more pronounced with a nonmonochromatic source, owing to secondary electrons created by the Bremsstrahlung radiation. Moreover, this kind of source is placed very close to the sample (about 1 cm), which generates a temperature rise of several tens °C.[17] With a focused X-ray beam local heating may be particularly crucial.[173]

The sensitivity to degradation varies appreciably according to the material. Degradation indices have been estimated on different polymers[57] by measuring the evolution of the concentration ratio between two elements as a function of time in very drastic conditions (source power 1.4 kW; 500 min). For instance, this index is 0, 10, 25, and 65 for PS, PTFE, PVC, and cellulose trinitrate, respectively. In practice, the anode power and the exposure time are much lower and so the degradation is often negligible. The degradation of cellulose is negligible over the period of a typical measurement ($\sim$15 min).[174] The correlation between the degradation rate and the X-ray source flux was examined in an interlaboratory study performed on PVC, nitrocellulose + cellulose acetate (NC + CA) and PTFE.[175] All the samples could be analyzed without significant degradation within 50–500 s with the different spectrometers used. PVC was found to be more stable than NC + CA or PTFE. A good correlation was observed between the rate constant of degradation of PVC or NC + CA and the X-ray flux, particularly when a nonmonochromatic source was used. The degradation behavior of PTFE was more complicated. The authors recommended to use the ratio of degradation rate constants between PVC and NC + CA to compare the performances of different instruments.

Electrons emitted by a metallic support with a high photoemission cross section like gold may provoke degradation of thin organic supported films such as self-assembled monolayers (SAM).[17] It has been demonstrated that photoinduced electrons were the principal source of damage in these monolayer systems, the X-rays being responsible for less than 10% of the degradation.[176]

A practical way to check that degradation of organic materials is not significant during the analysis is to start and to end the analysis with a detailed record of the C 1s peak. This is fast, compared to certain peaks that require accumulation. Degradation is often reflected in the shape of the C 1s peak. Moreover this allows the stability of surface charging to be checked at the same time (cf. the section on *Sample Charging and Charge Stabilization*). This quick test is recommended to be routinely included in the analysis procedure of organic samples.

Reduction of inorganic oxides can also be provoked during XPS analysis. For instance reduction of CuO has been demonstrated[177] to be due to both slow electrons generated from the X-ray window (nonmonochromatic source) and photoelectrons produced in the sample.

A black spot corresponding to the X-ray irradiated area is visible on the surface of certain samples after their analysis. This spot may correspond to some degradation; however, it is sometimes observed when no sample damage is detected. It could be explained by color centers that correspond to a lattice defect.[178] Sample damage can be drastically reduced by cooling the sample stage during the analysis of inorganic[179] as well as organic samples.[180]

Analytical Performances and Instrument Setting

Overview

Like any other method of analysis, XPS may be evaluated according to several criteria:

1. Selectivity, which determines the possibility to separate peaks characteristic of different elements or peak components characteristic of different functions.
2. Sensitivity, which is the signal obtained for a given concentration of the relevant element in the analyzed zone.
3. Precision, which must be analyzed considering the factors of variation between replicates: data treatment, instrument, specimen preparation, sampling, and time lapse between replicates.
4. Accuracy, which depends crucially on the evaluation of sensitivity factors (cf. the section on *Quantification Considering the Analyzed Zone as Homogeneous*), on the absence of peak overlapping, sample degradation, and artefacts.
5. Limit of detection (or quantification), which is the smallest concentration that can be detected (or quantified).

Artifacts

Beside X-ray satellites mentioned in the section *Source Satellites*, characteristic of a nonmonochromatic source, some ghost lines generated by a deteriorated or contaminated anode, by misalignment or by cross linking between Mg and Al anodes may appear.[33,181,182] X-ray fluorescence from a substrate may induce a secondary excitation in a thin film, creating ghost peaks.[183]

The increase of spectrometer sophistication to achieve high resolution and sensitivity involves a series of calibration parameters that must be checked and adjusted regularly with standard specimens. Subtle artifacts may also occur and need careful examination to be detected and solved. Figure 31 provides an illustration of some problems encountered.

1. Figure 31a and b shows that a lack of the energy scale linearity was responsible for an E_b shift of the F 1s peak of PTFE after E_b calibration with the C 1s peak.
2. Voltages on the different stages of the electrostatic lens must be carefully adjusted to avoid peak distortion, as illustrated in Fig. 31c for C 1s peak recorded on PET.
3. An unexpected satellite was observed on Ag $3d_{5/2}$ recorded on silver due to cross-linking deficiency between two channeltrons of the detector device. (Fig. 31d)
4. As already stated (cf. the section on *Sample Charging and Charge Stabilization*), charge stabilization is crucial for insulators as microorganisms and most

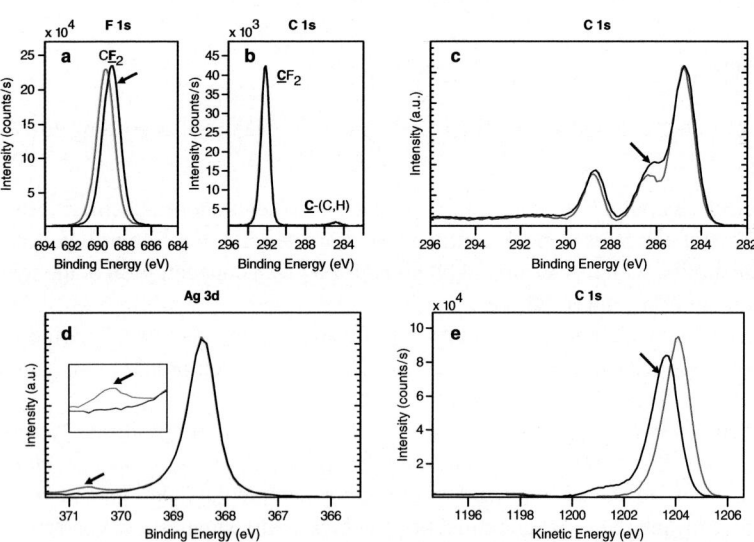

Fig. 31 Artifacts due to incorrect spectrometer setting; arrows indicate deficient spectra. Shift of (**a**) F 1s peak versus (**b**) C 1s peak of PTFE, due to a lack of energy scale linearity; (**c**) peak asymmetry on C 1s peak of PET due to bad electrostatic lens voltage adjustment; (**d**) satellite on Ag $3d_{5/2}$ peak of silver due to a detector problem; and (**e**) deformation of C 1s peak of PS due to incorrect charge stabilization. Kratos Axis Ultra spectrometer, monochromatized Al$_{K\alpha}$, pass energy 40 eV

biomaterials. If parameters of the charge stabilization device are misadjusted, recorded peaks are broadened or even distorted as illustrated in Fig. 31e for C 1s peak recorded on PS.

Charge stabilization and energy scale linearity must be checked regularly by the operator. The channeltron cross-linking deficiency was due to misadjustment of the channeltron high voltage, which could not be detected by the routine checks. Electrostatic lens voltage adjustment is a matter of technical expert.

Peak Width

The peak width is expressed by the full width at half the maximum intensity (FWHM) and is the key factor determining the selectivity. The FWHM is the convolution of three main contributions:

1. The *intrinsic line width* ΔE_n, which is inversely proportional to the lifetime (τ) of the core hole created by the photoemission. The line profile is of Lorentzian (Cauchy) character.
2. The width of the *excitation source* ΔE_x, which is about 0.26 eV for an Al$K\alpha$ monochromatized source. The X-ray line is assumed to be Lorentzian.
3. The width given by the *kinetic energy analyzer*, ΔE_p, which is determined by the pass energy and is selected by the operator. The energy analyzer function is of Gaussian character.

If all contributions are approximated by a Gauss curve:

$$\text{FWHM} = (\Delta E_n^2 + \Delta E_x^2 + \Delta E_p^2)^{1/2}. \qquad (44)$$

As mentioned before, the multiplet splitting effect (cf. the section on *Multiplet Splitting*) can also broaden the peak. Remember also that, for insulators, sample charging leads to peak broadening, even when a charge stabilization device is properly used (cf. the section on *Sample Charging and Charge Stabilization*).

The resolution can be improved by a deconvolution accounting for instrumental broadening. Monochromatized X-ray sources give directly a high resolution but the count rates and consequently the signal/noise ratios are generally lower.[169, 184]

Signal-to-Noise Ratio and Related Trade-Offs

The signal/noise ratio (S/N) is a critical parameter affecting the precision and the detection limit, as well as the reliability of peak decomposition.[123] The signal is the height of the peak maximum measured with respect to the background. The noise is measured on the background at a defined distance from the peak. It can be computed as peak-to-peak noise, as root-mean-square noise or as a standard deviation. A detailed discussion of the influence of the background intensity and of the X-ray beam monochromatization on the S/N ratio is presented in the literature.[185] The following example illustrates the S/N dependency on the pass energy and on the acquisition time.

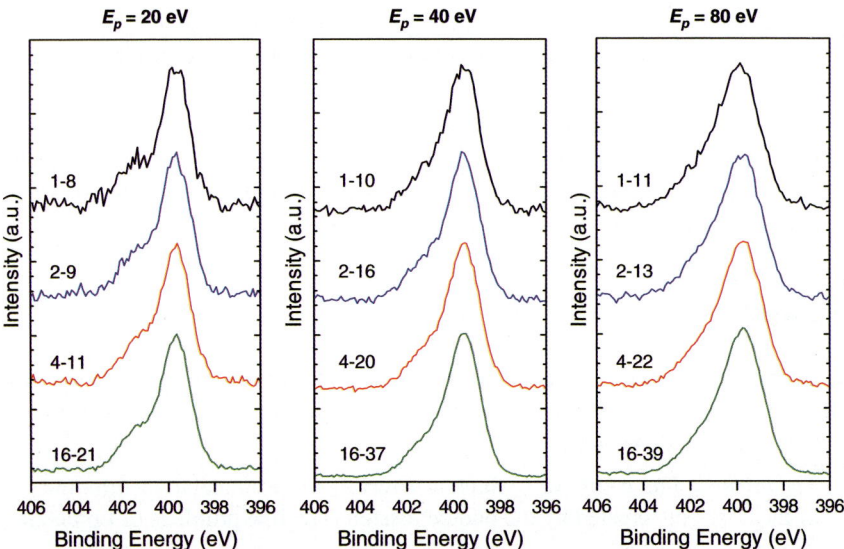

Fig. 32 N 1s peak recorded with different pass energies (E_p) and different acquisition times (1, 2, 4, 16 min) on a sample of *Bacillus subtilis*. Each record is labelled with acquisition time – signal to noise ratio. Kratos Axis Ultra spectrometer, monochromatized Al$_{K\alpha}$

The N 1s peak of *Bacillus Subtilis* vegetative cells was recorded with two spectrometers both equipped with a monochromatized Al$_{K\alpha}$ radiation – a SSX 100/206 (Surface Science Instruments, USA – 1988) and a Kratos Axis Ultra (Kratos Analytical, UK – 1999) – with different pass energies and acquisition times. The N 1s line was recorded to observe the variation of signal intensity and signal to noise ratio (S/N). Figure 32 presents the N 1s peaks recorded with the Kratos spectrometer in different conditions. It also gives the signal/noise (S/N) ratio. The peak to peak noise was computed using the Casa XPS software (Case Softwares Ltd.).[186] It shows that the S/N ratio increases with acquisition time. It also increases with the pass energy, but to the detriment of FWHM and of the separation between the components due to the peptide link (∼399.8 eV) and to protonated nitrogen (∼401.3 eV). For a small pass energy, the acquisition time may have to be increased to reach the desired S/N ratio but may have to be limited owing to the total time of spectrometer use, or the risk of sample degradation during analysis of sensitive materials. So there may be a trade-off between resolution, S/N ratio and sample integrity, in other words between pass energy, on the one hand, and X-ray beam intensity or, and acquisition time, on the other hand.[169]

The total intensity (peak + background) is directly proportional to the acquisition time. Figure 33 shows that the S/N ratio is proportional to the square root of the acquisition time as expected. It also shows that the S/N ratio tends to increase with the pass energy. With pass energies of 50 eV and 40 eV for the SSX and Kratos spectrometers respectively, which provide about the same FWHM, the S/N ratio at

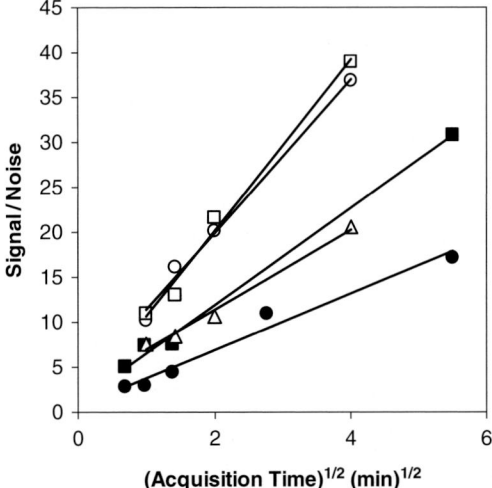

Fig. 33 Signal/noise ratio (peak to peak) plotted as function of the square root of acquisition time for the N 1s peak of a sample of *Bacillus subtilis*. SSX 100/206 spectrometer (*closed symbols*), monochromatized Al$_{K\alpha}$, pass energy of 50 (*filled circle*) and 150 eV (*filled square*); Kratos Axis Ultra spectrometer (*open symbols*), monochromatized Al$_{K\alpha}$, pass energy of 20 (*open triangle*), 40 (*open circle*), and 80 eV (*open square*)

the same acquisition time is about three times higher for the Kratos spectrometer than for the SSX spectrometer.

A comparison between six XPS instruments was conducted in 1989.[169] For comparing different spectrometers, a narrow peak width should be selected and the data acquisition time needed to obtain a useful S/N value should be determined. The variation of the peak width as a function of the pass energy, determined on nonconducting materials, provides information on the effectiveness of the charge stabilization system. If surface charging is responsible for a large FWHM, the X-ray power may be reduced, but this is to the detriment of S/N ratio and precision.

Accuracy

The influence of the adventitious contamination present on high surface energy solids, such as oxides or metals, has been discussed in the sections *Influence of Adsorbed Contaminants* and *Continuous Adlayer*. Note that these organic compounds also contain oxygen.[134]

The C 1s and O 1s peaks of model polymers [poly (bisphenol A carbonate) and poly (ethylene terephtalate)] have been recorded with two different spectrometers for which the O/C concentration ratio was determined in a different way: (a) ESCA 3 MkII spectrometer from Vacuum Generators, Mg$_{K\alpha}$, use of Wagner empirical sensitivity factors;[100, 101] and (b) SSX 100/206 spectrometer, monochromatized Al$_{K\alpha}$, sensitivity factors determined from cross sections[35] and considering IMFP proportional to $E_k^{0.7}$ (cf. the section on *Basic Methodology and Information*). They were decomposed with the constraint that all components of a given peak have the same FWHM and are described by the same function (Gaussian/Lorentzian proportion).[22] The O/C concentration ratio and the distribution of the various types of carbon and oxygen were in agreement (within 10%) with the expected values. A good agreement was also obtained for glucose if the portion of the carbon peak attributed to contamination was not taken into account.

The analysis of model biochemical compounds [glucose derivatives, poly(amino acids), etc.] with the SSX 100/206 spectrometer at a pass energy of 50 eV[134] has shown that two systematic errors may compensate each other. The presence of organic contamination at the sample surface reduced the apparent O/C ratio, because the contamination overlayer contained less oxygen than most of the analyzed standards. On the other hand, the assumption of a constant transmission function exagerated (by about 10%), the computed O/C concentration ratio. The two errors compensated each other and the apparent O/C concentrations ratios were in excellent agreement with the values expected from stoichiometry.

The problem of accuracy is further illustrated by Fig. 34 for a series of strains of *B. subtilis*. This presents a plot of the $(O+N)/C$ concentration ratios as a function of the proportion of carbon bound to oxygen or nitrogen (cf. the sections on *Semi-Quantitative Use of Angle Resolved Analysis* and *Peak Decomposition and Assignment*) i.e. $\{C \text{ total}-[\underline{C}-(C,H)]\}/C$ total, abbreviated by C_{ox}/C. The results obtained with two spectrometers are presented. The differences between the two spectrometers are more important along the ordinate scale than along the abscissa

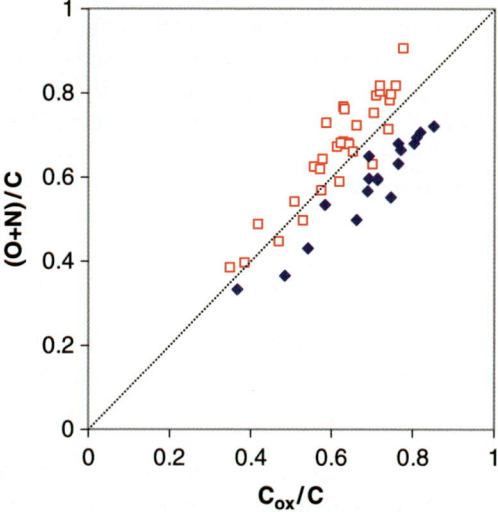

Fig. 34 Plot of the $(O+N)/C$ concentration ratio as a function of the proportion of carbon bound to oxygen or nitrogen, C_{ox}/C, for a series of strains of *Bacillus subtilis* analysed with two different spectrometers: SSX 100/206 (*open square*) and Kratos Axis Ultra (*closed diamond*)

scale. A unit slope is expected for the following functions: alcohol, primary amide, primary amine, and ester. The distribution of the experimental points slightly above the line of unit slope may be attributed to the presence of acetal of polysaccharide (C—O—C—O—C; cf. the section on *Peak Decomposition and Assignment*). The lower $(O+N)/C$ concentration ratios obtained for the Kratos spectrometer compared to the SSX 100/206 spectrometer are attributed to a systematic error on the oxygen sensitivity factor provided by the manufacturer. This conclusion is also supported by the analysis of metal oxides.

Limit of Detection and Quantification

The limit of detection has been stated as being the concentration at which the peak is three times as high as the noise of the background, i.e., the concentration for which the S/N ratio is greater than 3.[185] With recent spectrometers equipped with a monochromatized X-ray source, heavier elements such as lead or gold can be detected down to 0.02% mole fraction, but the figure is 2% for light elements such as boron or lithium.[26] The detection limit for the analysis of oxygen on PS was evaluated to be of the order of 0.1% mole fraction in routine conditions.[187] A value of about 0.1% was also found for sulfur, nitrogen, phosphorous, and potassium on samples of microorganisms[188,189] and for phosphorous, chlorine, and sulfur in corrosion investigations.[26]

The limit of quantification is a multiple of the limit of detection, the factor of multiplication depending on the precision required. The value zero in a table of concentrations is a nonsense and should be replaced either by the absence of any information (not determined) or by the notation "below detection limit" (bdl).

Determining the detection limit from a formal S/N ratio may be difficult for XPS peaks that show a complex shape. A practical way to determine the detection limit in a set of analyses is to consider a peak that is still clearly visible, to measure its area and to evaluate the factor by which the intensity should be decreased to loose the peak in the noise. The detection limit is then the concentration that was deduced from the peak considered, divided by the evaluated factor.

Surface Analysis of Microorganisms and Related Systems

Introduction

In an introductory review on the use of XPS for the analysis of organic and biological surfaces, dated 1984, Holm[190] made the following point: "Few attempts have so far been made to apply ESCA to biological tissues, bacteria cells, and so on. That is partly because such samples are considerably more difficult to prepare than are samples of polymers. Another reason is that protein molecules cannot be adequately characterized, and hence differentiated, by ESCA." He quoted a first review by

Millard dated 1978[191] and mentioned, in particular, the influence of the preparation technique, the lack of specificity of the signals of C, O, and N in the investigation of the surfaces of bacteria and the interest of the technique regarding the distribution of phosphorus (from teichoic acids) in Gram-positive and Gram-negative bacteria.

Marshall et al.[192] discussed the relevance of XPS for analysis of microbial cells surfaces and summarized a critical view as follows: "A number of steps in the process of preparing microorganisms for XPS examination may lead to the introduction of artefacts in the estimation of surface properties. These include the choice of organism, the growth conditions employed, the centrifugation of the cell suspension, the resuspension of the cells in distilled water, the freezing and drying processes, the ultrahigh vacuum conditions employed in XPS, and the possibility of X-ray damage to surface structures on the cells. Some of these processes may result in the leakage of internal cellular components or even to complete lysis of the cells prior to XPS examination, with major problems resulting from the adsorption to intact cell surfaces of released macromolecules. Reports of good correlations between XPS characterization of surfaces and other measured physical parameters need to be regarded with caution."

Therefore the present review emphasizes the information provided and its biological relevance, but also its limitation regarding sample preparation and analytical performances. A particular attention is devoted to the relationships between the surface chemical composition "as determined by XPS," and surface physico-chemical properties that are evaluated by other techniques (having also their own limitations) and that may play a major role in interfacial interactions: the surface electrical properties[193] and the surface hydrophobicity.[194] A final section gives a survey of applications of XPS to better understand interfacial phenomena involving microorganisms. Previous reviews may be found in references 22 to 25.

Chemical Functions

Peak Decomposition and Assignment

Figure 35 presents the O 1s, N 1s, and C 1s peaks of three different strains of *Bacillus subtilis*. The peak decomposition is also shown. For the C 1s and O 1s peak of microorganisms, it is recommended to perform the second step (cf. the section on *Peak Components of Biochemical Compounds*), imposing the same FWHM for all components of a given peak; this value may be let to be adjusted during the iterative process or may be the average of the values obtained in the first step.

The results of decomposing C 1s, O 1s, and N 1s peaks recorded on microbial cells may be summarized as follows.

The *carbon peak* can be decomposed into four contributions:

1. A component fixed at 284.8 eV, due to carbon bound only to carbon and hydrogen [$\underline{C}$—(C,H)]; this component is used as a reference to determine the binding energy of all the other components or peaks, i.e. to calibrate the binding energy scale

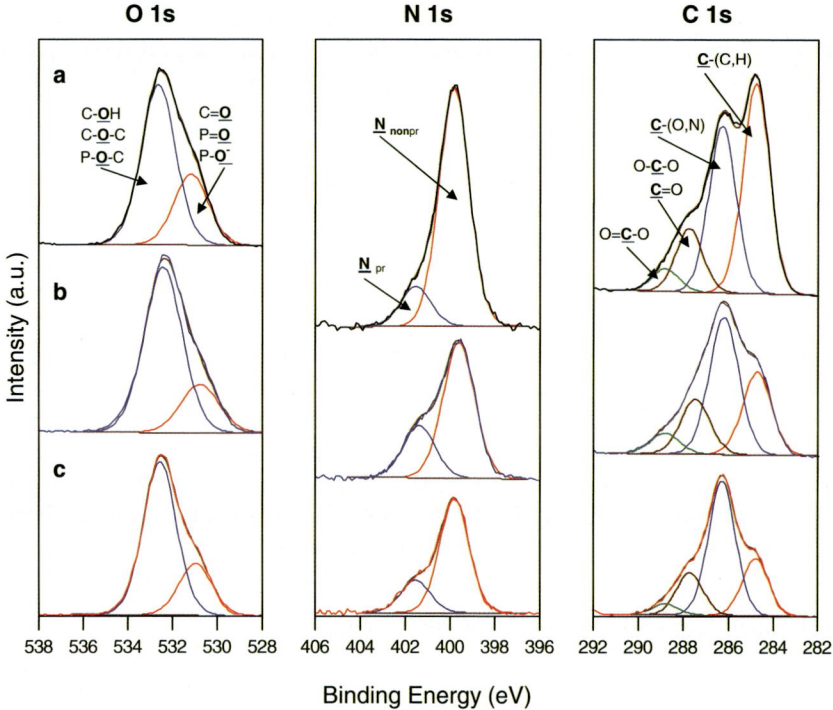

Fig. 35 Oxygen, nitrogen, and carbon 1s peaks of three strains (vegetative cells) of *Bacillus subtilis*: (**a**) ATCC 7058; (**b**) ATCC 15476; and (**c**) S499. Spectrometer Kratos Axis Ultra, monochromatized Al$_{K\alpha}$, pass energy 40 eV

2. A component at 286.1–286.3 eV, due to carbon singly bound to oxygen or nitrogen [$\underline{C}$—(O,N)], including alcohol, amide, acetal [C—O—C—O—C] or hemiacetal [C—O—C—OH]
3. A component at about 288.0 eV, due to carbon making one double bond [$\underline{C}$=O] or two single bonds with oxygen [O—$\underline{C}$—O], including amide, acetal and hemiacetal
4. Sometimes a weak component found near 289.0 eV and possibly due to carboxylic acid and ester

The *oxygen peak* can be decomposed into two contributions:

1. A component at about 531.2 eV, attributed to oxygen making a double bond with carbon [$\underline{O}$=C] in amide and to oxygen of phosphate [P=$\underline{O}$, P—$\underline{O}^-$]
2. A component at about 532.6 eV, attributed to hydroxide [C—$\underline{O}$H], acetal [C—$\underline{O}$—C—$\underline{O}$—C], hemiacetal, and phosphate ester [P—$\underline{O}$—C]

The main *nitrogen peak* appearing at about 399.8 eV is due to unprotonated amine or amide functions [N_{nonpr}]. Sometimes a weak component is found near 401.3 eV and attributed to protonated amine or ammonium [N_{pr}].

This summary results from works that started with yeast cells[195] and were extended to bacteria, both Gram-positive and Gram-negative, and to fungal spores, using successively three spectrometers with improved resolution: VG ESCA 3 MkII (Vacuum Generators), SSX 100/206 (Surface Science Instruments), Kratos Axis Ultra (Kratos Analytical).[22,23,25] The decomposition procedure described above was the most frequently used. Sulphur and iron peaks were observed on bacteria of interest in bioleaching and biobeneficiation.[196]

Quantitative Relationships

The consistency of the components and of their attribution are confirmed by quantitative relationships between data as discussed below. This consistency is remarkable in view of the difficulties pointed out in the section *Peak Components of Biochemical Compounds* to make reliable decomposition of peaks of simple compounds. It is due to the presence of a limited number of major chemical functions in living matter. It is also due to the reliability of the ratios of sensitivity factors used with the SSX 100/206 spectrometer (cf. the section on *Accuracy*).

Consistency Regarding Proteins and Polysaccharides

The peptidic link present in proteins, $\equiv$C—NH—(C=O)— must lead to a 1:1 ratio between the N and $\underline{C}$=O concentrations. Figure 36 presents results obtained for a soil bacterium *Azospirillum brasilense* at different culture times.[197] The correlation between the concentration of carbon responsible for the component at 287.9 eV

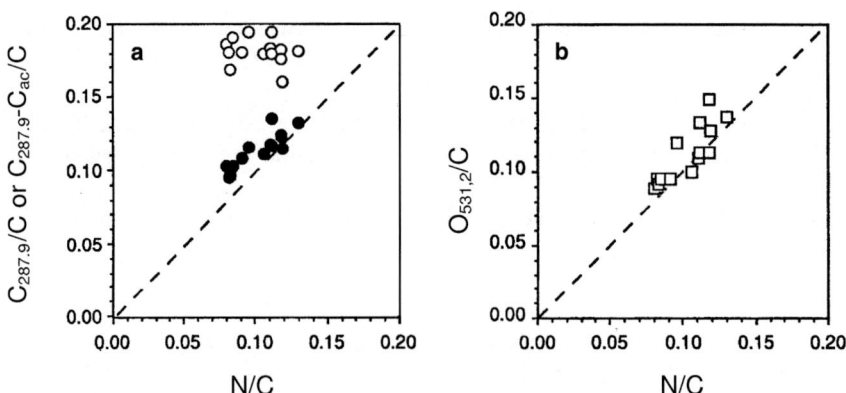

Fig. 36 Relations between concentration ratios (relative to total carbon) obtained for *Azospirillum brasilense* harvested at different culture times. Abscissa: nitrogen. Ordinate: (**a**) carbon responsible for the C 1s component at 287.9 eV before (*open circle*) and after (*closed circle*) subtraction of the contribution of acetal functions; and (**b**) oxygen responsible for the O 1s component at 531.2 eV[197]

($C_{287.9}$) and the nitrogen concentration (N) does not follow the 1:1 relationship expected for the peptidic link. The deviation may be explained by the contribution at 287.9 eV of carbon making two single bonds with oxygen, as present in the acetal links of polysaccharides. The concentration of acetal carbon C_{ac} may be evaluated by

$$C_{ac} = 0.2(C_{286.2} - N) \tag{45}$$

as one hexose unit in a polysaccharide contains one carbon atom of acetal type and 5 carbon atoms of C—OH type, and as the latter contribute to the $C_{286.2}$ component as well as carbon bound to nitrogen in the peptide links or in amines (C—N). Figure 36a shows that the difference $[C_{287.9} - C_{ac}]$, expected to be due to the peptidic link only (N—C=O), is indeed close to the nitrogen concentration.

A 1:1 relationship between $[C_{287.9} - C_{ac}]$ and N was also found for lactic bacteria[198] and for *Bacillus subtilis*.[375] In the case of the dormant spores of *Phanerochaete chrysosporium* (white rot fungus), the surface of which is very poor in polysaccharides, the raw $C_{287.9}$ concentration was very close to the N concentration.[199]

Consistency Regarding Proteins and Esterphosphates

The peptidic link must also lead to a 1:1 ratio between the N and O=C concentrations. Figure 36b shows that the relationship between N and $O_{531.2}$ concentrations is close to 1:1. This was also found for the spores of *Ph. chrysosporium*.[199] In contrast, for lactic bacteria the $O_{531.2}$ concentration was higher than the concentration of nonprotonated nitrogen (N_{nonpr}).[198] These bacteria also showed a high phosphorus concentration. A detailed study of correlations lead to the conclusion that the $O_{531.2}$ component contained a contribution of two oxygen atoms per phosphorus atom, i.e., P—O$^-$ and P=O of phosphate. This was confirmed recently by the study of a set of 9 strains of *B. subtilis* covering a range of P/C ratio from 0.007 to 0.055, which provided a 1:2 relationship between P and $[O_{531.2} - N_{nonpr}]$ or a 1:1 relationship between N_{nonpr} and $[O_{531.2} - 2P]$.

Consistency Regarding Polysaccharides

For each hexose unit of a polysaccharide, there are 5 oxygen atoms [C—OH or C—O—C], contributing to $O_{532.6}$, one acetal-type carbon O—C—O contributing to $C_{287.9}$ and 5 carbon atoms bound to hydroxide [C—OH]. The latter contributes to the $C_{286.2}$ component, together with carbon atoms bound to nitrogen. A 1:1 relationship was observed between the concentrations $O_{532.6}$ and $C_{286.2}$ for the sets of *A. brasilense* and of *B. subtilis* described above. Considering $[C_{286.2} - N]$ instead of $C_{286.2}$ shifts only slightly the regression line, due to the relatively low nitrogen concentration. Despite the relatively high concentration of phosphate of *B. subtilis* cells, the contribution of oxygen in the form of ester phosphate P—**O**—C to the $O_{532.5}$ component was too small to affect these correlations.

Consistency Between C 1s Peak Shape and Elemental Composition

A relationship close to 1:1 between the $(O+N)/C$ ratio and the proportion of carbon that is bound to a oxygen or nitrogen C_{ox}/C (Fig. 34) has been discussed in the section *Accuracy* in connection with the sensitivity factors. It has been observed for many microorganisms.[22,24,200] This is consistent with a dominance of alcohol, primary amide or amine, and ester functions. A high concentration of ether functions as in acetal links would lead to a ratio lower than 1:1. A high concentration of carboxyl, carboxylate, or ester phosphate would lead to a higher ratio.

Toward Molecular Composition

By combining experimental surface concentrations of elements and chemical functions with reasonable assumptions about the nature and composition of model molecular compounds, it is possible to figure out a surface molecular composition. For microorganisms, three main classes of basic constituents may be considered: proteins (Pr), polysaccharides (PS) with a general formula $(C_6H_{10}O_5)_n$, and hydrocarbonlike compounds (HC) with a general formula $(CH_2)_n$, which refers to lipids and other compounds that contain mainly carbon and hydrogen. In this approach, peptidoglycans are considered as a combination of proteins and polysaccharides, and (lipo)teichoic acids as a combination of hydrocarbonlike compounds and polysaccharides. The chemical composition of model constituents is given in Table 5.

Two sets of equations can be written, allowing the use of two independent sets of XPS data to evaluate the proportion of carbon associated with protein (C_{Pr}/C), polysaccharide (C_{PS}/C), and hydrocarbonlike components (C_{HC}/C).[23] One scheme is based on observed elemental concentration ratios (obs), where the model protein chosen is the major outer protein of a *Pseudomonas fluorescent* strain (Table 5).

$$[N/C]_{obs} = 0.279\,(C_{Pr}/C) \tag{46}$$

$$[O/C]_{obs} = 0.325\,(C_{Pr}/C) + 0.833\,(C_{PS}/C) \tag{47}$$

$$[C/C]_{obs} = (C_{Pr}/C) + (C_{PS}/C) + (C_{HC}/C) = 1. \tag{48}$$

A second scheme is based on the three main components of the carbon peak:

$$[C_{288}/C]_{obs} = 0.279\,(C_{Pr}/C) + 0.167\,(C_{PS}/C) \tag{49}$$

$$[C_{286.2}/C]_{obs} = 0.293\,(C_{Pr}/C) + 0.833\,(C_{PS}/C) \tag{50}$$

$$[C_{284.8}/C]_{obs} = 0.428\,(C_{Pr}/C) + 1\,(C_{HC}/C). \tag{51}$$

Table 5 Chemical composition of model constituents for the evaluation of the molecular composition of microbial surfaces

Model constituent	Concentration ratio (mol/mol)					Carbon concentration (mmol g^{-1})
	O/C	N/C	($\underline{C}$—(C,H))/C	($\underline{C}$—(O,N))/C	($\underline{C}$=O)/C	
Protein[a] (Pr)	0.325	0.279	0.428	0.293	0.279	43.5
Protein[b] (Pr)	0.33	0.27	0.40	0.32	0.28	42.4
Protein[c] (Pr)	0.347	0.286				43.1
Hydrocarbon (HC) $(CH_2)_n$	0	0	1.00	0	0	71.4
Polysaccharide (PS) $(C_6H_{10}O_5)_n$	0.833	0	0	0.833	0.167	37.0
Chitin $(C_8H_{13}O_5N)_n$	0.625	0.125	0.125	0.625	0.250	39.4

[a]Data computed for the major outer membrane protein of a *Pseudomonas fluorescens* strain used for the calculation of the molecular composition of *Azospirillum brasilense*[197]
[b]Data computed from the amino acid composition of different fungal wall proteins, used for the calculation of the molecular composition of *Phanerochaete chrysosporium* conidiospores[199]
[c]Data computed for the S-layer protein of *Lactobacillus helveticus*, used for the calculation of the molecular composition of lactic bacteria[198]

Solving one system of equations gives the proportion of the total carbon concentration (C_X/C) associated with each of the molecular constituents X. Dividing this proportion by the carbon concentration in the constituent (Table 5) gives the ratio of the weight of constituent X to the total concentration of carbon in the sample. Dividing the ratio characteristic of constituent X by the sum of the ratios of all constituents provides finally the weight fraction of X in the analyzed volume. The comparison of the results provided by the two schemes may be used to check internal consistency. The choice of the model protein composition is not crucial. According to (49) and (50), the evolution of the C 1s peak shapes in Fig. 35 reveals an increase of polysaccharide/polypeptide concentration ratio following the strain sequence a–b–c. This is in agreement with an increase of the O/N concentration ratio (3.8–5.1–6.0) and with the evolution of the O 1s peak shape.

Figure 37 presents results obtained for two lactic bacteria harvested at two culture stages[198] using the approach based on elemental concentration ratio and selecting the S-layer of *Lactobacillus helveticus* strain as the model protein (Table 5). It shows that the surface of these bacteria is poor in lipids (0–15% HC). The surface of *Lactococcus lactis* is rich in polysaccharides and its composition does not change much during the culture. The surface of *Lactobacillus helveticus* shows a more variable surface composition, with a polysaccharide concentration higher in stationary growth phase than in exponential growth phase; this increase of the polysaccharide concentration may be attributed either to an increased synthesis during growth or to a rupture of the protein S-layer present at the surface.

For spores and pellets of *Ph. chrysosporium*, alternative choices were examined for the model molecular compounds: proteins + glucose + hydrocarbons and proteins + chitin + hydrocarbons.[23,199] More sophisticated models may be used with

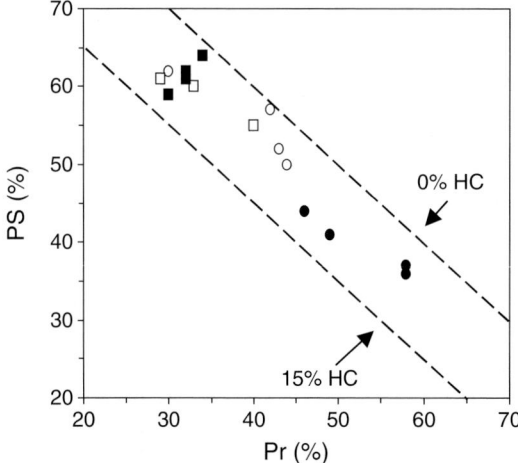

Fig. 37 Results of XPS analysis of lactic bacteria, expressed in terms of weight % of polysaccharides (PS), proteins (Pr), and hydrocarbons (HC). *Lactobacillus helveticus* (*circles*) and *Lactobacillus lactis* (*squares*), harvested in exponential (*closed symbols*) and in stationary growth phase (*open symbols*)[198]

a higher number of model constituents: chitin + glucan for fungal spores, consideration of teichoic acids,[201] consideration of nucleotides, etc. However, the results quickly loose significance as the number of unknowns and equations increases, due to the limited precision of the experimental data and the limited accuracy of the composition selected for the model constituents.

Microbial Sample Preparation

The high vacuum needed for XPS analysis requires a previous dehydration of the samples. A crucial question is then the representativity of the analyzed surface with respect to the real surface of the microorganisms. Any kind of chemical fixation being excluded for obvious reasons, freeze drying under controlled conditions has been used. One approach was based on depositing the washed and centrifuged cells in a trough, on quick freezing by immersion in $CHClF_2$ at liquid nitrogen temperature, and on evacuation below $-30\,°C$.[195] In subsequent works, the use of $CHClF_2$ was abandoned for environmental reasons and in order to prevent possible damage from the contact between the cell and a highly apolar solvent.[202]

The following protocol has been applied extensively,[189] using a freeze-dryer specially designed to allow the temperature of the shelf to be programmed at any value between $-80\,°C$ and room temperature by circulation of a cryogen. The cells are harvested by centrifugation at $4\,°C$ and washed with cold distilled water, by three successive suspensions and centrifugations. The last pellet is added with a small amount of water to obtain a concentrated suspension. A portion of the thick suspension (e.g., 2 mL of 10^{11} bacteria mL^{-1} or 1 mL of 3×10^9 yeasts mL^{-1}) is transferred into a container, typically of glass with an internal diameter and height of 2–4 cm, precooled and kept partially immersed in a liquid nitrogen bath. The container is covered by an aluminum foil and left in liquid nitrogen for 15 min. It is then immediately transferred into the freeze dryer (after piercing the aluminum foil) or stored in a freezer at $-80\,°C$ until freeze drying. For freeze drying, the containers are placed on the shelf at $-50\,°C$, the chamber is evacuated and the temperature program is started. This involves 3 h at $-50\,°C$, variation from $-50\,°C$ to $-5\,°C$ in 15 h, remaining at $-5\,°C$ until the pressure stabilizes at about 0.6 Pa. Then the shelf temperature is raised to $23\,°C$ in 3 h and maintained for a few hours, the vacuum chamber is vented with nitrogen and the flasks are quickly stoppered and stored in a dessicator.[189]

The influence of variations with respect to this procedure was examined with brewing yeasts.[189] The suspension must be cooled sufficiently fast to avoid migration of intracellular compounds to the surface. However, details regarding the freezing conditions were not found to be critical. No influence of cooling rate on the XPS data was observed with yeast in the range from a few thousands of $°C/s$ (droplets frozen by impact) to less than $1\,°C\,s^{-1}$ (samples frozen in a nonprecooled glass vial). On the other hand, a strict condition is that no melting takes place at any stage between freezing and complete dehydration. Figure 38 shows scanning electron micrographs for cells of *S. cerevisiae* freeze dried according to the above protocol (a), for cells obtained after freeze-drying with a bad temperature control allowing melting to occur (b), and for cells freeze-dried according to the standard protocol, rehydrated and again frozen and freeze dried (c). The XPS data obtained on the three preparations are presented in Table 6, together with those of a sample prepared by freezing and melting the suspension before final freeze drying (d), and of a sample prepared by crushing the cells before freeze drying (e). In comparison with properly freeze-dried cells (a), frozen cells submitted to melting, whether deliberately

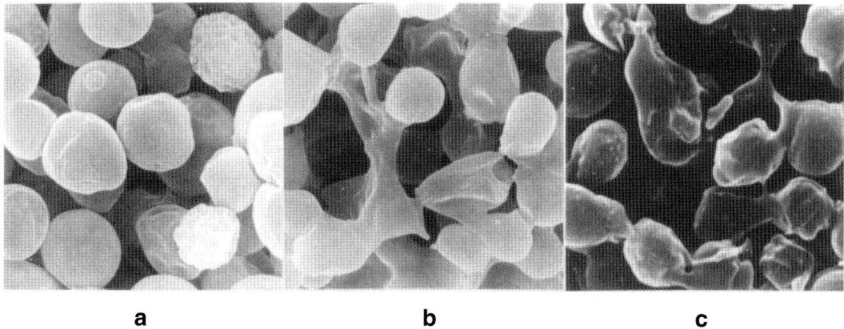

Fig. 38 Scanning electron micrographs of brewing yeast *Saccharomyces cerevisiae* (**a**) prepared according to the standard procedure; (**b**) freeze dried with a bad temperature control allowing melting to occur; and (**c**) freeze dried, rehydrated in water, and freeze dried again (**c**). Adapted from Ref. 189

Table 6 XPS analysis (mole fraction in %, excluding hydrogen) of brewing yeast (**a**) prepared according to the standard procedure; (**b**) freeze dried with a bad temperature control allowing melting to occur; and (**c**) freeze dried, rehydrated in water, and freeze dried again (see Fig. 38). Comparison of XPS data with (**d**) a sample resuspensed in water after a first freeze drying; or (**e**) crushed by grinding before freeze-drying

Sample history	$\underline{C}$—(C,H)	$\underline{C}$—(O,N)	$\underline{O}$—C	$\underline{O}$=C	N_{nonpr}	N_{pr}	P	K
(a) Standard procedure	31.5	29.2	27.3	2.0	1.5	<0.1	<0.1	<0.1
(b) Poor temperature control during freeze drying	44.9	19.4	14.6	7.0	3.7	0.6	0.7	0.3
(c) Cells freeze dried, rehydrated, and freeze dried again	47.6	18.8	15.6	5.4	2.4	0.9	0.7	0.5
(d) Suspension frozen and melted before freeze drying	51.5	16.8	14.5	4.8	3.0	0.6	0.8	0.4
(e) Cells crushed before freeze drying	54.4	14.4	9.7	7.3	4.9	0.5	0.5	0.3

(d) or by poor temperature control during freeze drying (b), or cells rehydrated after a first freeze drying (c) show an altered morphology with a clear evidence of a release of intracellular constituents. This is accompanied by an alteration of XPS data that become similar to those of crushed cells (e): increase of carbon only bound to carbon and hydrogen [$\underline{C}$—(C,H)], decrease of carbon making a single bond with oxygen or nitrogen [$\underline{C}$—(O,N)], decrease of oxygen making a single bond with carbon [$\underline{O}$-C], increase of oxygen making a double bond with carbon [$\underline{O}$=C] and of nonprotonated nitrogen [N_{nonpr}], increase of protonated nitrogen [N_{pr}], phosphate [P], and potassium [K]. This may be attributed to the release of lipids, proteins and possibly salts that are present in greater concentration at the surface of altered specimens, leading to a decrease of the carbohydrate concentration. Cells dehydrated by air-drying show XPS data similar to inadequately freeze-dried cells.

The analysis of yeast cells stored in a dessicator indicated a decrease of the proportion of [$\underline{C}$—(C,H)] in the C 1s peak and an increase of O/C and N/C ratios of the order of 10% after 3 months and 20% after 8 months.[189]

Besides sample damage due to improper freeze-drying, the representativity of the analyzed surface with respect to the surface of interest in hydrated conditions may be altered by a reorganization of the surface layer upon drying. To avoid a reorganization of cell wall polymers, it might be desirable to cool the sample so rapidly that the surface polymers are quenched by cryofixation or vitrification (i.e., solidifying all water in an amorphous state by very rapid cooling). This has not been explored. It would require (1) very high cooling rates (at least $10,000\,°C\,s^{-1}$), (2) a very small sample volume as no vitrification occurs at a depth larger than 25 μm, and (3) preservation of the cooled sample below its glass transition temperature.[189]

Analytical Performances

Sample Degradation

For yeast specimens and a spectrometer with nonmonochromatized radiation the P 2p signal increased by a factor of 2.5 over a period of 80 h of cycled analysis, while the N 1s signal decreased by about 20%.[195] A relative increase in the signal of carbon (0.4%/h), phosphate (4.0%/h), and potassium (4.2%/h), and a relative decrease in the signal of oxygen (-2.6%/h) and nitrogen (-2.0%/h) was measured.[202] The decrease of oxygen and nitrogen, on one hand, and the increase of phosphorus and potassium, on the other hand, suggest that organic matter was volatilized, concentrating inorganic compounds at the surface.

During a prolonged analysis of a strain of *S. cerevisiae*[188] with the SSX 100/206 spectrometer (monochromatized radiation), the irradiated area acquired a yellowish color; the relative contribution of $\underline{C}$—(C, H) and $\underline{C}$—(O, N) in the carbon peak increased and decreased respectively. The degradation was shown to be due to X-ray beam and not to the flood gun electrons. The degradation rate defined as the rate of decrease of the fraction of the C 1s peak due to carbon bound to oxygen or nitrogen C_{ox}/C (cf. the section on *Accuracy*) was found in the range of 5–12%/h over a period of 2–3 h, depending on spot size and yeast strain. The degradation rate measured in identical conditions was 1.2, 1.5, and 12%/h for sorbitol, starch, and yeast, respectively. A comparison between spectrometers with (SSX 100/206) and without (VG ESCA 3 MkII) a monochromator, using yeast cells, showed that the rate of degradation was about 3 times higher in the latter case, while the radiation dose per unit time and unit area was about 4 times larger.

Figure 39 (left) presents, for a collection of brewing yeasts, the relationship between the rate of degradation, measured as defined above, and the surface composition expressed in terms of model molecular constituents. It appears that the rate of degradation is directly correlated with the concentration of hydrocarbonlike compounds and inversely correlated with polysaccharide concentration. It appears thus that the relative rate of degradation of oxygen and nitrogen-containing functions increases as the concentration of polysaccharides decreases.[203]

Precision

A detailed study[188] of the precision of the analysis of a brewing yeast has shown that sample manipulation subsequent to freeze-drying, spectrum recording and data treatment were responsible for variation coefficients of less than 3% for the mole fraction of oxygen, of total carbon and of major types of carbon (mole fraction of several tens of %). The variation coefficients were below 10% for nitrogen (mole fraction of 1–2%), and reached 10–20% for minor elements (P, K; mole fraction close to the detection limit of the order of 0.1%) and for less abundant types of carbon. The biological material itself, and sample freezing and drying procedures were the main sources of variation. The variation coefficients taking these factors

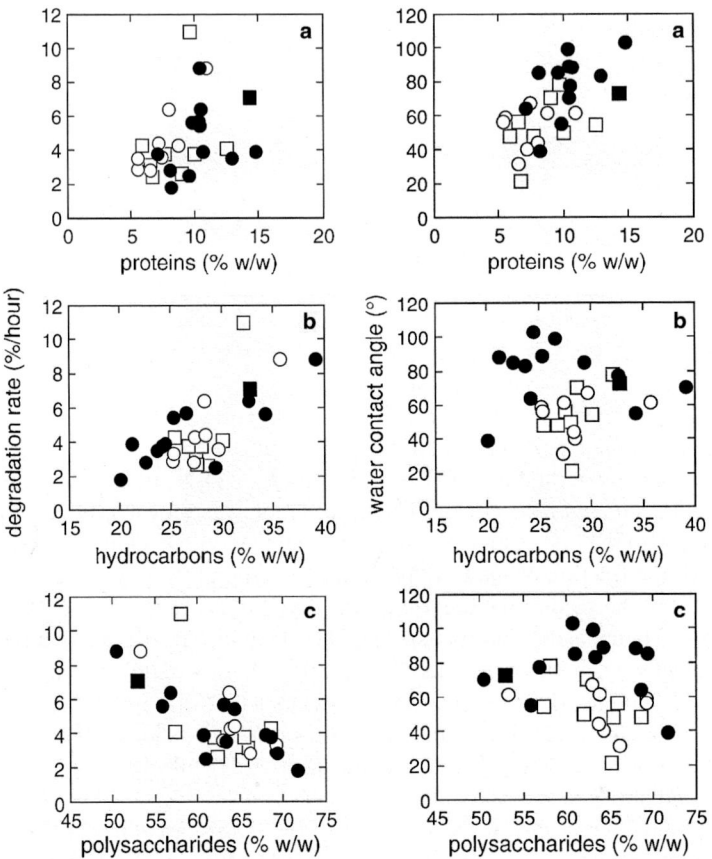

Fig. 39 Rate of degradation – left – (expressed by the relative decrease of the proportion, in the C 1s peak, of components due to carbon bound to O or N, divided by the time of exposure to X-rays) and water contact angle – right – of bottom- (*open square*, exponential; *open circle*, stationary cells) and top- (*closed squares*, exponential; *closed circles*, stationary cells) fermenting yeast strains as a function of the cell-surface concentration (weight %) of (**a**) proteins, (**b**) hydrocarbons, and (**c**) polysaccharides, as obtained by modeling XPS data[203]

into account were 3–7% for the major elements, about 10% for the main types of carbon, and 20% for nitrogen, phosphorus, and potassium were sometimes below the detection limit.

Surface Composition of Microorganisms

Variation of Surface Composition According to Species, Strain, and Physiological State

Gram-positive bacteria possess a well-defined rigid cell wall around the cytoplasmic membrane. The wall rigidity is caused by a thick peptidoglycan-based layer that makes about 10% of the total cell volume. Gram-negative bacteria have a thin (1–2 nm thick) peptidoglycan layer sandwiched between the cytoplasmic membrane and an outer membrane. Surface appendages, often containing proteins and lipoteichoic acids, as well as polysaccharide-rich capsules are often present on bacterial surfaces.[24,204] Yeasts, which are eukariotics, are characterized by a larger size than bacteria (order of 5–10 μm compared to 1–2 μm) and possess a thick wall. Generally, this constitutes 15–25% of the cell dry weight. Polysaccharides account for 80–90% of the wall, followed by smaller amounts of proteins and lipids. With few exceptions, glucans and mannans are the main polysaccharides, but a small amount of chitin is also present.[205]

Hierarchical cluster analyses on the basis of elemental surface concentration ratios N/C, O/C, and P/C revealed that, of 36 different streptococcal strains, only four *S. rattus* as well as nine *S. mitis* strains were located in distinct groups, presumably because of their relatively high N/C and O/C concentration ratios.[206] When the isoelectric points and water contact angles were included in the cluster analysis, the *S. rattus* strains remained grouped in one separated cluster, but the cluster comprised of the *S. mitis* strains was divided into two reasonably well-separated clusters by the tufted *S. sanguis* strains. "It is difficult, if not in principle impossible, to obtain a grouping of microbial strains on the basis of their cell-surface properties in accordance with their taxonomy, unless strains and species possess distinctly different surface compositions."[206]

Cluster analysis of 210 microbial strains revealed that 31 of the 33 staphylococci were located in a distinct group well separated from the other Gram-positive bacteria, from Gram-negative bacteria and from yeast strains, as a result of their generally high O/C surface concentration ratio.[24] This might indicate that these staphylococcal strains have capsular polysaccharide material at their outer surface, despite the fact that this was not always indicated by India ink staining. The yeast strains were in one distinct group, but the separation distance from the other microbial strains was less than for the staphylococci. Yeasts distinguish themselves from bacteria by a combination of low P/C and N/C and relatively high O/C surface concentration ratios. The Gram character of bacterial strains was not revealed by the cluster analysis.

Significant variations of surface composition are frequently observed when narrower sets of microorganisms are compared. The analysis of 27 genotypically characterized *Lactobacillus* species showed that the eight *L. acidophilus* strains were phenotypically different from other *Lactobacillus* strains, while it was not known whether there was an effect of growth phase.[207] The difference was caused by higher amounts of nitrogen-rich groups on the cell surface, responsible for an increased cell-surface hydrophobicity, isoelectric point, and adhesion to hexadecane.

Figure 39 (right) summarizes XPS data and the water contact angle obtained on a collection of brewing yeasts including 8 bottom-fermenting strains (*S. carlsbergensis*) and 12 top-fermenting strains (*S. cerevisiae*).[203] They were harvested in stationary culture phase and some of them also in exponential culture phase. The figure shows a direct correlation between cell-surface protein concentration and hydrophobicity (water contact angle measured on cell lawns). The inverse correlation with the polysaccharide concentration is less marked and there is no correlation between the water contact angle and the concentration of hydrocarbonlike compounds. Top strains are generally more hydrophobic than bottom strains; their surface possesses higher protein and lower phosphate concentrations. No systematic difference was found between yeast cells harvested in the exponential phase and in the stationary phase.

The culture medium of a *S. cerevisiae* strain used for alcohol production influenced the surface composition.[208] Phosphate depletion in the culture medium led to a lower phosphate concentration at the cell surface; depletion of other nutrients resulted in an increase of phosphate surface concentration due to a lower amount of biomass.

The phosphorus concentration at the surface of a plasmid-cured *Thermus* strain was three times lower compared to the wild strain, suggesting that the presence of plasmids influences cell-surface characteristics.[209]

The surface phosphate concentration of a given strain of *Bacillus licheniformis* decreased as a function of culture age and was influenced by aeration conditions and cultivation temperature.[210] The comparison between the phosphate concentration decrease in the culture medium, the increase of the cell concentration and the decrease of the phosphate concentration at the cell surface showed that the specific phosphate uptake is high at the beginning of the culture while the amount of phosphate assimilated is shared during the successive cell divisions. This holds not only for the total phosphate content but also for the phosphate present at the cell surface.

The surface composition of dormant spores of the white rot basidiomycete *Ph. chrysosporium* was modeled in terms of molecular compounds and estimated to be about 45% proteins, 20% polysaccharides, and 35% hydrocarbons. A significant increase of the polysaccharide concentration took place during germination and spore aggregation occurred concomitantly.[199] Atomic force microscopy (AFM) was used to probe, under water, the surface ultrastructure and molecular interactions of the spores. High-resolution images revealed that the surface of dormant spores was uniformly covered with rodlets having a periodicity of 10 ± 1 nm. In contrast, germinating spores had a very smooth surface partially covered with rough granular structures. Force–distance curve measurements demonstrated that the changes in spore surface ultrastructure during germination were correlated with profound modifications of molecular interactions: while dormant spores showed no adhesion with the AFM probe, germinating spores exhibited strong adhesion forces of 9 ± 2 nN magnitude. These forces were attributed to polysaccharide binding and suggested to be responsible for spore aggregation.[44]

When *A. brasilense* was cultured in Luria-Bertani rich medium, the surface composition varied during growth, as illustrated by Fig. 40. Modeling the composition in terms of molecular compounds indicated that the protein concentration increased, from 30 (exponential phase cells) to 50% (stationary phase cells), concomitantly with a decrease in the polysaccharide concentration, from 60 to 35%. These modifications were related to a change in cell-surface hydrophobicity, the water contact angle measured on cell lawns increasing from 20 to 60° (Fig. 40). No difference of electrophoretic mobility was detected between cells harvested in the exponential

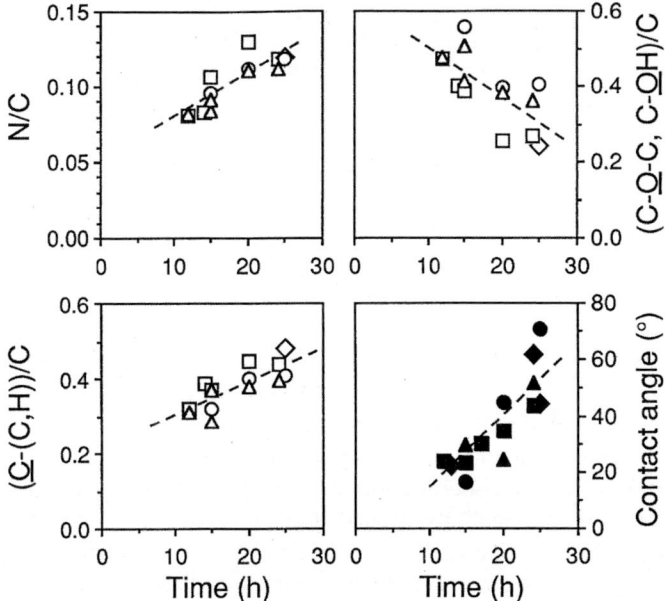

Fig. 40 Evolution of the surface composition, in terms of N/C, (C—$\underline{O}$H, C—$\underline{O}$—C)/C, and $\underline{C}$—(C,H)/C mole concentration ratios, and of the water contact angle of *A. brasilense* during growth. The different symbols correspond to four independent sets of determinations carried out on separate bacterial cultures; open and closed symbols originate from different cultures[197]

phase and cells harvested in the stationary phase. The increase of both cell-surface protein concentration and cell-surface hydrophobicity during growth was correlated with an increase of cell adhesiveness to model supports (cf. the section on *Material Surface Conditioning in Connection with Microorganisms and Biofilms*).[197]

In contrast with the growth in a complex medium, growth of *A. brasilense* in a synthetic medium led to a low reproducibility of XPS results, and did not show any significant variation of the cell-surface composition as a function of culture time.[211] Moreover, the surface composition was found to be generally lower in proteins, and higher in polysaccharides and in hydrocarbon-like compounds. The direct correlation between the phosphate concentration and the concentration of carbon bound only to carbon and hydrogen, $\underline{C}$—(C, H), indicated that hydrocarbon-like compounds detected at the surface of cells grown in synthetic medium are essentially phospholipids. The poor reproducibility of the XPS data obtained for cells grown in synthetic medium and the random observation of phospholipids at the outermost cell surface were not due to cell disruption or to migration of intracellular components during sample preparation. Exposure of phospholipids at the cell surface reflected either actual variations of the composition of the native surface induced by nutrient limitation or reorganization of cell-surface polymers upon freeze drying. Interestingly, the increase of the N/C surface concentration ratio during growth in the complex medium was accompanied by an increase of the P/C ratio. In contrast, the N/C surface concentration ratio of cells grown in the synthetic medium, which varied randomly, showed a narrow inverse correlation with the P/C ratio.

Relationship Between Chemical Composition and Surface Electrical Properties

The data accumulated for a large variety of microorganisms are very disperse regarding the relationship between electrical properties and surface composition. Better correlations are obtained when the comparisons are restricted to limited sets, particularly to taxonomically related cells.[23] The N/P concentration ratio was directly correlated with the isoelectric point (iep) for a collection of brewing yeasts[195] and inversely correlated with the electrophoretic mobility measured at pH 4 for a collection of strains including different yeasts and bacteria.[212,213]

The phosphate surface concentration of brewing strains was directly correlated with the electrophoretic mobility measured at pH 4. A 1:1 relationship was observed between the phosphate concentration and both the potassium concentration and the protonated nitrogen concentration, indicating that potassium and ammonium are the main counterions.[214] For a strain of *S. cerevisiae* used for alcohol production, cultured in different media, the isoelectric point decreased and the electrophoretic mobility at pH 4 increased with the surface phosphate concentration.[208] For *Escherichia coli* cells differing according to the strain and to the culture medium, the electrophoretic mobility at pH 4 increased with the phosphate surface concentration, which was close to the sum of K^+, Na^+, and NH_4^+ concentrations.[215] However, the examination of a collection of strains including different yeasts and bacteria showed that beyond a certain phosphate concentration (P/C about 0.005) and a certain P/N concentration ratio (about 0.1), the electrophoretic mobility showed no further variation.[216] This was presumably due to the effect of surface potential on the apparent acidity constant of remaining hydrogenophosphate moieties and to condensation of counterions on the surface. It seemed therefore that deprotonated carboxyl played a minor role in the development of the surface charge.

Direct correlations between the N/C surface concentration ratio and isoelectric point were found within sets of bacteria that showed only weak variations of the phosphate surface concentration: wild type and mutants of *Streptococcus salivarius*,[217] and a set of six strains of oral streptococci.[218,219] A more disperse but still clear correlation was obtained when these data were mixed with data concerning nine strains of *S. mitis* and four strains of *Peptostreptococcus micros*, another oral cavity microorganism.[220] The isoelectric point measured on 36 strains (including 7 different species) of oral streptococci was directly correlated with the N/C ratio and inversely correlated with the O/C ratio, as illustrated by Fig. 41.[206] The correlation was much narrower if the tufted *S. sanguis* strains were left apart. The data obtained for strains of *Serratia marcescens*,[221] urogenital and poultry lactobacilli,[222] and several anaerobic subgingival bacteria[223] fell within that correlation range. However, the distribution of data within each one of these last three sets did not show any

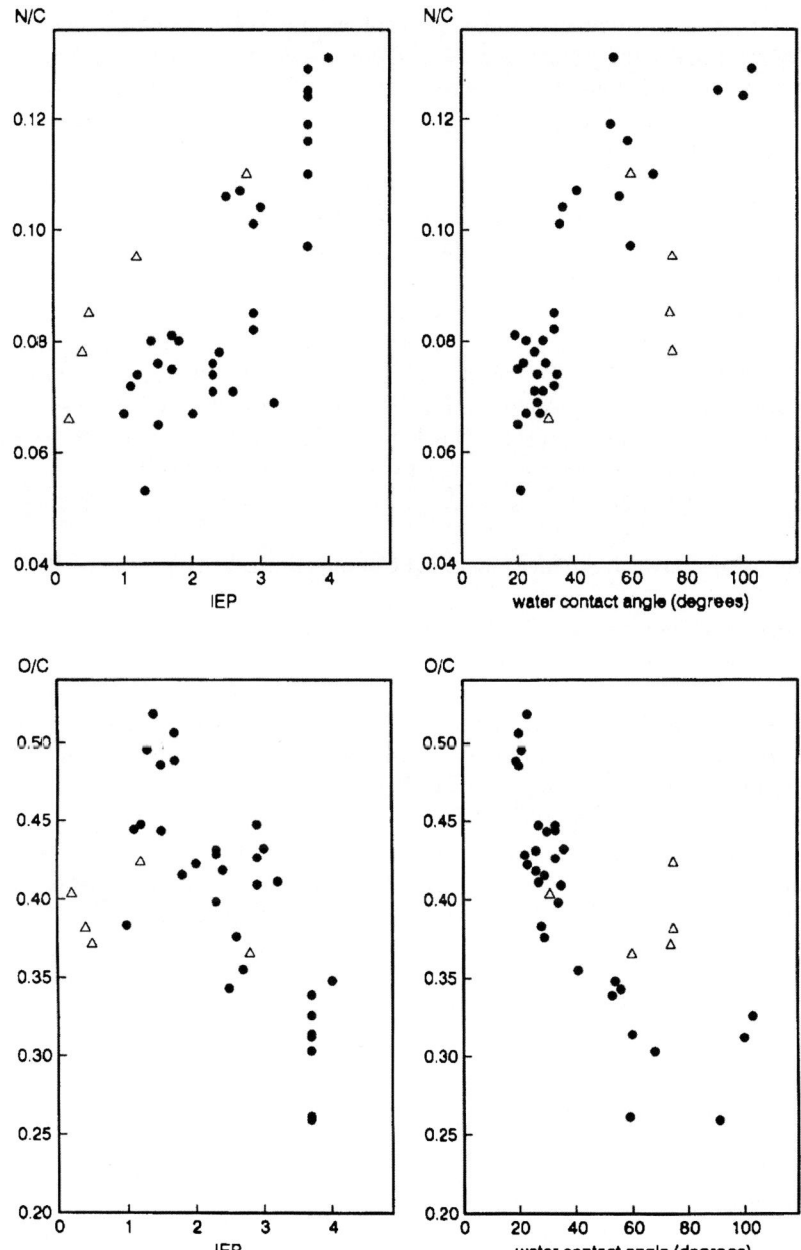

Fig. 41 Molar concentration ratio N/C and O/C as a function of the isoelectric point (IEP) and of the water contact angle, for peritrichous oral streptococci (*closed circle*) and tufted streptococci (*open triangle*). Reprinted from Ref. 206 with permission

clear correlation. A study of 27 strains of genotypically characterized *Lactobacillus* species showed that the group of eight *L. acidophilus*, which had a N/C surface concentration ratio about twice higher than the other strains (cf. the section on *Variation of Surface Composition According to Species, Strain, and Physiological State*), also had a high isoelectric point.[207]

The different profiles of electrophoretic mobility versus pH obtained for top and bottom yeast strains mentioned in the section *Variation of Surface Composition According to Species, Strain, and Physiological State* could be explained by evaluating the surface charge according to the surface chemical composition given by XPS. For bottom strains, the electrical properties were mainly controlled by phosphate, resulting in a low isoelectric point (pH 2 or below) and an electrophoretic mobility that did not become much more negative above pH 4. For the top strains, the electrical properties were mainly determined by the balance of protonated amino-groups and carboxylate groups in proteins, which gave a high isoelectric point (pH 4) and an electrophoretic mobility changing greatly with pH in the range of 2–7.[203]

Relationship Between Surface Chemical Composition and Hydrophobicity

A direct correlation between the N/C surface concentration ratio, i.e., the protein surface concentration, and the water contact angle measured on cell lawns was already mentioned in the section *Variation of Surface Composition According to Species, Strain and Physiological State* when comparing bottom and top brewing yeast strains (Fig. 39)[203] and for *A. brasilense* (Gram-negative bacteria) cultured for different times in a rich medium (Fig. 40).[197,224]

The N/C concentration ratio was also directly correlated with the water contact angle and inversely correlated with the polar contribution of the surface energy within various sets of Gram-positive bacteria: *Streptococcus mitis*,[24] urogenital and poultry lactobacilli,[222] small sets of staphylococci,[225] and oral streptococci.[218,219,226] Data obtained for two bacteria relevant to fouling in the food industry, *S. thermophilus* and *Leuconostoc mesenteroides*, fit also the latter correlation.[227] The opposite trend was found with the O/C concentration ratio, but the correlation was often less clear owing to the smaller range of relative variation of O/C than of N/C. A disperse correlation between N/C and the water contact angle was obtained for a large collection of strains including oral streptococci and *P. micros*.[220] As observed for the iep (cf. the section on *Relationship Between Surface Chemical Composition and Electrical Properties*) and illustrated by Fig. 41, the direct correlation between the water contact angle measured on 36 strains (including 7 different species) of oral streptococci and the N/C ratio, and the inverse correlation with the O/C ratio, were narrower if the tufted *S. sanguis* strains were left apart.[206] The group of *L. acidophilus* characterized by a high N/C ratio and a high isoelectric point (cf. the section on *Relationship Between Surface Chemical Composition and Electrical Properties*), also showed a higher water contact angle $(40-80°)$ compared to the 3 groups of other species of lactobacilli $(20-50°)$.[207] An inverse correlation was observed between the P/C ratio of two pairs of coagulase negative staphylococcal isolates with different plasmid profiles, and the surface hydrophobicity as evaluated by adhesion to hexadecane.[228]

Correlations between water contact angles and XPS data were not followed for *E. coli* and *Actinobacillus actinomycetem comitans* presumably owing to their Gram-negative character or to the presence of a complex surface structure.[24] The strong difference of surface hydrophobicity between two strains of *Acinetobacter calcoaceticus* was not accompanied by a significant difference of surface chemical composition.[229]

Surface Analysis vs Bulk Analysis

Certain studies revealed a relationship between the infrared absorption (IR) spectrum of whole bacterial cells dispersed in a KBr pellet and XPS data. A set of encapsulated and nonencapsulated coagulase-negative staphylocci[225] showed a direct correlation between concentration ratios of elements or functions and intensity ratios of infrared bands:

1. N/C vs. amide II band at $1,542\,\text{cm}^{-1}/$CH stretching band, the amide II band being due to N-H bending in proteins
2. $(\underline{C}=O)/C$ vs. amide I band at $1,654\,\text{cm}^{-1}/$CH stretching band, the amide I band being due to C=O stretching in proteins
3. P/C vs. phosphate band at $1,237\,\text{cm}^{-1}/$CH stretching band
4. $[\underline{C}-(O,N)]/C$ vs. phosphate-sugar band at $1,070\,\text{cm}^{-1}/$CH stretching band

A narrow correlation was observed for a collection of 6 different *Streptococcus* strains between N/C concentration ratio and amide II/CH stretching intensity ratio, and between $(\underline{C}=O)/C$ concentration ratio and amide I/CH stretching intensity ratio.[218] The second correlation was also followed for a collection of five tufted *S. sanguis* strains, in contrast with the first correlation.[226] A narrow correlation was observed between N/C concentration ratio and amide II/CH stretching intensity ratio for *S. salivarius* strains but no such correlation was found for *E. coli* strains, which are Gram-negative.[24]

As IR analysis provides information on the bulk composition, the correlation with XPS data may be explained by the fact that in certain sets of Gram-positive strains, a variation of surface composition reflects the variation of the cell wall composition and that the cell wall represents a large proportion of the cell dry weight. This explanation is supported by the study of 5 different species of Gram-positive bacteria.[200] The polypeptide surface concentration of the isolated cell walls deduced from XPS analysis was in excellent agreement with the polypeptide concentration determined by biochemical analysis. On the other hand, *Bacillus brevis* whole cells showed a twice higher surface concentration of polypeptides compared to that of four coryneform bacteria, which was attributed to the presence of an S-layer. Consistently the polypeptide concentration of the cell wall of *B. brevis* was intermediate between that of whole cells of *B. brevis* and of cell walls and whole cells of the coryneform bacteria. This supports both the surface specificity of the XPS analysis of whole cells and the relevance of the deduced surface composition expressed in terms of model biochemical compounds.

A narrow inverse correlation was observed between the N/C surface ratio of six isolates of *S. salivarius* and the concentration of lipoteichoic and teichoic acids at the surface (LTA-TA), measured by antibodies.[217] This suggests that LTA-TA were diluted or shielded by proteinaceous moieties at the surface. Four of these isolates were analyzed by secondary ion mass spectroscopy (SIMS), which revealed an excellent correlation between the peak intensity ratio PO_3^-/CNO^-, and the P/N concentration ratio measured by XPS.[230]

Stripping the S-layer of a strain of *Lactobacillus helveticus* by LiCl treatment led to a decrease of the protein surface concentration evaluated by XPS, and to an increase of the surface concentration in LTA-TA and polysaccharides–peptidoglycanes.[201] This was revealed by the variation of elemental concentration ratios (*N/C*, *O/C*, *P/C*) and of C 1s and O 1s peak profiles, and was in agreement with biochemical analysis of LiCl-treated isolated cell walls. The XPS analysis of the precipitated extract confirmed that this was almost pure protein.

The XPS analysis of dormant and germinating basidiospores and its consistency with AFM observation and spore aggregation confirmed the potentiality of the method for helping to understand interfacial phenomena involving microorganisms (cf. the section on *Variation of Surface Composition According to Species, Strain and Physiological State*). This was also nicely illustrated by the comparison of a fibrillated strain of *Streptococcus salivarius* HB and a nonfibrillated mutant HBC12. The loss of proteinaceous fibrillar surface antigen of the mutant was accompanied by a variation of the N/C surface concentration ratio from 0.104 to 0.053 and a shift of the isoelectric point from 3.0 to 1.3; a similar evolution was provoked by protease treatment.[217] Direct measurement of the cell-surface softness was performed more recently by AFM observations in aqueous solutions.[231] Upon approach of the fibrillated cells in water, the AFM tip experienced a long-range repulsion force, starting at ~ 100 nm, attributed to the compression of the soft layer of fibrils present at the cell surface. In 0.1 M KCl, repulsion was only experienced when the tip was closer than ~ 10 nm, reflecting a stiffer cell surface due to collapse of the fibrillar mass. Force–distance curves indicated that the nonfibrillated strain, probed both in water and in 0.1 M KCl, was much stiffer than the fibrillated strain in water, and a repulsion force was experienced by the tip at close approach only (20 nm in water and 10 nm in 0.1 M KCl). Differences in cell-surface softness were further supported by differences in cell-surface morphology; the fibrillated strain imaged in water showed characteristic topographical features attributable to fibrils.

An attempt of depth profiling (ARXPS) was performed by examining *S. salivarius* HB and four fibrillar mutants at different angles of photoelectron collection.[232] Note that a freeze-dried powder of bacteria is far from having the smoothness suitable to get clear-cut results by ARXPS. The angle dependence of the O/C and P/C surface concentration ratios was difficult to interpret. However, the angle dependence of the N/C surface concentration ratio provided the following results: N/C did not vary with sampling depth on the bald strain (HBC12) and on a strain with a dense array of fibrils of uniform length; N/C decreased as sampling depth increased in case of a sparsely fibrillated strain and eventually reached the value observed for the bald strain; a high N/C at small sampling depth was observed for HB strain that possessed protruding protein-rich fibrils. No further significant observation could be obtained, owing to variations of results between independent cultures.

Depth profiling was also performed in the early days of XPS analysis of microorganisms, by combination with oxygen plasma treatment.[233] The phosphorus signal increased upon etching. However, this may be due to accumulation of nonvolatile residues (ashes) from oxidation rather than a variation of concentration as a function of depth.

Application of XPS to the Study of Interfacial Phenomena Involving Microorganisms

We present below a survey of applications of XPS to elucidate processes occurring at interfaces and involving microorganisms. For sake of clarity, a distinction is made between the following categories:

1. Influence of cell-surface chemical composition on microbial behavior at interfaces (adhesion, aggregation, flotation)
2. Interaction between microbial surfaces and the liquid environment
3. Material surface conditioning in connection with microorganisms and biofilms
4. Material alteration or generation by microorganisms.

The works that refer directly to biomedical applications will be considered in the section *XPS Study of Systems Related to Biomedical Applications*.

Influence of Cell-Surface Chemical Composition on Microbial Behavior at Interfaces

The differences of surface composition, isoelectric point, and hydrophobicity between top and bottom brewing yeasts, as outlined in the earlier sections, are in accordance with their behavior at the end of fermentation. The higher hydrophobicity of the top brewing yeast may explain their top-cropping behavior at the end of beer fermentation; the flocs associate with CO_2 bubble and rise to the top of the fermented wort. On the contrary, the less hydrophobic aggregates of the bottom strains settle down.[203] In contrast, no systematic difference of surface composition observed by XPS, of surface electrical properties or of surface hydrophobicity was found between flocculating and nonflocculating strains, or between cells from the exponential and stationary growth phases, even for strains where flocculation occurred during the transition from one growth phase to the other. The tendency to flocculate and the occurrence of flocculation are governed by mechanisms that differ according to the strain, e.g., appearance of lectins at the surface for bottom strains and variation of steric repulsion for some top strains.[234,235] However, no relationship can be established with data obtained by XPS.

The flotation performances of two strains of *S. cerevisiae* were examined in relation with cell separation in continuous fermentation.[236] For batch cultures, the more hydrophobic strain (water contact angle 69°, compared to 27°), which showed a higher degree of flotation, showed a surface concentration higher in nitrogen and carbon of hydrocarbon type, and lower in oxygen. For continuous cultures, the relationship between surface chemical composition, surface electrical properties, surface hydrophobicity, and flotation was less clear.

Phosphate limitation upon growing *Corynebacterium glutamicum* led to a lower phosphate concentration and a lower hydrophobicity of the cell surface, with no significant change of O/C and N/C ratios. Phosphate-depleted cells required a higher DEAE-dextran concentration to aggregate and showed a lower tendency to adhere to glass treated by an aminosilane or by adsorption of DEAE-dextran.[237]

The XPS analysis of marine bacterial strains indicated that the strain showing a higher protein surface concentration adhered to SS, glass, or Teflon™ surfaces in a much higher quantity compared to two other strains.[238]

Interactions Between Microbial Surfaces and the Liquid Environment

A decrease of O/C and P/C concentration ratios at the surface of *Streptococcus sobrinus* was reported as a result of preincubation with polyclonal antibodies. A combination of the pH-dependent zeta potential and the XPS data suggested that polyclonal antibody adsorption occurred through blocking of surface phosphate.[239] XPS measurements provided evidence for IgG attachment to *Pseudomonas aeruginosa*. The antibody-modified cells showed a reduced adhesion to glass under flow in a nutrient medium.[240]

Sorption of lead cations by acetone-washed *Saccharomyces uvarum* resulted in a shift of the oxygen peak that, together with infrared data, supported the hypothesis that lead uptake occurred mainly through binding to carboxylate. The nitrogen peak decreased after lead sorption suggesting that nitrogen containing groups were also involved in the biosorption process.[241]

XPS at low temperature showed that exposure of *Clostridium* sp. to solutions of uranyl acetate led to the immobilization of uranium onto the cell wall by reduction to sparingly soluble U^{IV}.[242] The speciation of elements trapped in a biomass generated by *Desulfovibrio* sp. was also examined after culture in the presence of solutions of Fe^{III}, Cr^{III}, Ni^{II}, and molybdate.[243] The aim was to simulate exposure of sulfate-reducing bacteria (SRB) to the microenvironment next to SS. The presence of sulfur in different oxidation stages was observed (S^{2-}, S^0, SO_3^{2-}, SO_4^{2-}), and the extent of reduction appeared to be influenced by the nature of the metal added. The metal ion species were found to exist in several chemical states, and in all cases this included their sulfides, the formation and stability of which was facilitated by the microbial production of hydrogen sulfide gas. Molybdate was also shown to be reduced to Mo^V in a deaerated solution by reaction with exopolymers containing proteins.[242]

Material Surface Conditioning in Connection with Microorganisms and Biofilms

The occurrence of quick surface conditioning when a material is exposed to a natural aqueous medium was pointed out in the section *Quantification Considering the Analyzed Zone as Homogeneous Illustration* and Fig. 18. The presence of a conditioning film (also called primary film) or of a biofilm was also revealed by XPS analysis of SS exposed to natural seawater.[244]

It was shown by XPS that proteins were adsorbed from a suspension of *A. brasilense*, even in absence of contact of the cells with the surface.[224] Adsorption of proteins released by the cells enhanced the density of cells adhering on model substrates (glass, PS). This explained the influence of temperature and of cell aging on cell adhesion. It was also related to the observation of a higher density of cells adhering from a suspension in phosphate buffer saline (PBS) compared to water. Thus the presence of salts influenced bacterial adhesion in a more subtle way than double layer electrostatic interactions between the cells and the support. In the next stage, in situ secretion of proteins during the prolonged contact between the cells and the support strengthened adhesion. A more recent AFM study showed that the presence of macromolecules adsorbed from *A. brasilense* suspensions caused a significant change of substratum properties, as revealed by modifications of force–distance curves.[224, 245–247]

The role of extracellular polymer substances (EPS) in the initial adhesion of *P. aeruginosa* to substrata differing according to hydrophobicity was investigated. XPS indicated that the bacterial footprints consisted of uronic acids, the prevalence of which increased with the number of deposition and detachment cycles.[248]

The release of proteins and nucleic acids by *Lactococcus lactis* was found to be much larger for cells suspended in PBS compared to water, owing to cell lysis. The XPS analysis of PS and glass supports used to test the adhesion of the bacteria showed that organic extracellular substances (proteins and carboxylate-rich compounds) were adsorbed from the aqueous phase. They may have an effect on cell adhesion by bringing macromolecules at the interface.[187]

Adhesion of *Staphylococcus aureus* and *Listeria monocytogenes* to a SS surface increased as the amount of adsorbed milk proteins decreased, as indicated by the N 1s peak. Alpha-lactalbumin exhibited sparse coverage (AFM and XPS analysis) compared to both skim milk and kappa-casein and was less effective at reducing bacterial adhesion.[249]

SS and ceramics were subjected to repeated treatments that involved soiling using a milk powder inoculated with *P. aeruginosa* and *S. aureus* and spraying with water. XPS showed that surfaces were conditioned rapidly to saturation with organic material within one cycle, whereas fouling by microorganisms was less rapid. The microbiological and chemical methods of analysis described provide a way of testing the cleanability of surfaces found in food processing facilities and for screening novel cleaning procedures and/or surface materials.[250]

In a study of yeast adhesion to the surface of champagne bottles, inorganic surfaces (glass, bentonites) were shown by XPS to adsorb wine constituents. A low concentration of nitrogen was observed, indicating the adsorption of proteins. The main increase of carbon concentration concerned the component at 286.1 eV. Considering these results in the light of the present knowledge on XPS data interpretation (cf. the sections on *Chemical Functions* and *Toward Molecular Composition*) indicates that the compounds adsorbed in contact with wine were mainly of polysaccharidic nature.[251]

Material Alteration or Generation Under the Action of Microorganisms

Nanoparticles were synthesized extracellularly by a silver-tolerant yeast in the presence of a silver solution. Several techniques, including XPS, confirmed that the nanoparticles were elemental silver.[252]

Enzyme adsorption by cellulose films was observed by several methods, including XPS, and was followed by enzymic degradation of the films.[253] Silicate glass surfaces were found to be depleted in Fe and Al as a result of exposure to *Bacillus* sp.[254] XPS revealed that reduced sulfur forms in aqueous solutions of coal-derived products were not oxidized by dibenzothiophene-degrading bacteria, supporting the indication that extensive degradation of the carbon structure was concurrent with the loss of sulfur.[255] In the context of the extraction of iron from a hornblende by *Arthrobacter* sp., the formation of siderophore–Fe surface complexes was examined by an XPS study of Fe with and without the siderophore on gold surfaces, and of the siderophore on hornblende crystal surfaces.[256]

Bioleaching is an application of mineral alteration under the action of microorganisms. The combination of molecular biology, reflectance microscopy, and XPS demonstrated that reduction of $Fe^{(III)}$ oxides under the influence of *Geobacter sulfurreducens* could be related to the expression of specific genes by individual bacterial cells and cell aggregates associated with the mineral surface.[257] Mineral transformation by *G. pelophilus* was also demonstrated to be mediated by the surface associated population.[258] During bioleaching of manganese dioxide minerals, Mn^{IV} was progressively reduced to Mn^{II} and signals due to microorganisms appeared in the C 1s and N 1s peaks.[259] Uranium complexes seemed to be associated in larger quantity with hematite surfaces colonized by SRB than with bacteria-free surfaces. At least a portion of the U^{VI} that accumulated in the presence of the SRB was exterior to the cells, possibly associated with the extracellular biofilm matrix.[260] Upon bio-oxidation of arsenic-bearing sulfides in gold concentrates, sulfides are progressively oxidized from the surface of minerals to the core, leading to a layered structure with sulfate at the surface, which is found in the grains of the bio-oxidation residue.[261] Additional references on the study of bacteria–minerals interactions may be found in a review.[256]

XPS was also used to get deeper insight into microbiologically influenced corrosion (MIC). The changes produced by exposure of 304 and 317 L austenitic SS to sulfate-reducing bacteria were examined.[242,263,264] After exposure the samples were transferred under argon into the spectrometer. To prevent sulfur loss from the samples and instrument contamination, the samples was cooled below $-30\,°C$. The analysis revealed significant sulfidation of Fe, Cr, and Ni, with sulfides being present throughout the anodic film. It was demonstrated that a combination of XPS and electrochemical analysis can be used to develop an accelerated test for alloy susceptibility to microbiological corrosion using a suitable culture. A similar examination performed on molybdenum metal revealed than the surface stability of molybdenum was compromised owing to MoS_2 formation during exposure to SRB. Moreover, molybdate formation induced by bacterial H_2S production was inhibitory to sulfate reduction by forming protective Mo^V—S complexes with intermediate S-containing amino groups and proteins.[265]

XPS results verified the formation of biofilm containing extracellular polymers on 304 SS exposed to *Burkholderia* sp. Changes in the relative Fe concentration and Fe 2p peak shape indicated also that iron had accumulated in the film.[266] Time-of-flight secondary ion mass spectroscopy (ToF-SIMS) and XPS were employed to investigate the interactions between EPS from SRB of the genus *Desulfuvibrio*, and Fe ions released from steel.[267]

MnO_2 was identified by XPS in biofilms of *Leptothrix discophora* grown on 316 L SS and ennobling the open circuit potential. A mixture of different manganese minerals was revealed by XPS on field-exposed samples, and it was shown that electrochemical reduction of MnO_2 to Mn^{II} proceeded through MnOOH.[268] XPS analysis also contributed to the examination of bacteria-induced corrosion of ancient bronze mirrors found in ground.[269]

A review on biocorrosion and biofouling of metals and alloys of industrial usage was published recently.[270] Perspectives consider the potential of innovative approaches in microscopy and electrochemistry and in spectroscopic techniques used for the study of corrosion products and biofilms.[270]

XPS Study of Systems Related to Biomedical Applications

Introduction

Interfacial phenomena play an essential role in many biomedical applications. The reaction of the body towards implants largely depends on the surface properties of the latter.[271] The corrosion or the degradation of materials placed in contact with biological fluids is initiated at the material–fluid interface. The successful design of biosensors or of supports for cell culture relies on the appropriate modification of a material surface and on the interaction of that modified surface with macromolecules in solution or with cells. An overview of key constituents and processes that are

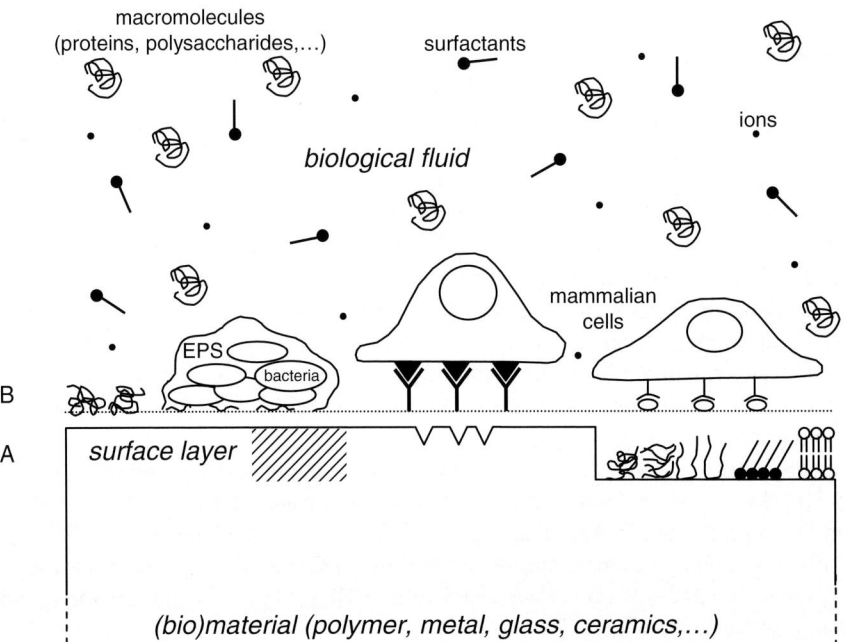

Fig. 42 Schematic representation of interfaces of interest in biomedical applications: bulk of the material, surface layer of the material (possibly modified), physically adsorbed or grafted compounds, cells, biological fluid. The separation between zones A and B illustrates the fact that any of the modified material surfaces depicted in zone A (chemical or topographical alteration, model surface obtained by self-assembly or LB deposition, immobilized macromolecules) may be in contact with any of the immobilized compounds (macromolecules) or cells (mammalian cells, bacteria) depicted in zone B

encountered at interfaces typical of biomedical applications is presented in Fig. 42. Given its surface sensitivity, XPS is the most important tool to study the chemical composition of these interfaces; it may also contribute to a better understanding of their spatial organization.

The aim of this section is to illustrate the use of XPS for biomedical applications. While we attempted to cover the diversity of such uses in a structured manner and to highlight the capabilities and the limitations of the method, we could of course not provide an exhaustive review. We focus first on the examination of the surface of materials that are developed in view of biomedical applications. Secondly, the use of XPS to investigate the adsorption of macromolecules, especially of proteins, at the material surface is detailed. A third part is centered on the study of the modification of material surfaces following contact with biological fluids or cells in culture (in vitro) or following explantation. In the fourth part, the direct analysis of the surface of mammalian cells is discussed. Note that applications of XPS related to microorganisms are dealt with in the section *Surface Analysis of Microorganisms and Related Systems*. Finally, a last part is dedicated to the study of model systems that are used to increase our understanding of interfacial phenomena in biological systems.

Materials Designed for Biomedical Purpose

Native Materials

Metals and polymers are the two most important classes of materials used as biomaterials. In many papers, XPS is used to check or to evaluate the chemical composition of the surface layer of these materials, with the aim to correlate this chemical composition with properties such as resistance to or easiness of degradation, inhibition or promotion of protein adsorption, increase or decrease of cell adhesion. The properties that are searched for largely depend on the application that is envisioned.

Titanium implants are widely used for bone reconstruction. Given the important chemical shift occurring when titanium is oxidized as TiO_2 (E_b Ti $2p_{3/2}$ = 458.7 eV) compared to the metallic form Ti^0 (E_b Ti $2p_{3/2}$ = 453.9 eV),[26] XPS is a suitable tool to evaluate the thickness of the TiO_2 layer covering metallic implants. In a comparative study of screw-shaped titanium implants, XPS showed, on the one hand, that the surface is made of TiO_2, and, on the other hand, that variable amounts of carbon and nitrogen-containing contaminants are present.[272] Further implantation for 12 weeks monitored by histological evaluation of bone formation did however not reveal significant differences resulting from the slightly different chemical compositions of implant surfaces. Actually, for improving such titanium implants, the efforts are usually focused on altering the surface topography in order to increase the surface area. However, an extensive study of 34 different commercially available dental titanium implants showed a clear correlation between surface composition, assessed by XPS, and surface topography.[273] The samples were divided into four groups according to their surface finish: machined, sandblasted, acid-etched, or plasma-sprayed. The carbon content was decreased and the titanium content was increased on acid-etched and plasma-sprayed surfaces compared to machined surfaces (Fig. 43). Clearly, evaluation of the influence of surface chemistry must accompany studies of the influence of surface topography on bone healing. In particular, attention must be paid to the influence of material treatment on the

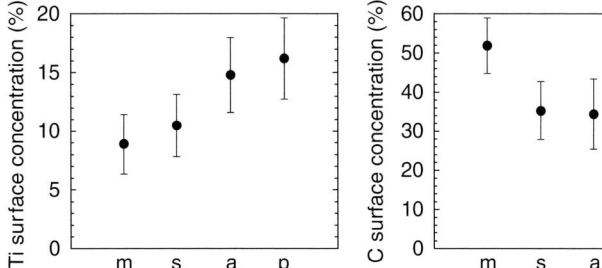

Fig. 43 Mean surface concentration of titanium (*left*) and carbon (*right*) determined by XPS on four groups of dental implants: m, machined; s, sandblasted; a, acid etched; p, plasma sprayed. The error bars show the 95% confidence intervals. Redrawn from Ref. 273 with permission from Quintessence Publishing Co Inc.

thickness of the titanium oxide layer and of surface contamination by organic compounds (cf. the sections on *Influence of Adsorbed Contaminants* and *Data Interpretation Through Simulation*).

Biodegradable polymers are used both for surgery and for drug-delivery applications. Their surface composition is critical to ensure biocompatibility and to control the degradation. XPS was used to compare the surface composition to the bulk composition of a variety of polyesters, including poly(lactic acid) (PLA) and poly(hydroxy butyrate).[274] After subtraction of a low level of hydrocarbon contamination from the C 1s peak, the experimental O/C concentrations ratios agreed well with the ratios expected from the stoichiometry of these polymers.

A strategy that is often used to obtain materials with both appropriate mechanical properties and controlled surface characteristics is to combine materials. Block, statistic, or grafted copolymers, but also polymer blends, are hence frequently found in biomedical applications. Polyetherurethanes (PEU) are block copolymers containing hard and soft segments, made of polyurethane and polyether, respectively. Because of their appropriate mechanical properties and their acceptable hemocompatibility, they are used for the design of blood-contacting devices such as catheters and vascular prostheses. XPS was used to analyze the surface composition of PEU, and the $\underline{C}$—(C,H)/$\underline{C}$—O concentration ratio measured on a variety of PEU was directly correlated to the decrease of platelet retention measured using an arterio–venous shunt inserted in a baboon (Fig. 44).[275] In contrast, the C—(C,H)/C—O ratio of different PEU containing polyethers was found to correlate directly with a platelet retention index determined using a column packed with PEU-coated beads.[276] This raised the question whether the surface layer probed by XPS was homogeneous or whether enrichment of one of the segments could occur at the extreme surface. Therefore, ARXPS studies were performed,[275,277] and the effect of the ratio of soft to hard segments was also examined.[278] In some cases, a high angle dependence was observed, pointing at surface enrichment with soft segments. In other cases, the absence of angle-dependence suggested that a mixture of hard and soft segments was present at the outermost surface. It should be kept in mind that the surface enrichments observed with XPS under vacuum may be nonrepresentative of the surface state in aqueous solutions (cf. the section on *Surface Reorganization and Low Temperature Analysis*).

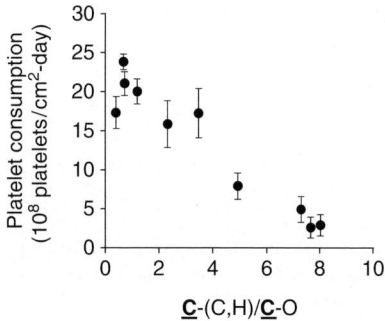

Fig. 44 Platelet consumption by polyurethanes, measured using an arterio–venous shunt inserted in a baboon, as a function of the concentration ratio of carbon in a hydrocarbon-type environment [$\underline{C}$—(C,H)] to carbon in an ether-like environment [$\underline{C}$—O] measured by XPS on these polyurethanes. Redrawn from Ref. 275 with permission from Elsevier

Poly(lactic acid-*co*-lysine) copolymers were synthesized to benefit from the biodegradability of PLA while using the amino-groups of lysine as grafting sites for RGD peptides, which are known to be recognized by mammalian cell receptors. XPS was used to determine the lysine and the peptide densities at the surface. This was made possible by the incorporation of diiodotyrosine into the peptide. The RGD-containing polymer was later found to promote the spreading of bovine aortic endothelial cells.[279]

PS/poly(2-hydroxyethyl methacrylate) (PS/PHEMA) composite microspheres were prepared with the aim to selectively adsorb fibrinogen compared to albumin. Selective adsorption of a given protein may lead to advances in biosensors design and may help modulating the biocompatibility of a material. XPS showed that the surface of PS/PHEMA microspheres was strongly enriched with PHEMA. The XPS results, combined with adsorption studies, allowed the optimum surface composition leading to selective fibrinogen adsorption to be found at about 20% of PHEMA.[280] In another study, a material was developed to selectively adsorb albumin compared to fibrinogen.[281] Therefore, poly(vinyl chloride)-graft-(ω-stearyl-polyethylene oxide) was synthesized. XPS showed that the poly(ethylene oxide) (PEO) concentration was much higher at the surface compared to the bulk.

PEO chains were also grafted to poly(methyl methacrylate) (PMMA) in order to improve the hemocompatibility of the material. XPS, used in the angle-dependent mode, showed that PMMA tends to be enriched at the surface, except when the PEO side-chains are long and the PEO content is high. The hemocompatibility of these materials, assessed using the recalcification time of platelet-rich plasma, was shown to increase with the PEO concentration at the surface.[282]

Compounds containing phosphatidylcholine (PC) moieties are an alternative to PEO in view of obtaining surfaces with protein-resistant properties and improved hemocompatibility. Copolymers of *n*-butyl methacrylate and of different methacrylates containing a PC group were prepared. XPS was applied, in the angle-resolved mode, to both vacuum-dried and freeze-dried copolymers. It was shown that the freeze-dried polymers were always enriched with PC at the surface. In some cases, this was also true for the vacuum-dried samples. The presence of PC at the surface was related to a good hemocompatibility. The materials characterized by a PC enrichment even after vacuum-drying did not need pre-hydration to show this good hemocompatibility, which may be advantageous for some applications.[283] Similarly, polysulfone was blended with a PC-containing polymer to improve the properties of hollow fibers used for blood purification. Again, XPS showed that the surface was enriched with PC units. This resulted in less protein adsorption and less platelet adhesion on the membrane.[284]

Collagen, a structural protein, and chitosan, an aminopolysaccharide derived from chitin, have been cross-linked together with the aim to produce an artificial matrix for liver cells. The XPS spectrum of the collagen/chitosan matrix was a linear combination of those of pure collagen and pure chitosan, suggesting that chitosan was successfully cross-linked with collagen.[285]

Surface-Modified Materials

The materials that show the adequate mechanical properties and that can easily be processed in order to serve for biomedical purpose often do not have the needed surface properties. Surface modification is thus a strategy that is commonly followed to develop materials presenting enhanced properties in terms of interaction with biological media, while keeping the bulk properties intact. Apart from modifications of the topography, which were mentioned in the preceding section, the surface of polymers and of metals may be modified by a chemical reaction involving the material itself or by the retention of (macro)molecules, using adsorption, grafting, or coating. The purpose of such modifications is usually either to allow the control of the behavior of specific cell types, or to prevent protein adsorption as much as possible in order to decrease bacterial and/or mammalian cells adhesion (so-called anti-fouling or protein-resistant surfaces).

XPS is often used to detect the changes of surface composition brought by the treatment. In many cases, each step of the modification procedure is monitored by XPS. The useful information may be limited to the appearance or the disappearance of given elements from the survey spectra. More often, careful peak decomposition allows the quantification of new chemical functions brought at the surface. In some cases, the molecules used for the surface treatment bear a tag (e.g. fluorine atoms), which facilitates their detection. In some advanced studies, the results are modeled, taking into account the formation of successive layers with different properties.[286, 287] The two following examples illustrate the typical approach that is used when analyzing surface-modified materials by XPS.

Merret et al.[288] studied the immobilization of the transforming growth factor beta-2 (TGF-β2) via a bifunctional poly(ethylene glycol) (PEG) spacer onto poly(dimethyl siloxane) (PDMS). The objective was to improve the control of the growth of corneal epithelial cells, in view of developing an artificial cornea. The surface chemical composition of the unmodified PDMS was close to that expected. It was then first aminated using plasma polymerization of allyl amine. This resulted in a significant decrease of the Si 2p signal, in the appearance of N 1s signal, and in the modification of the shape of the C 1s peak. The PEG layer grafted on the aminated PDMS was then analyzed: a higher oxygen content and a lower carbon content were recorded, in accordance with the presence of PEG. Adsorption of TGF-β2 on a nonreactive PEG layer did not lead to the expected increase of the nitrogen signal. Since other data obtained using radiolabeling showed that TGF-β2 could adsorb on that material, the most probable explanation is that TGF-β2 molecules were embedded in the PEG layer. On the other hand, modification of the surface with TGF-β2 through binding to a reactive PEG layer resulted in an increase of the nitrogen signal and in a modification of the shape of the C 1s peak. The comparison of the results obtained with a nonreactive PEG and those obtained with a reactive PEG suggested that a significant amount of the growth factor was covalently coupled to the PEG in the second case, although physical adsorption (as in the first case) certainly also played a role. This study illustrates the use of XPS to follow each step of a surface modification procedure. It also emphasizes the difficulties encountered when covalent grafting must be distinguished from physical adsorption. The use of complementary analytical techniques is generally required to solve this kind of issue.

Kingshott et al.[289] have examined the influence of the attachment strategy of PEG on the reduction of bacterial adhesion: a layer of branched poly(ethylene imine) (PEI) was first immobilized on SS or on carboxylated poly(ethylene terephthalate) (PET—COOH); PEG was then grafted on that PEI layer. The PET—COOH substrate was prepared from native PET, using an intermediate reaction producing PET-OH. The C 1s spectra of PET, PET—OH, and PET—COOH were almost the same. Therefore, derivatization of PET—OH and PET—COOH using fluorine-containing molecules was performed, allowing the presence of the —OH and the —COOH reactive groups on the PET to be assessed (chemical reactions in Fig. 45). It should be noted that the study included a series of controls in order to take into account the physical adsorption of the molecules used as tags (table in Fig. 45). PEI was, on the one hand, adsorbed on SS, and, on the other hand, grafted to PET—COOH. This resulted in a significant increase of the nitrogen signal. The thickness of the PEI layer was calculated, from the XPS data, to be about 0.8 nm. Finally, grafting of PEG led to an O/C concentration ratio close to that of pure PEG, although weak signals from the PEI layer and the substrate were still present. The presence of a majority of ether functions in the C 1s peak further confirmed the high degree of PEG immobilization. The surfaces were used for protein adsorption and bacterial adhesion experiments. The PEG layer grafted on PEI-grafted PET significantly reduced bacterial adhesion, while the PEG layer grafted on PEI physisorbed on SS did not. Protein adsorption could not be totally prevented, even in the best cases, which may account for the residual level of bacterial adhesion that was still observed with the best surface.

The two studies cited here above illustrate the fact that demonstrating grafting by XPS requires the use of appropriate control samples. It is often challenging to distinguish physically adsorbed molecules from chemically grafted ones. Grafting can only be evidenced if the difference between the signal obtained after grafting and the signal measured on the control sample is large compared to the variability of the data collected on controls. If this is not the case, the grafted amount is below the limit of quantification.

An overview of surface modifications by retention of (macro)molecules, performed in view of biomedical applications, which have been investigated by XPS is presented in Table 7. Peptides or proteins are mainly used to provide recognition signals for cells that will be brought in contact with the material surface, thereby controlling the adhesion, the proliferation and even the differentiation of the cells. Whole proteins are, in many studies, immobilized by physical adsorption; this will be dealt with in the next section. In some cases, however, attempts are made to covalently bind proteins to the substrate.[293, 294] Sugars are important molecules in the frame of the hemocompatibilization of materials, as demonstrated by the intensively used heparin. Some cell types, including hepatocytes, are also able to recognize sugars via specific surface receptors. Surface modifications using other polymers often aim at reducing protein adsorption. This is mainly performed with PEG. On the other hand, the immobilized polymer may provide reactive sites for further surface modification, as it is the case with poly(acrylic acid) (PAA).

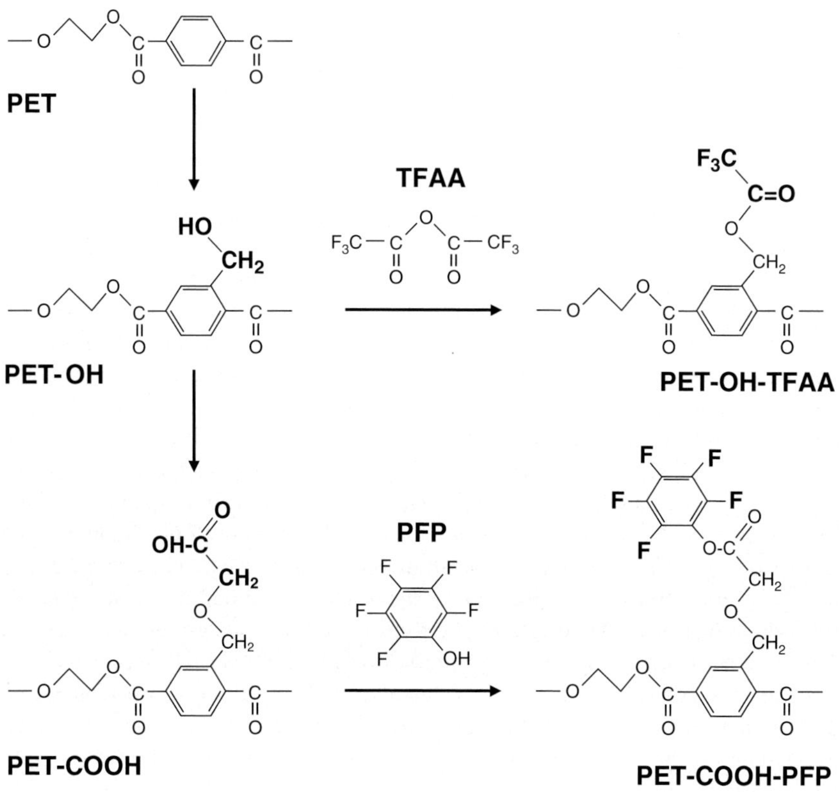

sample	% C	% O	% F
PET (expected)	71.4	28.6	bdl
PET	70.8	29.2	bdl
PET-OH	69.5	30.5	bdl
PET+TFAAads	72.7	26.8	0.5
PET-OH-TFAAgraft	67.7	26.2	6.0
PET-COOH	68.8	31.3	bdl
PET+PFPads	68.6	29.5	1.8
PET-OH+PFPads	68.9	30.1	1.0
PET-COOH-PFPgraft	65.8	28.1	6.1

bdl: below detection limit

Fig. 45 Scheme of the modification of PET into PET—COOH through a PET—OH intermediate, and of the derivatization of PET—OH using trifluoroacetic anhydride (TFAA) and of PET—COOH using pentafluorophenol (PFP). Table of the XPS elemental composition of these materials as well as of control samples obtained by adsorption of TFAA or PFP. The suffix graft designates the materials derivatized with TFAA or PFP, while ads designates the materials on which TFAA or PFP were adsorbed. Adapted from Ref. 289 with permission from Langmuir, copyright (2003) American Chemical Society

Table 7 Examples of surface modifications by immobilization of (macro)molecules investigated by XPS

Immobilized (macro) molecule[a]	Substrate[a]	Field of application	Reference
Peptides or proteins			
GRGD	Chitosan	Promoting endothelialization	290
GRGD	PU-PEG	Promoting endothelialization	291
Peptidomimetic of RGD	PET	Promoting cell adhesion	292
TGF-β2	PDMS	Controlling the growth of corneal epithelial cells (artificial cornea)	288
Fibronectin	PVDF	Osseointegration	293
Collagen	Titanium	Osseointegration	294
Oligo- or polysaccharides			
Galactose, lactose	PS	Culture of hepatocytes	295
Heparin	Polypyrrole, SS	Enhance hemocompatibility	296, 297
Heparin (+RGD)	Poly(carbonate-urea) urethane	Vascular grafts	298
Chitosan	PLA	Controlling the morphology of fibroblasts and hepatocytes	299
Alginic acid, hyaluronic acid	PS	Anti-fouling surfaces	287
Dextran	FEP, PTMSP	Various biomedical applications including contact lenses	300
Dextran + RGD or IKVAV	Silicon, polyimide, gold	Neural implants	301
Dextran sulfate	SS	Enhance hemocompatibility	297
Other oligo- or polymers			
Tetraglyme	FEP	Anti-fouling surfaces	302
PEO	Polyaniline	Anti-fouling surfaces	303
PEO	Polyurethane-urea	Anti-fouling surfaces	304
PLL-PEG	Metal oxides	Anti-fouling surfaces	286
PEI + PEG	SS, PET-COOH	Anti-fouling surfaces	289
PAA	PET, PVDF, Titanium	Allowing further protein immobilization	293, 294, 305
Miscellaneous			
Phosphatidyl choline	Glass, PHEMA, PVA	Enhance hemocompatibility	306
Alkyltrichlorosilanes	Silicon	Biocompatibilisation of silicon microdevices	307

[a] A full list of the abbreviations may be found in Appendix

The surface modification of biomaterials is also frequently achieved using plasmas. The physico-chemical properties of the material surface may be modified using plasma discharge in different gases, or a polymer coating may be deposited using plasma polymerization (cf. the section on *Data Interpretation Through Simulation*). A well-known example of the first approach is tissue-culture polystyrene (TCPS), which is commonly used to culture cells in vitro (Fig. 19). XPS shows that plasma oxidation of PS introduces a whole range of oxygen-containing functions at the PS surface.[171] There is a vast literature related to the surface chemical modifications of polymers submitted to plasma discharges, a frequent aim being to improve adhesion; this will not be reviewed here in detail. The combined use of O_2 and NH_3 to treat the surface of polytetrafluoroethylene (PTFE) by plasma was shown to improve the attachment of endothelial cells compared to plasma discharge in a single gas. This was explained by the simultaneous presence of oxygen and nitrogen-containing groups, as detected by XPS.[308]

Although the plasma modification or the plasma polymer coating of the surface may in itself improve the characteristics of a material in view of biomedical applications, it is also often used to bring reactive sites at the surface of inert materials, thereby allowing further surface modification. This is illustrated by the introduction of amine functions at the surface of poly-1-trimethylsilyl-1-propyne (PTMSP) and of perfluorinated ethylene-propylene copolymer (FEP), in order to immobilize polysaccharides on these unreactive polymers. Amine functions were introduced either by submitting the polymers to a plasma discharge in NH_3, or by depositing a thin polymer film at their surface by plasma polymerization of *n*-heptylamine. The introduction of amine groups was assessed by XPS.[300]

A particularly important challenge for biomaterials is the possibility to sterilize them without affecting their functional properties. XPS is a valuable tool to study the effect of sterilization on the surface properties of the materials. In the case of metals, the fate of the oxide layer, which is related to the biocompatibility, is of interest.[309] For polymers, surface oxidation possibly brought by the sterilization must be examined.[310]

For different reasons, modifying surfaces in a heterogeneous manner may be advantageous. This is for example the case to built multiarray sensors, to orient the growth of cells, to present recognition sites for cells in a geometrically defined manner, or to trigger synergetic effects of different chemical functions or specific molecules. XPS may be useful to check the modification procedures on plain surfaces before performing the patterning method,[311] or to obtain the average composition of the patterned surface. The lateral resolution of XPS in the imaging mode is limited to 3–10 µm (cf. the section on *XPS Imaging*). Images showing domains with different chemical compositions may thus only be obtained if the patterns are above that size. Tracks of oxidized PS drawn by a combination of plasma treatment and photolithography (Fig. 46a,b) were imaged by XPS, using the intensity of the O 1s peak. These tracks could be used to selectively adsorb extracellular matrix proteins (collagen, fibronectin), by taking advantage of the influence of surface hydrophobicity on adsorption in competition with a polymer surfactant (Pluronic F68), and to confine the adhesion of different mammalian cell types (Fig. 46c,d).[312] In another study, XPS analysis along a line crossing 100 µm dots of Nb in a matrix of Al showed the alternation of the chemically distinct domains, by means of the O 1s, C 1s, Nb 3d, and Al 2p intensities as a function of the distance (Fig. 47).[313] These patterned metal surfaces were further used to study the behavior of osteoblast cells.

XPS Analysis of Biosystems and Biomaterials 277

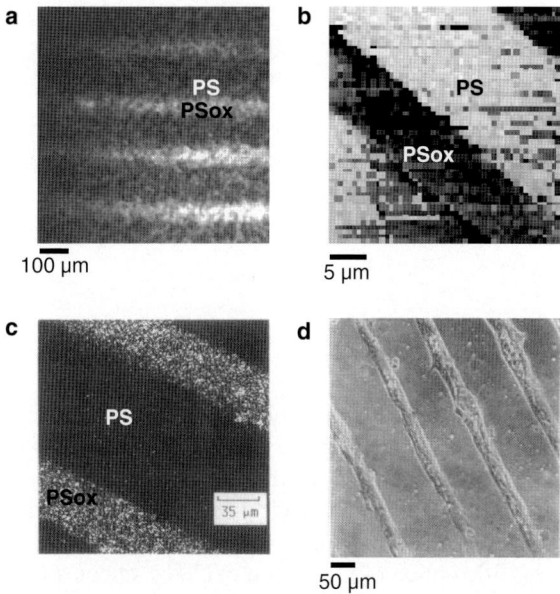

Fig. 46 Data obtained on PS/PSox patterned surfaces produced by photolithography and oxygen plasma oxidation of polystyrene: (**a**) XPS mapping of the O 1s peak intensity (obtained with a Kratos Axis Ultra spectrometer); (**b**) adhesion map obtained by AFM in water with a silicon probe (vertical scale = 25 nN), revealing the hydrophobicity contrast in the pattern; (**c**) ToF-SIMS image recorded with the signal of CNO^- ions on a patterned surface conditioned with a solution of fibronectin and Pluronic F68, revealing the selective adsorption of the extracellular matrix protein on the oxidized tracks; and (**d**) micrograph of rat hepatocytes on a patterned substrate conditioned with a solution of type I collagen and Pluronic F68, showing the selective adhesion of the cells on the oxidized tracks. Adapted from Refs. 25 and 312

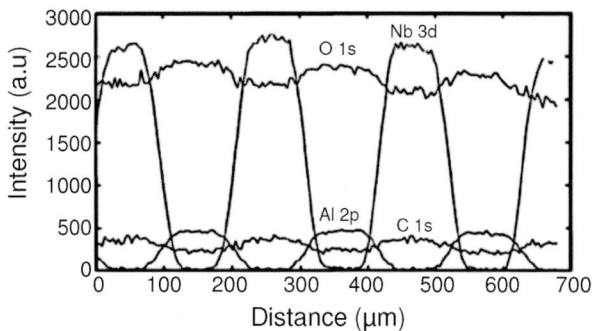

Fig. 47 Line scans of the XPS intensities of O 1s, C 1s, Al 2p, and Nb 3d along a line crossing 100 μm dots on a Nb/Al patterned surface. The results were obtained on a PHI Quantum 2000 spectrometer. From Ref. 313 with permission from Biomaterials, copyright (2002) Elsevier Science Ltd.

Adsorbed Biological Molecules

When a material is placed in contact with a biological fluid, adsorption of biomolecules immediately occurs. Further interactions with the biological environment, including living cells, are determined by the amount, the nature, the conformation, and the orientation of the biomolecules within the adsorbed phase. Characterizing such adsorbed layers is thus a key step in the development of materials for biomedical applications.

Adsorption of Model Compounds

Certain XPS studies centered on amino acids or homopeptides have already been presented in the section *Peak Components of Biochemical Compounds*. Information regarding the adsorption phenomenon and the resulting organization of the adsorbed layer was collected for different amino acids on a variety of substrates.[314–318] Such studies provided a better understanding of protein adsorption while looking at less complex molecules. Adsorption of glycine on oxidized silicon was shown to result in the orientation of the adsorbed molecules, with COO^- groups pointing away from the surface while NH_3^+ groups were oriented towards the surface.[314] Adsorption of cysteine on titanium was examined as a function of pH.[318] Two maxima of thickness of the adsorbed layer were observed, around pH 3.5 and 10. This was explained on the basis of complex formation between amphoteric species.

Adsorption from Single Protein Solutions

Many studies are dealing with the adsorption of one protein from a defined solution (buffer). Although this is far from the complexity of biological fluids, it allows a better understanding of the adsorption phenomenon to be achieved. In 1985, Paynter and Ratner[319] have reviewed the use of XPS for the study of adsorbed proteins. They presented spectra for thick films of a variety of proteins. Since the elemental and functional compositions of different proteins are quite similar, it is usually impossible to use XPS to discriminate one protein from another. However, due to the presence of nitrogen and to the involvement of carbon in C—(O,N) and O=C—N bonds, adsorbed proteins can be easily detected on most substrates.

In a first approach, XPS may be used to evaluate the thickness of the adsorbed layer, assuming a constant thickness and a complete surface coverage. This can be achieved by measuring the ratio of nitrogen to an element originating from the substrate, and then taking into account the decay of the photoelectron intensity with depth. A more reliable determination may be obtained by acquiring XPS data at different angles of photoelectron collection (ARXPS; cf. the sections on *Semi-Quantitative Use of Angle Resolved Analysis* and *Data Interpretation Through Simulation*) or using different X-ray sources (energy-resolved XPS; based on the modification of λ occurring when the energy of the X-ray source is changed). This was done for different proteins adsorbed on mica.[320] Angle-resolved profiling was shown to be more accurate than energy-resolved profiling. The obtained thicknesses were compatible with the adsorption of a monolayer of protein molecules.

It is however quite rough to consider that adsorbed layers have a constant thickness and cover the substrate in a homogeneous manner. In many studies, the possibility that the adsorbed protein forms patches or islands at the surface is taken into account.[167,319,321–323] In that way, XPS becomes a tool to investigate the organization of the adsorbed layer. From a mathematical point of view, this could be achieved by recording spectra at different angles θ of photoelectron collection and then computing the thickness of the adsorbed layer (t) and the degree of surface coverage (φ) from a set of two (or more) equations with two unknowns (cf. the section on *Tentative Consideration of the Heterogeneity of the Analyzed Zone*). However, owing to uncertainties related to both the experimental data and the model, this is unpractical. This is why XPS data are usually coupled to information brought by another technique (adsorbed amount determined by radiolabeling, thickness of the layer or surface coverage estimated by AFM, ellipsometry, surface plasmon resonance, or quartz crystal microbalance) when the organization of the adsorbed layer must be determined.

In a coupled AFM/XPS study of patterned collagen layers obtained by adsorption and dewetting on a PS substrate, AFM provided information on the thickness of the collagen pattern ($t = 7.5$ nm), which showed a net-like structure. This allowed a surface coverage of 13% to be computed from the N/C ratio measured by XPS. This value could be compared with the surface coverage observed on the AFM images, which was of the order of 28% but was overestimated due to the broadening effect of the AFM probe.[45]

Figure 28 (cf. the section on *Materials with Adsorbed Layers*) was presented to illustrate the influence of the drying rate on the organization of adsorbed collagen layers obtained on PS by incubation for different times in a solution of low concentration ($7 \mu g$ mL^{-1}). Slow drying resulted in the formation of discontinuous layers at short adsorption times, as evidenced by AFM. The N/C concentration ratio measured by XPS and the adsorbed amount were modeled as described in the section *Data Interpretation Through Simulation* (37, 39, 40). Figure 48 shows, on the one hand (a), a chart of the computed variations of the adsorbed amount as a function of N/C concentration ratio for fixed t or φ values, and on the other hand (b), experimental data, the adsorbed amount being measured by radiolabeling.[167]

From the comparison of Figs. 48 a and b, it can be concluded that for the fast-dried samples, the progressive increase of N/C and of the adsorbed amount with the adsorption duration is mainly due to a progressive increase of the surface coverage. Adsorption does not proceed through the accumulation of layers (t increase) of closely packed molecules ($\varphi = 1$). The thickness of about 8 nm must be taken with caution owing to the meaning of the model, on the one hand, and to inherent simplifications, on the other hand. It is not the physical thickness of the layer but a cumulated thickness of the collagen molecules forming the layer. It is an apparent value while the reality must be a distribution of different thicknesses. However, it may be compared with the dimensions of the collagen molecule (length

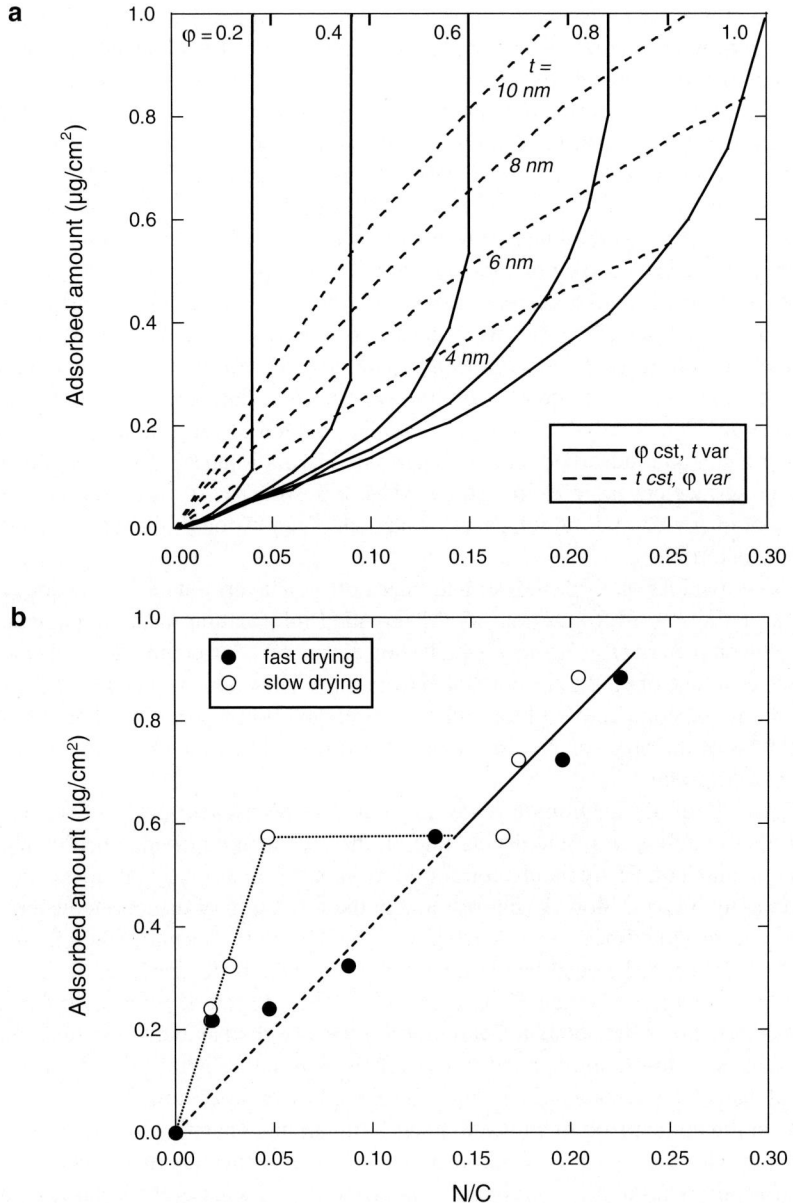

Fig. 48 Variation of the amount of collagen adsorbed on PS as a function of the N/C ratio given by XPS: (**a**) Chart of the computed variations at the indicated constant surface coverage φ (*plain lines*) and at the indicated constant thickness t (broken lines); and (**b**) experimental data for collagen adsorption from a $7\,\mu\text{g}\,\text{mL}^{-1}$ solution for different periods of time followed by rinsing and fast (*closed circle*) or slow (*open circle*) drying. Mean values are presented for the sake of clarity, except for the results obtained after 2 h of adsorption, which showed a low reproducibility. The broken and the dotted lines in (**b**) highlight the different trends found at low adsorbed amount for fast and slow drying, respectively; the plain line shows that a similar behavior is then observed at high adsorbed amount. Adapted from Ref. 167

~300 nm, diameter = 1.4 nm). Again this indicates that the fast-dried adsorbed layer may be regarded not as made of closely-packed molecules but as constituted by the accumulation of aggregated molecules. For the slow-dried samples, the surface coverage remained lower than 25% at short adsorption times, in line with the observation of discontinuous structures on the AFM images. The difference of collagen layer organization between fast- and slow-drying was reflected in the water contact angle (Fig. 28).

A similar approach was followed by Lhoest et al.,[322] who concluded, for the adsorption of fibronectin on PS and plasma-oxidized PS (PSox), that the thickness of the adsorbed layer increased with the concentration of the protein solution, while the surface coverage remained about constant (about 70% on PS and 80% on PSox).

The fraction of the surface covered by haemoglobin or fibronectin adsorbed on different fluoropolymers was determined using angle-dependent XPS analysis coupled with radiolabeling.[323] In that case, the data obtained at different angles of photoelectron collection were used to strengthen the set of results, the determined surface coverage being the mean of values obtained at different angles. The degree of surface coverage was highly dependent on the protein nature, the substrate chemical composition, and the protein concentration in solution.

XPS data on adsorbed proteins may also complement data obtained using radioassays owing to the fact that XPS only detects those proteins that are adsorbed at the external surface of a material, while radioassays also quantify molecules that would have diffused into the material. The adsorption and the penetration of lysozyme and albumin onto and into carboxymethylated PHEMA hydrogels were examined that way. The comparison of the amounts of protein detected by XPS and radiolabeling showed that lysozyme molecules mainly penetrate into the hydrogels, while a substantial fraction of albumin molecules remains located at the surface.[324]

While XPS may help to characterize the spatial distribution of adsorbed proteins, in most cases, it is not the appropriate tool to examine the conformation or the orientation of adsorbed proteins, owing to (1) the explored depth (2–10 nm), which is too large compared to the scale of interest, (2) the occurrence of the same elements, involved in similar bonds, in most of the amino acids, (3) the poor lateral resolution, and (4) the alterations that may occur upon dehydration. However, this can be made possible by the presence or the addition of elements differing from C, O, and N and acting as markers. Margalit and Vasquez[325] reported a study on the orientation of myoglobin on metal surfaces. Myoglobin, which contains a Fe atom, was derivatized with pentaamineruthenium, which was attached to a specific histidine. The distance between the Fe and the Ru atoms was known, and was comparable to the photoelectron attenuation length. The relative intensities of Fe and Ru as determined by XPS were thus sensitive to the orientation of the protein. On In—Sn oxide surfaces as well as on aluminum, the results indicated a preferential orientation of the Ru—Fe axis, with the Ru furthest from the substrate.

Some biomaterials are designed to reduce protein adsorption (cf. the section on *Surface-Modified Materials*). XPS may be used to check whether this goal is reached. In 1986 already, Gölander et al.[326] showed that the modification of PVC or sulfated PE with PEO was effective to reduce albumin adsorption. The nitrogen content detected on the untreated surfaces after albumin adsorption was of the order of 7%, while it was 0.8% on the best PEO-modified surface. Since the ultimate aim to be reached is the design of surfaces that would completely prevent protein adsorption, it is interesting to evaluate the detection limit of XPS for adsorbed proteins. This was done recently by Wagner et al..[327] Therefore, they compared the XPS signal collected for fibrinogen adsorbed on mica, PTFE, allyl amine, or heptyl amine plasma-deposited polymers with the absolute adsorbed amount measured using radioassays. Not surprisingly, they concluded that the detection limit depends on the chemical composition of the substrate. While the detection limit was estimated to be 10 and 10–25 ng cm^{-2} on mica and PTFE, respectively, it was much higher (up to 200 ng cm^{-2}) on both nitrogen-containing polymers. The organization of the adsorbed layer (continuous layer vs patches) also affects the sensitivity of XPS. On very effective protein-resistant surfaces, XPS was shown to fail detecting adsorbed lysozyme or fibronectin, while time-of-flight secondary ion mass spectroscopy (ToF-SIMS) allowed the detection of minute amounts of these proteins.[328] These surfaces were thus not entirely protein-resistant. This highlights the limitations of XPS when low levels of proteins must be detected.

Adsorption from More Complex Solutions

The usefulness of XPS to study layers obtained by adsorption of mixed protein solutions is limited since different proteins in the adsorbed layers can usually not be distinguished. However, XPS may still bring some information regarding the organization (thickness, surface coverage) of these mixed protein layers. This can be coupled to the information brought by other techniques in view of obtaining a detailed description of the adsorbed layer.

Wagner et al. reported the study of the adsorption from binary and ternary mixtures of albumin, fibrinogen and immunoglobulin G on mica by XPS, radiolabeling, and ToF-SIMS.[329] The measured N/Al concentration ratios were compared to N/Al ratios computed on the basis of the mass fraction of each of the protein in the layer, obtained by radiolabeling, and of the N/Al ratio measured on single protein adsorbed layers. This is based on the assumption that the thickness of a mixed protein film is the average of the thicknesses of single protein films of each component, and that the adsorbed layers are continuous. Agreement between the measured and expected N/Al values should occur if both components maintain the same conformation and orientation in the mixed layers compared to single protein layers. A good agreement was found in the case of fibrinogen and albumin adsorption, indicating that fibrinogen was adsorbed in the same orientation whatever the fibrinogen/albumin mass ratio in the solution used for the adsorption. The high N/Al ratio and the high fibrinogen adsorbed amount suggested that fibrinogen molecules were adsorbed in end-on orientation. This paper emphasized on the need of a multitechnique approach to understand the complexity of mixed protein films. The same authors also reported a study of protein adsorption from plasma or serum,[330] using the same techniques.

The limitations of XPS regarding the investigation of multicomponent films were highlighted.

The adsorption of proteins together with other macromolecules, possibly of biological origin, may also be relevant and has been the object of XPS studies. Bartzoka et al. reported the study of albumin and silicone films on glass substrates.[331] Their aim was to understand better the interactions between protein and silicone, the latter being extensively used for medical applications. They prepared samples either by first immobilizing albumin on glass by covalent coupling and then adsorbing silicones, or by first coating glass with silicone, then proceeding to albumin adsorption. ARXPS examination of these silicone-on-protein and protein-on-silicone samples showed similar results in terms of Si and N content, although the ultimate layer was supposedly different. This is explained by the diffusion of silicone molecules, resulting in the formation of a mixed protein/silicone layer rather than the formation of two layers separated by a defined boundary.

The adsorption of collagen in competition with Pluronic F68, a nonionic polymer surfactant (triblock PEO—PPO—PEO copolymer), on PS and TCPS was examined by XPS.[332] The C 1s peaks recorded on the untreated substrates as well as on the substrates treated with Pluronic and/or collagen are presented in Fig. 49. On both substrates, adsorption of Pluronic leads to the expected appearance of an ether contribution ($\underline{C}$—O), while adsorption of collagen leads to a C 1s peak typical for proteins, with $\underline{C}$—(O,N) and O=$\underline{C}$—N components. Adsorption from a mixed collagen/Pluronic solution is different on PS, where a peak similar to that observed for single Pluronic adsorption is found, compared to TCPS, where the typical components of collagen are present, with however a reduced intensity compared to adsorption from a single collagen solution. In presence of Pluronic, the adsorption of collagen is thus inhibited on the hydrophobic PS, while collagen adsorption remains high on TCPS. Similar results were obtained for the adsorption of fibronectin in presence of Pluronic.[333] These studies were at the origin of the selective adsorption of extracellular matrix proteins on PSox tracks drawn in a PS matrix (cf. the section on *Surface-Modified Materials* and Fig. 46), thereby allowing selective cell adhesion.

Modifications Induced by Biological Exposure

Materials After Exposure In Vitro

Owing to the difficulties inherent to in vivo studies, the surface modifications of materials placed in biological environments are often examined in vitro. Besides the adsorption of biomacromolecules, which inevitably occurs when materials are placed in contact with biological solutions, some authors have examined the effect of such solutions and of cells on the surface chemistry of the material itself. This approach is made rather difficult owing to the necessity to separate the effect of adsorbed biomolecules from that of the substrate alteration. The protocols used

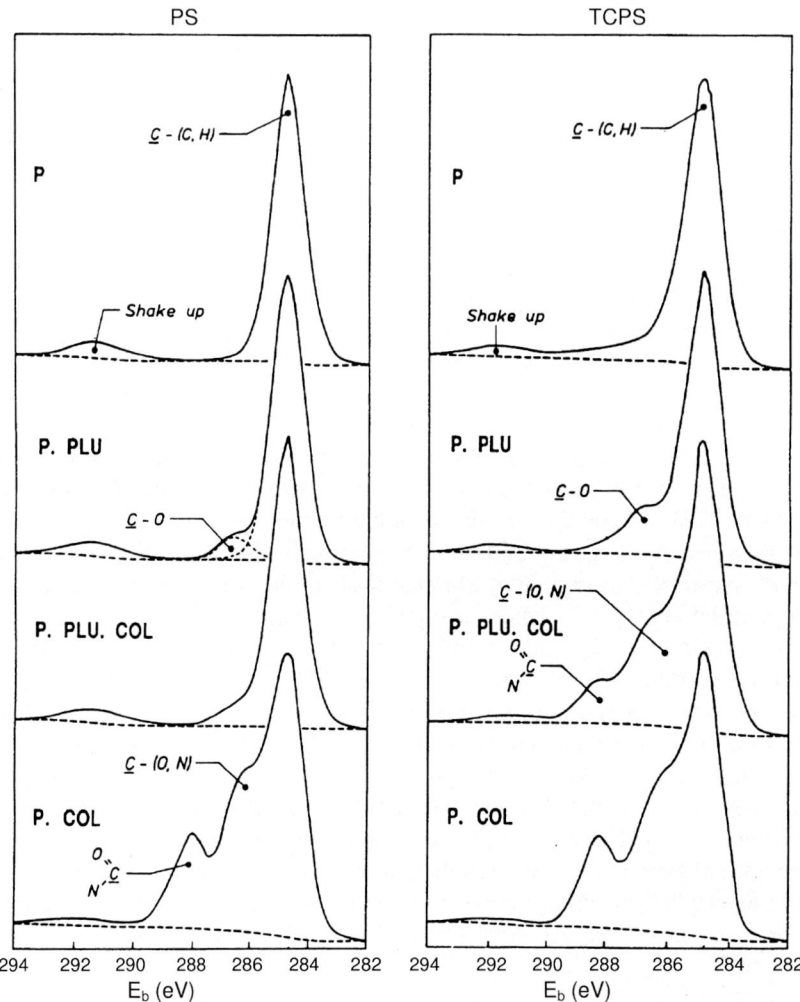

Fig. 49 C 1s peaks of PS and TCPS conditioned with phosphate buffer (reference samples; P), Pluronic F68 (P.PLU), a mixed solution of Pluronic F68 and collagen (P.PLU.COL), or collagen (P.COL). From Ref. 332

for removing the adsorbed layer are not always explicit, their efficacy may be sometimes criticized, and they could themselves be at the origin of chemical modification of the material surface.

The biocorrosion of thin copper films under the effect of bacterial culture supernatant was investigated with XPS.[334] This revealed that Cu was oxidized and partially removed from the substrate. A poly(ester urethane) polymer developed for vascular applications was examined by XPS after contact with enzyme solutions and cleaning. A significant decrease of the carbonate group content was recorded, pointing at the removal or the modification of these groups by the enzymes.[335,336] The resistance to corrosion of metal alloys, used for dental applications, when placed in artificial saliva was associated with the formation of a passive layer at the surface and related to the Cr and Mo content of that layer, as evidenced by XPS.[337]

In presence of activated macrophages, an increase of the surface oxides of a Co—Cr—Mo alloy was observed. This was attributed to oxidation and nitration reactions induced by reactive chemical species (H_2O_2, O_2^-, NO) released by the cells. This could contribute to a better resistance to corrosion of the alloy when used in vivo.[338] Several works were dedicated to the study of titanium surfaces before and after cell culture.[339–341] Osteoblast-like cells were shown to enhance the precipitation of P- and Ca-containing compounds at the titanium surface, which was not the case with fibroblasts. This allowed a better understanding of implant-cells interactions to be achieved.[340] The interactions of silicate minerals with cells have been extensively studied by Seal et al..[342,343] Their aim was to investigate the pathogenicity of such minerals, especially under the form of fibers. The coordination and the concentration of Si, Al, Mg, and Fe in the silicates were shown to be altered by the cells. This was accompanied by an increase of the Fe content of the cells, as shown by atomic absorption spectroscopy[344,345] and was attributed to direct chemical interaction occurring between the silicates and the cells.

Materials After Use or Explantation

In a limited number of studies, XPS was used to compare the surface composition of a material before and after a more or less prolonged implantation. SS plates were implanted for one year in the tibiae of rabbits. XPS analysis after explantation revealed the formation of a passive layer strongly enriched with Cr.[346] Electroactive aniline polymer plates were implanted beneath the dorsal skin of rats for 19–90 weeks. While examination of the explanted samples by infrared spectroscopy showed that the bulk composition of the polymers was intact, XPS revealed that the outermost surface underwent hydrolysis as well as carbon oxidation during implantation.[347]

Compared to experiments conducted in a defined medium, XPS analysis after in vivo experiments is even more challenging given the complexity (nature and amount) of the compounds encountered by implanted surfaces. Again, studies that aim at understanding the modifications of the material itself are perturbed by the inevitable presence of biomolecules at the material surface after explantation. Paynter et al.[348] carefully examined different procedures to remove the adsorbed layer from polyurethane vascular prostheses explanted after up to 6 months from dogs, in order to evaluate the degradation of the material during implantation. The presence of proteins at the surface of the explanted grafts was evidenced by the increased nitrogen content and the modification of the C 1s peak shape. Sample cleaning using sodium bicarbonate followed by sodium hypochlorite was more efficient than a laboratory detergent or than isopropanol to remove the adsorbed layer. Although visual observation of the grafts showed evident material degradation, no significant changes in the XPS spectra were noted, taking into account the probable effects of sample cleaning. This indicated that the degraded material was ablated away from the surface. In a more recent study, polyurethane vascular grafts were implanted in human patients.[349] Some of the implants failed and were then explanted and examined by different techniques, including XPS. The carbonate content of the surface was significantly lower on the explanted compared to the virgin material. Although this might be partly attributed to incomplete removal of adsorbed proteins, the authors concluded that chemical degradation and hydrolysis of the material had occurred in the body.

In the studies presented in this section so far, the concern was the integrity of the material after implantation, and efforts were made to avoid the interference of adsorbed biomolecules in the XPS data.

Cleaning reusable medical devices raises questions regarding possible residual contamination by prions. In this context, XPS was used to evaluate the effectiveness of detergent formulations to remove proteins from SS surface soiled with a brain homogeneate from human origin.[350]

XPS was used, with SEM and electron microprobe analysis, to examine conditioning films, biofilms, and encrustations on ureteral stents recovered from patients or incubated in vitro.[351,352]

A study of PHEMA-containing contact lenses was performed in order to evaluate the adsorption of biomolecules (proteins, lipids, polysaccharides). Therefore, contact lenses were analyzed by XPS after having been worn for 10 min or 1 h by human patients.[353] Proteins were adsorbed on all the lenses tested, as well as mucins or lipids in some cases. Clearly, the composition of the adsorbed layer depended on the chemical nature of the lenses and on the duration of the contact with the eye.

Two kinds of soft hydrogel contact lenses were cleaned with different lens care solutions (LCS) and analyzed by XPS. The elemental composition did not vary strongly between the different systems but the $\underline{O}\!=\!C/\underline{O}\!-\!C$ concentration ratio varied by a factor of 5. It was inversely correlated with the initial deposition rate measured in adhesion tests of *P. aeruginosa* performed in a laminar flow chamber. This revealed the adverse effect of compounds rich in $\underline{O}$-C moieties adsorbed from the LCS.[354] After the contact lenses were worn, the presence of nitrogen-rich compounds and a much higher $\underline{O}\!=\!C/\underline{O}\!-\!C$ concentration ratio were observed. The initial deposition rate of bacteria was lower on worn than on unworn contact lenses.[355]

Direct Analysis of Mammalian Cell Surfaces

The major concern that arises when performing XPS analysis on cells is related to the relevance of results obtained in ultrahigh vacuum, far from the physiological conditions. The sample preparation procedure is certainly critical for the success of such studies. The works performed on microbial cells were reviewed in the section *Surface Analysis of Microorganisms and Related Systems*.

A few studies of mammalian cells by XPS were reported in the seventies. This was reviewed by Millard.[191] XPS combined with etching allowed to detect the increase of calcium concentration, especially in the cytoplasm subjacent to the plasma membrane, following activation of the DNA synthesis of hepatoma cells.[356] Similarly, XPS study combined with etching of red blood cells allowed the presence of thallium located at a depth of about 10 nm to be revealed. Thallium may be an activator of enzymes located in and below the phospholipids bilayer. Iron was only detected at a depth of about 50 nm.[357]

The XPS analysis of fibroblasts, epithelial cells, and smooth muscle cells washed in the absence or in the presence of serum proteins showed that they adsorb hardly any serum proteins in contrast with many materials mentioned in this section and in the section *Data Interpretation and Evaluation*.[358] To our knowledge, no recent studies of the surface of mammalian cells by XPS were reported.

Characterization of Model Materials

Besides material surfaces, which are specifically tailored for biomedical applications, model systems have also been developed to perform basic studies, allowing a better understanding of interfacial phenomena in biological systems to be achieved. XPS is often used to characterize such model surfaces, in terms of chemical composition as well as organization.

Phospholipid mono- or bilayers obtained by the Langmuir–Blodgett (LB) technique are used to mimic the cell membrane. Such layers may be useful to understand the interactions of cell membranes with proteins or with drugs. Solletti et al.[359] reported a XPS/AFM study of dipalmitoyl phosphatidylethanolamine (DPPE) and DPPC LB films on mica and highly oriented pyrolytic graphite (HOPG). Deposition of a monolayer of DPPE on mica provoked the increase of the C 1s signal while the apparent concentration of mica constituents markedly decreased as expected. Compared to the values expected from the stoichiometry of DPPE, the $\underline{C}$—(C, H) component of the C 1s peak was in excess while the $\underline{C}$—O and O=$\underline{C}$—O components were weaker, indicating that the hydrocarbon chains of DPPE were oriented away from the mica. Using ARXPS, the thicknesses of mono- and bilayers of DPPE on mica were evaluated on the basis of the C 1s/K 2p intensity ratio, considering flat and uniform layers. A value of 2.07 was found for the ratio of the thickness of the bilayer to that of the monolayer. However, AFM images of the bilayer in air actually showed domains on top of a homogeneous layer, with a step height compatible with the thickness of a bilayer. Combination of AFM with XPS thus pointed to the reorganization in air of the bilayer into a monolayer covered by bilayer domains. Since bilayers are only stable under liquid, the usefulness of XPS to probe such systems is limited, unless analysis can be performed in the frozen-hydrated state (cf. the section on *Surface Reorganization and Low Temperature Analysis*).

Deleu et al.[152] examined mixed surfactin/DPPC monolayers obtained by the LB technique using XPS and AFM. Surfactin is a surface-active lipopeptide produced

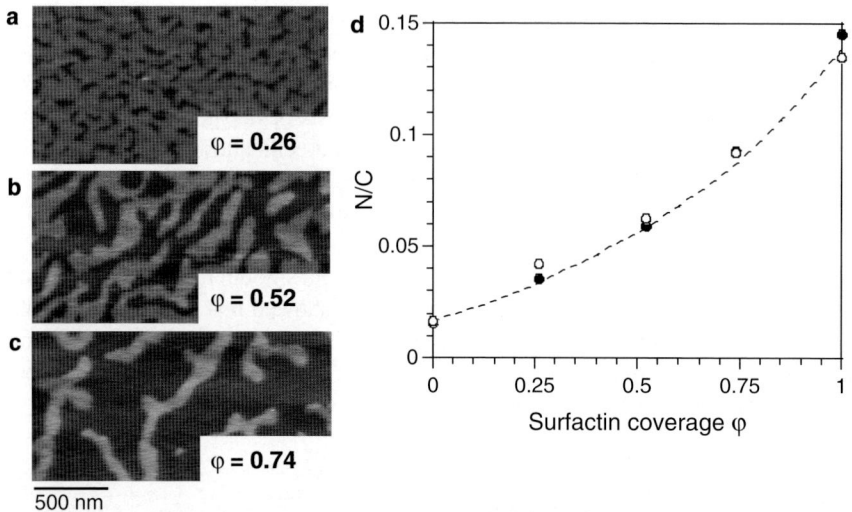

Fig. 50 Study of mixed surfactin/DPPC monolayers on mica. AFM topographic images at (**a**) 0.1, (**b**) 0.25, and (**c**) 0.5 surfactin/DPPC molar ratios. The DPPC domains are 1.2 nm thicker than the surfactin domains. The surfactin surface coverage φ is indicated for each molar ratio. (**d**) Variation of N/C XPS ratio as a function of φ measured by AFM. *Circles*, experimental data (two independent sets); *broken line*, curve computed by using experimental data for the pure monolayers and the respective surface coverages. Adapted from Ref. 152 with permission from Biophysical Journal, copyright (1999) Biophysical Society

by *Bacillus subtilis* strains, with antiviral, antibacterial, and hemolytic activities. Understanding its interactions with cell membranes is important. The N/C concentration ratio measured on a pure surfactin layer was compatible with random orientation of surfactin, while the N/C ratio of a pure DPPC layer fitted with polar headgroups oriented towards the mica substrate, as demonstrated using models taking into account the polar head and the hydrocarbon tail thicknesses. For mixed surfactin/DPPC monolayers, the experimental N/C ratio followed exactly the variation foreseen by considering a combination of two pure monolayers with the respective degrees of surface coverage determined by AFM (Fig. 50). These two studies of LB films of phospholipids show again the excellent complementarity of AFM and XPS to probe the organization of thin films.

SAMs of alkylsilanes on hydroxylated surfaces or of alkanethiols on gold[360] is another example of model systems used to investigate biological interactions with material surfaces. The advantage of SAMs lies in the well-defined and stable surface chemical composition that can be achieved, allowing the role of specific chemical functions on protein adsorption or cell adhesion to be investigated.[361,362] The characterization of SAMs, including the use of XPS, was extensively reviewed.[363,364]

Mixed SAMs of 11-mercapto-1-undecanol ($C_{11}OH$) and 1-hexadecanethiol ($C_{15}CH_3$) on gold were prepared to study albumin adsorption as a function of surface hydrophilicity.[365] XPS showed a significant incorporation of $C_{11}OH$ in the

SAM only when the thiol solution contained 70% of $C_{11}OH$. A gradual increase of the O/C concentration ratio was then observed for solutions containing from 70 to 100% of $C_{11}OH$; this was correlated with a linear decrease of the water contact angle measured on the monolayer. While albumin adsorption was lower on the pure $C_{11}OH$ compared to the pure $C_{15}CH_3$ layer, no significant decrease of albumin adsorption with the increased hydrophilicity of mixed SAMs was observed. However, the study of exchanges between adsorbed albumin and albumin or fibrinogen in solution indicated that a surface content of about 65% of $C_{11}OH$ may favor albumin adsorption in detriment of fibrinogen adsorption.

Pure SAMs of HS—$(CH_2)_{10}$-R with different end-terminal groups [R being $CH_3, CH_2OH, COOH, CH_2(OCH_2CH_2)_3OH$, or $CH_2OPO_3CH_2CH_2N(CH_3)_3$] were prepared to investigate leucocyte adhesion, an important phenomenon related to the inflammatory response to biomaterials.[366] A good agreement was found between expected and experimental elemental and functional compositions. For instance, $\underline{C}$—O/C concentration ratios of 0.099 and 0.394 were found for R being CH_2OH and $CH_2(OCH_2CH_2)_3OH$, respectively, the expected ratios being 0.091 and 0.412. Leucocyte adhesion was high on the CH_3- and COOH-terminated layers, and lower on the other surfaces. Preincubation of the surfaces with fibrinogen decreased cell adhesion in all cases.

Another example of model materials are the so-called chemical gradient surfaces,[367] which present a surface chemical composition changing gradually along a given direction. Such gradients allow the relationship between surface properties (wettability, electrical charge) and biological interactions (protein adsorption, cell adhesion) to be investigated in one single experiment. The preparation, characterization, and application of chemical gradient surfaces were reviewed.[368] Since the modifications of the surface properties are usually spread over millimeters, XPS may be useful to probe these modifications, either by analyzing separate spots or by using the imaging mode. The O 1s signal was shown to increase and the C 1s signal was shown to decrease from the hydrophobic to the hydrophilic side of a gradient obtained by silanization of glass with dichlorodimethylsilane, using a method based on the diffusion of the silane from one solvent to another.[368]

Conclusion

The use of XPS is still expanding, as shown by Fig. 1. This concerns not only inert materials but also biosystems and even less common areas such as art or archeology.

XPS is a major tool to scrutinize the *spatial distribution* of the constituents of a complex solid or solid-like system. Figure 51 presents a comparison with other techniques, regarding the chemical information and the spatial resolution along the plane of the sample surface (lateral resolution) and perpendicular to that plane (explored depth). XPS provides the best compromise in terms of qualitative information, quantification, and explored depth, in order to collect the chemical information which is needed to understand important interfacial (bio)processes (adsorption, adhesion,

Technique	Prim. part.[a]	Phenomenon - measurement	Lateral resolution	Explored depth	Qualitative information	Quantif.[b]
ToF-SIMS[c]	Ions	Erosion - mass spectrum	~ 1 μm	Surface, ~ 1 nm	Molecular fragments	SQ
XPS	hν	Ionisation - electron emission	~ 5 μm	Quasi-surface, ~ 5 nm	Elements - functions	Q
Auger spectrosc.	e	Ionisation, relaxation - electron emission	~ 100 nm	Quasi-surface, ~ 5 nm	Elements	Q
Infrared spectrosc.	hν	Absorption, reflection - light intensity	~ 1 mm	Sub-surface[d], ~ 1 μm Ads. layer[e], ~ 1 nm	Chemical functions[f]	SQ
Electron microprobe	e	Ionisation, relaxation - X-ray emission	~ 1 μm	Sub-surface, ~ 1 μm	Elements	Q

[a] Primary particles used.
[b] Q, possibility of quantification; SQ, quantification limited to semiquantitative comparisons.
[c] Time-of-Flight Secondary Ion Mass Spectroscopy, or static SIMS.
[d] By attenuated total reflectance.
[e] By reflection of polarized light with a metal substratum.
[f] Molecular compounds, with certain limitations.

Fig. 51 Overview of techniques providing a chemical analysis of "surfaces"

aggregation, flotation,etc.). Time of Flight Secondary Ion Mass Spectroscopy probes a smaller thickness and is thus still more delicate regarding contamination and representativity of the analyzed surface. It does not provide a direct quantitative analysis; on the other hand, advanced methods of data treatment open the way to information on the conformation of proteins in the adsorbed state.[369] Auger spectroscopy is limited to conductive samples. Detailed information on the nano-organization of dehydrated layers of adsorbed biomacromolecules can be obtained by combining XPS with the measurement of the adsorbed amount and, or with AFM. As the lateral resolution is concerned, progresses in XPS are expected through the use of tunable synchrotrons as X-ray sources, which open the way to a surface spectromicroscopy.[370]

The *chemical information* given by XPS concerns not only the elements but also their speciation, in particular the chemical functions in the case of organic materials. This is highly valuable in the context of the analysis of biosystems and biomaterials. The chemical composition can be expressed in terms of concentrations of relevant model molecular compounds. The guidelines established to interpret the spectra of microorganisms were successfully used for the spectra of food and food ingredients, namely model cakes and their constituents: flour, sugar, oil (analyzed at low temperature), and freeze-dried egg white, egg yolk, and whole egg.[371] Relationships similar to those presented in Fig. 36 were also obtained. For phospholipid-rich systems (egg yolk, cake), a 1:1 relationship was found between protonated nitrogen and phosphorous. The acquisition of multispectral data sets and their treatment by multivariate techniques is expected to develop, with the aim to improve quality of the data regarding chemical information but also signal/noise ratio and therefore spatial resolution.[372]

An important constraint of XPS analysis is that the sample must be dehydrated. This is no problem for most classical materials; however, it must be kept in mind that materials made of amphiphilic compounds may expose a different surface to an aqueous medium, on the one hand, and to air or vacuum, on the other hand. For adsorbed phases, dehydration may be a complication as it may alter the distribution of adsorbed compounds; on the other hand, it may be a way to create novel surface architectures.[373]

In the study of *systems which are intrinsically hydrated*, such as hydrogels, microorganisms and biofilms, the dehydration may generate artefacts and the results must be considered carefully, taking into account both the sample preparation procedure and the aim of the analysis. Moreover, in the case of microorganisms, the XPS analysis is subject to the same perturbation as all other methods involving culture, centrifugation, and resuspension.[374] For brewing yeast, profound surface alteration was shown to be absent, provided freeze drying was performed in adequate conditions. The same may be expected for other yeasts and for Gram-positive bacteria; the situation may be more difficult for Gram-negative bacteria. The possible reorganization of cell wall or cell-surface polymers as a result of exposure to air or vacuum could be investigated further by using cryofixation, on the one hand, and by performing the XPS analysis on samples maintained at low temperature after freeze-drying, on the other hand. Nevertheless, the review presented above shows that meaningful information can be obtained on microbial surfaces. The chemical composition, as seen by XPS, is indeed correlated with properties (electrical properties, nanomechanical properties observed by AFM in water, macroscopic behavior involving interfacial processes), which are governed by the state of surfaces in contact with water. While the correlations demonstrate the relevance of the XPS data, they suggest but do not demonstrate a direct dependence of these properties on the chemical features revealed by XPS. In any case, generalization to other situations in terms of strains, conditions, and properties must be avoided.

Acknowledgments C.C. Dupont-Gillain is a Research Associate of Belgian National Foundation for Scientific Research (F.N.R.S.). The contribution of numerous young researchers is gratefully acknowledged. The authors thank R.A. Jacob and F. Bouvy for their help. The support of F.N.R.S., of the Federal Office for Scientific, Technical and Cultural Affairs (Interuniversity Poles of Attraction Program), of the Research Department of the Communauté Française de Belgique (Concerted Research Actions) and of Région wallonne is also acknowledged.

References

1. V.R. Gowariker, N.V. Viswanathan and J. Sreedhar, *Polymer Science*, Wiley, New York, 1986.
2. D.J. Gallant, B. Bouchet and P.M. Baldwin, *Carbohydr. Polym.* **32**, 177–191 (1997).
3. H. Lodish, A. Berk, S.L. Zipursky, P. Matsudaira, D. Baltimore and J. Darnell, *Molecular Cell Biology, 4th ed.*, W.H. Freeman and Company, New York, 2000.
4. P.C. Hiemenz and R. Rajagopalan, *Principles of Colloid and Surface Chemistry, 3rd ed.*, Marcel Dekker, New York, 1997.
5. B. Alberts, D. Bray, J. Lewis, M. Raff, K. Roberts and J.D. Watson, *Molecular Biology of the Cell*, Garland Publishing, New York, 1983.

6. S.J.B. Reed, *Electron Microprobe Analysis, 2nd ed.*, Cambridge University Press, 1997.
7. J.D. Andrade, in *Surface and Interfacial Aspects of Biomedical Polymers, Vol. 1*, J.D. Andrade (ed.), Plenum Press, New York, 1985, Chap. 5, pp. 105–195.
8. G. Binnig, C.F. Quate and C. Gerber, *Phys. Rev. Lett.* **56**, 930–933 (1986).
9. K. Siegbahn, C. Nordling, A. Fahlman, R. Nordberg, K. Hamrin, J. Hedman, G. Johansson, T. Bergmark, S.-E. Karlsson, I. Lindgren and B. Lindberg, *ESCA. Atomic, Molecular and Solid State Structure Studied by Means of Electron Spectroscopy*, Almqvist & Wiksells Boktryckeri AB, Uppsala, 1967.
10. K. Siegbahn, *J. Electron Spectrosc. Relat. Phenom.* **137–140**, 3–42 (2004).
11. E. Paparazzo, *J. Electron Spectrosc. Relat. Phenom.* **134**, 9–24 (2004).
12. H.P. Bonzel and C. Kleint, *Prog. Surf. Sci.* **49**, 107–153 (1995).
13. D.A. Shirley and C.S. Fadley, *J. Electron Spectroc. Relat. Phenom.* **137–140**, 43–58 (2004).
14. H. Seyama and M. Soma, *Anal. Sci.* **19**, 487–497 (2003).
15. A.M. Venezia, *Catal. Today* **77**, 359–270 (2003).
16. K. Asami and K. Hashimoto, *Langmuir* **3**, 897–904 (1987).
17. D. Briggs, *Surface Analysis of Polymers by XPS and Static SIMS*, Cambridge University Press, Cambridge, 1998.
18. B.D. Ratner and B.J. McElroy, in *Spectroscopy in the Biomedical Sciences*, R.M. Gendreau (ed.), CRC Press, Boca Raton, 1986, Chap. 5, pp. 107–140.
19. K. Merrett, R.M. Cornelius, W.G. McClung, L.D. Unsworth and H. Sheardown, *J. Biomater. Sci.: Polym. Ed.* **13**, 593–621 (2002).
20. G. Buckton, R. Bulpett and N. Verma, *Int. J. Pharm.* **72**, 157–162 (1991).
21. C. Bosquillon, P.G. Rouxhet, F. Ahimou, D. Simon, Ch. Culot, V. Préat and R. Vanbever, *J. Control. Release* **99**, 357–367 (2004).
22. P.G. Rouxhet and M.J. Genet, in *Microbial Cell Surface Analysis: Structural and Physicochemical Methods*, N. Mozes, P.S. Handley, H.J. Busscher and P.G. Rouxhet (eds.), VCH Publishers, New York, 1991, Chap. 8, pp. 173–220.
23. P.G. Rouxhet, N. Mozes, P.B. Dengis, Y.F. Dufrêne, P.A. Gerin and M.J. Genet, *Colloids Surf. B: Biointerfaces* **2**, 347–369 (1994).
24. H.C. van der Mei, J. de Vries and H.J. Busscher, *Surf. Sci. Rep.* **39**, 1–24 (2000).
25. P.G. Rouxhet, C.C. Dupont-Gillain, M.J. Genet and Y.F. Dufrêne, in *Biofilms in Medicine, Industry and Environmental Biotechnology*, P. Lens, A.P. Moran, T. Mahony, P. Stoodley and V.O'Flaherty (eds.), IWA Publishing, London, 2003, Chap. 16, pp. 259–284.
26. D. Briggs and M.P. Seah (eds.), *Practical Surface Analysis: Auger and X-ray Photoelectron Spectroscopy, Vol. 1*, Wiley, Chichester, 2nd ed., 1990.
27. T.L. Barr, *Modern ESCA. The Principles and Practice of X-ray Photoelectron Spectroscopy*, CRC Press, Boca Raton, 1994.
28. B.D. Ratner and D.G. Castner, in *Surface Analysis – The Principal Techniques*, J.C. Vickerman (ed.), Wiley, Chichester, 1997, Chap. 3, pp. 43–98.
29. D. Briggs and J.T. Grant (eds.), *Surface Analysis by Auger and X-ray Photoelectron Spectroscopy*, IM Publications, Chichester, 2003.
30. J.F. Watts and J.W. Wolstenholme, *An Introduction to Surface Analysis by XPS and AES*, Wiley, Chichester, 2003.
31. C.J. Powell, *J. Vac. Sci. Technol. A* **21**, S42–S53 (2003).
32. T.A. Koopmans, *Physica* **1**, 104–113 (1933).
33. J.F. Moulder, W.F. Stickle, P.E. Sobol and K.D. Bomben, *Handbook of X-ray Photoelectron Spectroscopy*, J. Chastain (ed.), Perkin-Elmer, Eden Prairie, 1992.
34. D.R. Lide (ed.), *Handbook of Chemistry and Physics, 76th ed.*, CRC Press, Boca Raton, USA, 1992–1996.
35. J.H. Scofield, *J. Electron Spectrosc. Relat. Phenom.* **8**, 129–137 (1976).
36. R. Wilson, in *Surface Analysis – The Principle Techniques*, J.C. Vickerman (ed.), Wiley, Chichester, 1997, Chap. 2, pp. 9–39.
37. J. Schweppe, R.D. Deslattes, T. Mooney and C.J. Powell, *J. Electron Spectrosc. Relat. Phenom.* **67**, 463–478 (1994).

38. G. Beamson, S.R. Haines, N. Moslemzadeh, P. Tsakiropoulos, P. Weightmanaud and J.F. Watts, *Surf. Interface Anal.* **36**, 275–279 (2004).
39. S. Delsarte, V. Serin, A.-M. Flank, F. Villain and P. Grange, *J. Solid State Chem.* **163**, 163–177 (2002).
40. M.J. Remy, M.J. Genet, P.P. Notté, P.F. Lardinois and G. Poncelet, *Microporous Mat.* **2**, 7–15 (1993).
41. C.D. Wagner and A. Joshi, *J. Electron Spectrosc. Relat. Phenom.* **47**, 283–313 (1988).
42. G. Margaritondo, in *Surface Analysis by Auger and X-ray Photoelectron Spectroscopy*, D. Briggs and J.T. Grant (eds.), IM Publications, Chichester, 2003, Chap. 26, pp. 733–748.
43. A. Curtis and C. Wilkinson, *Biomaterials* **8**, 1573–1583 (1997).
44. Y.F. Dufrêne, C.J.P. Boonaert, P.A. Gerin, M. Asther and P.G. Rouxhet, *J. Bacteriol.* **181**, 5350–5354 (1999).
45. C.C. Dupont-Gillain and I. Jacquemart, *Surf. Sci.* **539**, 145–154 (2003).
46. C.C. Dupont-Gillain, J.A. Allaerts and J.L. Dewez, *Bio-Med. Mater. Eng.* **14**, 281–291 (2004).
47. A. Touhami, B. Nysten and Y.F. Dufrêne, *Langmuir* **19**, 4539–4546 (2003).
48. A. Oelsner, O. Schmidt, M. Schicketanz, M. Klais and G. Schönhense, *Rev. Sci. Instr.* **72**, 3968–3974 (2001).
49. A. Nambu, J.M. Bussat, M. West, B.C. Sell, M. Watanabe, A.W. Kay, N. Mannella, B.A. Ludewigt, M. Press, B. Turko, G. Meddeler, G. Zizka, H. Spieler, H. van-der-Lippe, P. Denes, T. Ohta, Z. Hussain and C.S. Fadley, *J. Electron Spectrosc. Relat. Phenom.* **137–140**, 691–697 (2004).
50. H.J. Busscher, H.C. van der Mei, M.J. Genet, J.F. Perdok and P.G. Rouxhet, *Surf. Interface Anal.* **15**, 344–346 (1990).
51. C.C. Bryson, *Surf. Sci.* **189–190**, 50–58 (1987).
52. G. Vereecke and P.G. Rouxhet, *Surf. Interface Anal.* **26**, 490–497 (1998).
53. M.J. Genet, C. Jacques, N. Mozes, C. Van Hove, A. Lejeune and P.G. Rouxhet, *Surf. Interface Anal.* **33**, 601–606 (2002).
54. M.A. Kelly, *J. Chem. Educ.* **81**, 1726–1733 (2004).
55. ISO 15472, *Surface Chemical Analysis – X-ray Photoelectron Spectrometers – Calibration of Energy Scales*, ISO, Geneva, 2001.
56. NIST, *X-ray Photoelectron Spectroscopy Database*, NIST Standard Reference Database 20, Version 3.4 (Web Version) http://srdata.nist.gov/xps/index.htm.
57. G. Beamson and D. Briggs, *High Resolution XPS of Organic Polymers: The Scienta ESCA300 Database*, Wiley, Chichester (1992).
58. S. Tougaard, *Surf. Interface Anal.* **26**, 249–269 (1998).
59. S. Tougaard, in *Surface Analysis by Auger and X-ray Photoelectron Spectroscopy*, D. Briggs and J.T. Grant (eds.), IM Publications, Chichester, Chap. 12, 2003, pp. 295–343.
60. A. Proctor and D. Hercules, *Appl. Spectrosc.* **38**, 505–518 (1984).
61. P. Boulanger, J.J. Pireaux, J. Verbist and J. Delhalle, *J. Electron Spectrosc. Relat. Phenom.* **63**, 55–73 (1993).
62. R. Foerch, G. Beamson and D. Briggs, *Surf. Interface Anal.* **17**, 842–846 (1991).
63. T. Murakami, S. Kuroda and Z. Osawa, *J. Colloid Interface Sci.* **200**, 192–194 (1998).
64. C.D. Wagner, W.M. Riggs, L.E. Davis, J.F. Moulder and G.E. Muilenberg (eds.), *Handbook of X-ray Photoelectron Spectroscopy*, Perkin Elmer, Eden Prairie, 1979.
65. L. Kövér, in *Surface Analysis by Auger and X-ray Photoelectron Spectroscopy*, D. Briggs and J.T. Grant (eds.), IM Publications, Chichester, 2003, Chap. 16, pp. 421–464.
66. D. Briggs, in *Surface Analysis by Auger and X-ray Photoelectron Spectroscopy*, D. Briggs and J.T. Grant (eds.), IM Publications, Chichester, 2003, Chap. 2, pp. 31–56.
67. T.A. Carlson, *Surf. Interface Anal.* **4**, 125–134 (1982).
68. T. Murakami, S. Kuroda and Z. Osawa, *J. Colloid Interface Sci.* **202**, 37–44 (1998).
69. P.C. Schamberger, J.I. Abes and J.A. Gardella Jr., *Colloids Surf. B: Biointerfaces* **3**, 203–215 (1994).
70. A. Galtayries and J.-P. Bonnelle, *Surf. Interface Anal.* **23**, 171–179 (1995).

71. T.L. Barr, *Crit. Rev. Anal. Chem.* **22**, 567–635 (1991).
72. E. Desimoni, G.I. Casella, A. Morone and A.M. Salvi, *Surf. Interface Anal.* **15**, 627–634 (1990).
73. G. Beamson, D.T. Clark, J. Kendrick and D. Briggs, *J. Electron Spectrosc. Relat. Phenom.* **57**, 79–90 (1991).
74. D.G. Castner and B.D. Ratner, *Surf. Interface Anal.* **15**, 479–486 (1990).
75. M.P. Seah, *Surf. Interface Anal.* **9**, 85–98 (1986).
76. R.F. Reilman, A. Msezane and S.T. Manson, *J. Electron Spectrosc. Relat. Phenom* **8**, 389–394 (1976).
77. C.J. Powell and A. Jablonski, *Electron Effective Attenuation Length Database*, Version 1.0 (SRD 82), US Department of Commerce, National Institute of Standards and Technology, Gaithersburg, 2000.
78. I.M. Band, Y.I. Kharitonov and M.B. Trzhaskovskaya, *At. Data Nucl. Data Tables* **23**, 443–505 (1979).
79. ISO 18115, *Surface Chemical Analysis – Vocabulary*, ISO, Geneva, 2001.
80. A. Jablonski and C.J. Powell, *Surf. Sci. Rep.* **47**, 35–91 (2002).
81. C.R. Brundle, *J. Vac. Sci. Technol.* **11**, 212–224 (1974).
82. R.M. Friedman, *Silicates Ind.* **9**, 247–253 (1974).
83. M.P. Seah and W.A. Dench, *Surf. Interface Anal.* **1**, 2–11 (1979).
84. J.C. Ashley and C.J. Tung, *Surf. Interface Anal.* **4**, 52–55 (1982).
85. W. Hanke, H. Ebel, M.F. Ebel, A. Jablonski and K. Hirokawa, *J. Electron Spectrosc. Relat. Phenom.* **40**, 241–257 (1986).
86. C.J. Powell, *J. Vac. Sci. Technol. A* **3**, 1338–1342 (1985).
87. J. Szajman, J. Liesegang, J.G. Jenkin and R.C.G. Leckey, *J. Electron Spectrosc. Relat. Phenom.* **23**, 97–102 (1981).
88. C.D. Wagner, L.E. Davis and W.M. Riggs, *Surf. Interface Anal.* **2**, 53–55 (1980).
89. P.J. Cumpson, *Surf. Interface Anal.* **31**, 23–34 (2001).
90. M.P. Seah, S.J. Spencer, F. Bodino and J.J. Pireaux, *J. Electron Spectrosc. Relat. Phenom.* **87**, 159–167 (1997).
91. A.E. Hughes and C.C. Phillips, *Surf. Interface Anal.* **4**, 220–226 (1982).
92. J. Osterwalder, M. Sagurton, P.J. Orders, C.S. Fadley, B.D. Hermsmeier and D.J. Friedman, *J. Electron Spectrosc. Relat. Phenom.* **48**, 55–99 (1989).
93. M.P. Seah and M.T. Anthony, *Surf. Interface Anal.* **6**, 230–241 (1984).
94. M.P. Seah, M.E. Jones and M.T. Anthony, *Surf. Interface Anal.* **6**, 242–254 (1984).
95. L.T. Weng, G. Vereecke, M.J. Genet, P. Bertrand and W.E.E. Stone, *Surf. Interface Anal.* **20**, 179–192 (1993).
96. M.P. Seah, *J. Electron Spectrosc. Relat. Phenom.* **71**, 191–204 (1995).
97. http://www.npl.co.uk/npl/cmmt/sis/index.html
98. M.P. Seah, *Surf. Interface Anal.* **20**, 243–266 (1993).
99. G.C. Smith, *J. Electron Spectrosc. Relat. Phenom.* **148**, 21–28 (2005).
100. C.D. Wagner, L.E. Davis, M.V. Zeller, J.A. Taylor, R.H. Raymont and L.M. Gale, *Surf. Interface Anal.* **3**, 211–225 (1981).
101. C.D. Wagner, *J. Electron Spectrosc. Relat. Phenom.* **33**, 99–102 (1983).
102. M.P. Seah, *Surf. Interface Anal.* **8**, 85–98 (1986).
103. H. Piao and N.S. MacIntyre, *Surf. Interface Anal.* **33**, 591–594 (2002).
104. M. Callewaert, P.A. Gerin, M.J. Genet, L. Boulangé-Petermann and P.G. Rouxhet, *Surf. Interface Anal.*, submitted.
105. E. Desimoni, G.I. Casella, A.M. Salvi, T.R.I. Cataldi and A. Morone, *Carbon* **30**, 527–531 (1992).
106. J.P. Bojon, N. Hilleret and K. Weiss, *Vacuum* **53**, 247–251 (1999).
107. D.R. Cousens, B.J. Wood, J.Q. Wang and A. Atrens, *Surf. Interface Anal.* **29**, 23–32 (2000).
108. D.J. Miller, M.C. Biesinger and N.S. McIntyre, *Surf. Interface Anal.* **33**, 299–305 (2002).
109. J.E. Castle, A.M. Salvi and M.R. Guascito, *Surf. Interface Anal.* **27**, 753–760 (1999).
110. M.F. Ebel, M. Schmid, H. Ebel and A. Vogel, *J. Electron Spectrosc. Relat. Phenom.* **34**, 313–316 (1984).

111. G. Vereecke and P.G. Rouxhet, *Surf. Interface Anal.* **27**, 761–769 (1999).
112. J.L. Dewez, A. Doren, Y.J. Schneider and P.G. Rouxhet, *Biomaterials* **20**, 547–559 (1999).
113. A. Savitsky and M.J.E. Golay, *Anal. Chem.* **36**, 1627–1639 (1964).
114. D.A. Shirley, *Phys. Rev. B* **5**, 4709–4714 (1972).
115. A. Proctor and P.M.A. Sherwood, *Anal. Chem.* **54**, 13–19 (1982).
116. S. Tougaard, *Solid State Commun.* **61**, 547–549 (1987).
117. S. Tougaard, *Surf. Interface Anal.* **25**, 137–154 (1997).
118. M.F. Koenig and J.T. Grant, *J. Electron Spectrosc. Relat. Phenom* **33**, 9–22 (1984).
119. N.S. McIntyre, A.R. Pratt, H. Piao, D. Maybury and S.J. Splinter, *Appl. Surf. Sci.* **144–145**, 156–160 (1999).
120. N. Fairley, in *Surface Analysis by Auger and X-ray Photoelectron Spectroscopy*, D. Briggs and J.T. Grant (eds.), IM Publications, Chichester, 2003, Chap. 15, pp. 397–420.
121. D. Briggs and M.P. Seah (eds.), *Practical Surface Analysis: Auger and X-ray Photoelectron Spectroscopy, 1st ed.*, Wiley, Chichester, 1983.
122. J.M. Conny and C.J. Powell, *Surf. Interface Anal.* **29**, 856–872 (2000).
123. R. Hesse, T. Chassé, P. Streubel and R. Szargan, *Surf. Interface Anal.* **36**, 1373–1383 (2004).
124. http://www.acg.nist.gov/std
125. W.F. Egelhoff Jr., *Surf. Sci. Rep.* **6**, 253–415 (1987).
126. D.M. Hercules, *J. Chem. Educ.* **12**, 1751–1766 (2004).
127. J.L. Dewez, A. Doren, Y.J. Schneider, R. Legras and P.G. Rouxhet, in *Interfaces in New Materials*, P. Grange and B. Delmon (eds.), Elsevier, London, 1991, pp. 84–94.
128. M.L. Miller and R.W. Linton, *Anal. Chem.* **57**, 2314–2319 (1985).
129. http://www.surfacespectra.com
130. D.T. Clark, J. Peeling and L. Colling, *Biochim. Biophys. Acta* **453**, 533–545 (1976).
131. R.C. Cothern, W.E. Moddeman, R.G. Albidge, W.J. Sandress, P.L. Kelly, W.S. Handley and L. Field, *Anal. Chem.* **48**, 162–166 (1976).
132. A. Dilks, *ACS Symp. Ser.* **162**, 293–317 (1981).
133. W.J. Landis and J.R. Martin, *J. Vac. Sci. Technol. A* **2**, 1108–1111 (1984).
134. P.A. Gerin, P.B. Dengis and P.G. Rouxhet, *J. Chim. Phys.* **92**, 1043–1065 (1995).
135. P.A. Gerin, M.J. Genet and P.G. Rouxhet, *Surf. Sci. Spect.* **4**, 28–32 (1997).
136. S. Bartiaux, J.B. Lhoest, M.J. Genet, P. Bertrand and P.G. Rouxhet, *Surf. Sci. Spect.* **3**, 342–347 (1997).
137. R.W. Paynter, *Surf. Interface Anal.* **27**, 103–113 (1999).
138. M.P. Seah, *J. Vac. Sci. Technol. A* **22**, 1564–1571 (2004).
139. S. Tanuma, C.J. Powell and D.R. Penn, *Surf. Interface Anal.* **25**, 25–35 (1987).
140. C.J. Powell and A. Jablonski, *Electron Inelastic Mean Free Path Database*, Version 1.1 (SRD71), US Department of Commerce, National Institute of Standards and Technology, Gaithersburg, 2000.
141. M. Mantel and J.P. Wightman, *Surf. Interface Anal.* **21**, 595–605 (1994).
142. J.L. Dewez, V. Berger, Y.J. Schneider and P.G. Rouxhet, *J. Colloid Interface Sci.* **191**, 1–10 (1997).
143. R.W. Paynter, *Surf. Interface Anal.* **33**, 14–22 (2002).
144. R.W. Paynter, *J. Electron Spectrosc. Relat. Phenom.* **135**, 183–190 (2004).
145. R.W. Paynter, *Surf. Interface Anal.* **35**, 932–939 (2003).
146. C. Changui, A. Doren, W.E.E. Stone, N. Mozes and P.G. Rouxhet, *J. Chim. Phys.* **84**, 275–281 (1987).
147. Y.F. Dufrêne, T.G. Marchal and P.G. Rouxhet, *Appl. Surf. Sci.* **144–145**, 638–643 (1999).
148. V.M. De Cupere, J. Van Wetter and P.G. Rouxhet, *Langmuir* **19**, 6957–6967 (2003).
149. C.C. Dupont-Gillain, B. Nysten and P.G. Rouxhet, *Polym. Int.* **48**, 271–276 (1999).
150. V.M. De Cupere and P.G. Rouxhet, *Surf. Sci.* **491**, 395–404 (2001).
151. F.A. Denis, P. Hanarp, D.S. Sutherland, J. Gold, Ch. Mustin, P.G. Rouxhet and Y.F. Dufrêne, *Langmuir* **18**, 819–828 (2002).
152. M. Deleu, M. Paquot, P. Jacques, P. Thonart, Y. Adriaensen and Y.F. Dufrêne, *Biophys. J.* **77**, 2304–2310 (1999).

153. A. Cimino, D. Grazzoli and M. Valigi, *J. Electron Spectrosc. Relat. Phenom.* **104**, 1–29 (1999).
154. P.L.J. Gunter, O.L.J. Gijzeman and J.W. Niemantsverdriet, *Appl. Surf. Sci.* **115**, 342–346 (1997).
155. C. Defossé, P. Canesson, P.G. Rouxhet and B. Delmon, *J. Catal.* **51**, 269–277 (1978).
156. S. Tanuma, C.J. Powell and D.R. Penn, *Surf. Interface Anal.* **17**, 911–926 (1991).
157. S. Tanuma, C.J. Powell and D.R. Penn, *Surf. Interface Anal.* **21**, 165–176 (1994).
158. R. Payling and J. Szajman, *J. Electron Spectrosc. Relat. Phenom.* **43**, 37–51 (1987).
159. D.E. Gray (ed.), *American Institute of Physics Handbook*, McGraw-Hill, New-York, 2nd ed., (1963).
160. O.V. Krylov, *Catalysis by Non Metals*, Academic Press, New York, 1970.
161. W.H. Strehlow and E.L. Cook, *J. Phys. Chem. Ref. Data* **2**, 163–199 (1973).
162. J. Szajman and R.C.G. Leckey, *J. Electron Spectrosc. Relat. Phenom.* **23**, 83–96 (1981).
163. D.P. Norton, *Mat. Sci. Eng. R.* **43**, 139–247 (2004).
164. S.R. Morisson, *Electrochemistry at Semiconductor and Oxidized Metal Electrodes*, Plenum, New York, 1980.
165. J.C. Ashley, *J. Electron Spectrosc. Relat. Phenom.* **28**, 177–194 (1982).
166. S. Tanuma, in *Surface Analysis by Auger and X-ray Photoelectron Spectroscopy*, D. Briggs and J.T. Grant (eds.), IM Publications, Chichester, 2003, Chap. 11, pp. 259–294.
167. I. Jacquemart, E. Pamula, V.M. De Cupere, P.G. Rouxhet and C.C. Dupont-Gillain, *J. Colloid Interface Sci.* **278**, 63–70 (2004).
168. B.D. Ratner, P.K. Weathersby, A.S. Hoffman, M.A. Kelly and L.H. Sharpen, *J. Appl. Polym. Sci.* **22**, 643–664 (1978).
169. B.D. Ratner and D.G. Castner, *Colloids Surf. B: Biointerface* **2**, 333–346 (1994).
170. E. Occhiello, M. Morra, G. Morini, F. Garbassi and P. Humphrey, *J. Appl. Polym. Sci.* **42**, 551–559 (1991).
171. C.C. Dupont-Gillain, Y. Adriaensen, S. Derclaye and P.G. Rouxhet, *Langmuir* **16**, 8194–8200 (2000).
172. M.J. Genet, P. Cogels, Y.A. Adriaensen and P.G. Rouxhet, *Surf. Interface Anal.* **40**, 338–342 (2008).
173. D.R. Baer, D.J. Gaspar, M.H.Engelhard and A.D. Lea, in *Surface Analysis by Auger and X-ray Photoelectron Spectroscopy*, D. Briggs and J.T. Grant (eds.), IM Publications, Chichester, 2003, pp. 211–233.
174. L.S. Johansson and J.M. Campbell, *Surf. Interface Anal.* **36**, 1018–1022 (2004).
175. K. Yoshihara and A. Tanaka, *Surf. Interface Anal.* **33**, 252–258 (2002).
176. R.L. Graham, C.D. Bain, H.A. Biebuyck, P.E. Laibinis and G.M. Whitesides, *J. Phys. Chem.* **97**, 9456–9464 (1993).
177. Y. Iijima, N. Niimura and K. Hiraoka, *Surf. Interface Anal.* **24**, 193–197 (1996).
178. C. Kittel, *Introduction to Solid State Physics, 6th ed.*, Wiley, New York, 1986.
179. E.M. Gaigneaux, M.J. Genet, P. Ruiz and B. Delmon, *J. Phys. Chem. B* **104**, 5724–5737 (2000).
180. L. Ruangchuay, J. Schwank and A. Sirivat, *Appl. Surf. Sci.* **199**, 128–137 (2002).
181. R.L. Chaney, *Surf. Interface Anal.* **10**, 36–47 (1987).
182. K. Prabhakaran, Y. Kobayashi and T. Ogino, *J. Electron Spectrosc. Relat. Phenom.* **63**, 283–288 (1993).
183. M.F. Ebel, R. Svagera, R. Ashury, H. Ebel, P.C. Zalm and C. Van der Marel, *J. Electron Spectrosc. Relat. Phenom.* **131–132**, 145–151 (2003).
184. M.F. Koenig and J.T. Grant, *J. Electron Spectrosc. Relat. Phenom.* **36**, 213–215 (1985).
185. M.F. Koenig and J.T. Grant, *Surf. Interface Anal.* **7**, 217–222 (1985).
186. http://www.casaxps.com
187. C.J.P. Boonaert, Y.F. Dufrêne, S.R. Derclaye and P. G. Rouxhet, *Colloids Surf. B: Biointerfaces* **22**, 171–182 (2001).
188. P.B. Dengis, P.A. Gerin and P.G. Rouxhet, *Colloids Surf. B: Biointerfaces* **4**, 199–211 (1995).
189. P.B. Dengis and P.G. Rouxhet, *J. Microbiol. Methods* **26**, 171–183 (1996).

190. R. Holm, in *Analysis of Organic and Biological Surfaces*, P. Echlin (ed.), Wiley, New York, 1984, Chap. 5, pp. 37–72.
191. M.M. Millard, in *Contemporary Topics in Analytical and Clinical Chemistry, Vol. 3*, D.M. Hercules, G.M. Hieftje, L.R. Snyder and M.A. Everson (eds.), Plenum, New York, 1978, pp. 1–55.
192. K.C. Marshall, R. Pembrey and R.P. Schneider, *Colloids Surf. B: Biointerfaces* **2**, 371–376 (1994).
193. A.M. James, in *Microbial Cell Surface Analysis: Structural and Physicochemical Methods*, N. Mozes, P.S. Handley, H.J. Busscher and P.G. Rouxhet (eds.), VCH Publishers, New York, 1991, Chap. 9, pp. 221–262.
194. H.C. van der Mei, M. Rosenberg and H.J. Busscher, in *Microbial Cell Surface Analysis: Structural and Physicochemical Methods*, N. Mozes, P.S. Handley, H.J. Busscher and P.G. Rouxhet (eds.), VCH Publishers, New York, 1991, Chap. 10, pp. 263–287.
195. J.L. Van Haecht, C. Defossé, B. Van Den Bogaert and P.G. Rouxhet, *Colloids Surf.* **4**, 343–358 (1982).
196. P.K. Sharma and K.H. Rao, *Miner. Metall. Process.* **22**, 31–37 (2005).
197. Y.F. Dufrêne and P.G. Rouxhet, *Can. J. Microbiol.* **42**, 548–556 (1996).
198. C.J.P. Boonaert and P.G. Rouxhet, *Appl. Environ. Microb.* **66**, 2548–2554 (2000).
199. P.A. Gerin, Y. Dufrêne, M.N. Bellon-Fontaine, M. Asther and P.G. Rouxhet, *J. Bacteriol.* **175**, 5135–5144 (1993).
200. Y.F. Dufrêne, A. van der Wal, W. Norde and P.G. Rouxhet, *J. Bacteriol.* **179**, 1023–1028 (1997).
201. N. Mozes and S. Lortal, *Microbiology.* **141**, 11–19 (1995).
202. D.E. Amory, M.J. Genet and P.G. Rouxhet, *Surf. Interface Anal.* **11**, 478–486 (1988).
203. P.B. Dengis and P.G. Rouxhet, *Yeast* **13**, 931–943 (1997).
204. I.C. Hancock, in *Microbial Cell Surface Analysis: Structural and Physicochemical Methods*, N. Mozes, P.S. Handley, H.J. Busscher and P.G. Rouxhet (eds.), VCH Publishers, New York, 1991, Chap. 8, pp. 21–59.
205. G.H. Fleet, in *The Yeasts, 2nd ed.*, A.H. Rose and J.S. Harrison (eds.), Academic Press, London, 1991, Chap. 5, Vol. 4, pp. 199–277.
206. H.C. van der Mei and H.J. Busscher, *Adv. Dent. Res.* **11**, 388–394 (1997).
207. K.W. Millsap, G. Reid, H.C. van der Mei and H.J. Busscher, *Can. J. Microbiol.* **43**, 284–291 (1997).
208. N. Mozes, L.L. Schinckus, C. Ghommidh, J.M. Navarro and P.G. Rouxhet, *Colloids Surf. B: Biointerfaces* **3**, 63–74 (1994).
209. K.M. Nordström and N. Mozes, *Colloids Surf. B: Biointerfaces* **2**, 67–72 (1994).
210. P.F.G. Herben, N. Mozes and P.G. Rouxhet, *Biochim. Biophys. Acta* **1033**, 184–188 (1990).
211. Y.F. Dufrêne and P.G. Rouxhet, *Colloids Surf. B: Biointerfaces* **7**, 271–279 (1996).
212. D.E. Amory, N. Mozes, M.P. Hermesse, A.J. Léonard and P.G. Rouxhet, *FEMS Microbiol. Lett.* **49**, 107–110 (1988).
213. N. Mozes, A.J. Léonard and P.G. Rouxhet, *Biochim. Biophys. Acta* **945**, 324–334 (1988).
214. D.E. Amory and P.G. Rouxhet, *Biochim. Biophys. Acta* **938**, 61–70 (1988).
215. H. Latrache, N. Mozes, C. Pelletier and P. Bourlioux, *Colloids Surf. B: Biointerfaces* **2**, 47–56 (1994).
216. N. Mozes, D.E. Amory, A.J. Léonard and P.G. Rouxhet, *Colloids Surf.* **42**, 313–329 (1989).
217. H.C. van der Mei, A.J. Léonard, A.H. Weerkamp, P.G. Rouxhet and H.J. Busscher, *J. Bacteriol.* **170**, 2462–2466 (1988).
218. H.C. van der Mei, A.J. Léonard, A.H. Weerkamp, P.G. Rouxhet and H.J. Busscher, *Colloids Surf.* **32**, 297–305 (1988).
219. H.C. van der Mei, M.J. Genet, A.H. Weerkamp, P.G. Rouxhet and H.J. Busscher, *Arch. Oral Biol.* **34**, 889–894 (1989).
220. M.M. Cowan, H.C. van der Mei, P.G. Rouxhet and H.J. Busscher, *J. Gen. Microbiol.* **138**, 2707–2714 (1992).
221. H.C. van der Mei, M.M. Cowan, M.J. Genet and P.G. Rouxhet, *Can. J. Microbiol.* **38**, 1033–1041 (1992).

222. P.L. Cuperus, H.C. van der Mei, G. Reid, A.W. Bruce, A.E. Khoury, P.G. Rouxhet and H.J. Busscher, *J. Colloid Interface Sci.* **156**, 319–324 (1993).
223. M.M. Cowan, H.C. van der Mei, P.G. Rouxhet and H.J. Busscher, *Appl. Environ. Microb.* **58**, 1326–1334 (1992).
224. Y.F. Dufrêne, C.J-P. Boonaert and P.G. Rouxhet, *Colloids Surf. B: Biointerfaces* **7**, 113–128 (1996).
225. H.C. van der Mei, P. Brokke, J. Dankert, J. Feijen, P.G. Rouxhet and H.J. Busscher, *Appl. Environ. Microb.* **55**, 2806–2814 (1989).
226. H.J. Busscher, P.S. Handley, P.G. Rouxhet, L.M. Hesketh and H.C. van der Mei, in *Microbial Cell Surface Analysis: Structural and Physicochemical Methods*, N. Mozes, P.S. Handley, H.J. Busscher and P.G. Rouxhet (eds.), VCH Publishers, New York, 1991, Chap. 12, pp. 317–338.
227. H.J. Busscher, M.N. Bellon-Fontaine, N. Mozes, H.C. van der Mei, J. Sjollema, A.J. Léonard, P.G. Rouxhet and O. Cerf, *J. Microbiol. Methods.* **12**, 101–115 (1990).
228. H.J. Busscher, H. Bialkowska-Hobrazanska, G. Reid, M. van der Kuijl-Booij and H.C. van der Mei, *Colloids Surf. B: Biointerfaces* **2**, 73–82 (1994).
229. H.C. van der Mei, M.M. Cowan and H.J. Busscher, *Curr. Microbiol.* **23**, 337–341 (1991).
230. B.J. Tyler, *Ann. NY Acad. Sci.* **831**, 114–126 (1997).
231. H.C. van der Mei, H.J. Busscher, R. Bos, J. de Vries, C.J.P. Boonaert and Y.F. Dufrêne, *Biophys. J.* **78**, 2668–2674 (2000).
232. H.C. van der Mei, P.S. Handley and H.J. Busscher, *Cell Biophys.* **20**, 99–110 (1992).
233. M.M. Millard, R. Scherrer and R.S. Thomas, *Biochem. Biophys. Res. Commun.* **72**, 1209–1217 (1976).
234. P.B. Dengis and P.G. Rouxhet, *J. Inst. Brew.* **103**, 257–261 (1997).
235. C.J.P. Boonaert, C.C. Dupont-Gillain, P.B. Dengis, Y.F. Dufrêne and P.G. Rouxhet, in *Encyclopedia Bioprocess Technology: Fermentation, Biocatalysis and Bioseparation*, M.C. Flickinger and S.W. Drew (eds.), Wiley, New York, 1999, pp. 531–548.
236. R. Tybussek, F. Linz, K. Schügerl, N. Mozes, A.J. Léonard and P.G. Rouxhet, *Appl. Microbiol. Biotechnol.* **41**, 13–22 (1994).
237. J. Büchs, N. Mozes, C. Wandrey and P.G. Rouxhet, *Appl. Microbiol. Biotechnol.* **29**, 119–128 (1988).
238. C.M. Pradier, C. Rubio, C. Poleunis, P. Bertrand, P. Marcus and C. Compere, *J. Phys. Chem. B* **109**, 9540–9549 (2005).
239. M. van Raamsdonk, H.C. van der Mei, J.J. de Soet, H.J. Busscher and J. de Graaff, *Infect. Immun.* **63**, 1698–1702 (1995).
240. K.A. Poelstra, H.C. van der Mei, B. Gottenbos, D.W. Grainger, J.R. Van Horn, H.J. Busscher and G. Anthony, *J. Biomed. Mater. Res.* **51**, 224–232 (2000).
241. R. Ashkenazy, L. Gottlieb and S. Yannai, *Biotechnol. Bioeng.* **55**, 1–10 (1997).
242. C.R. Clayton and G.P. Halada, *Electrochem. Soc. Proc.* **2001-5**, 102–112.
243. J.T. Kearns, C.R. Clayton, G.P. Halada, J.B. Gillow and A.J. Francis, *Mater. Perform.* **31**, 48–51 (1992).
244. D. Costa, P. Marcus, M.N. Bellon-Fontaine, B. Rondot, M. Walls, O. Vidal, P. Lejeune and C. Compere, *Electrochem. Soc. Proc.* **97-26**, 450–461 (1998).
245. B. van der Aa and Y.F. Dufrêne, *Colloids Surf. B: Biointerfaces* **23**, 173–182 (2002).
246. Y.J. Dufrêne, H. Vermeiren, J. Vanderleyden and P.G. Rouxhet, *Microbiology (UK)* **142**, 855–865 (1996).
247. Y.F. Dufrêne, V.B. Wiertz and P.G. Rouxhet, *Biofouling* **9**, 307–315 (1996).
248. C. Gomez-Suarez, J. Pasma, A.J. van der Borden, J. Wingender, H.-C. Flemming, H.J. Busscher and H.C. van der Mei, *Microbiology* **148**, 1161–1169 (2002).
249. L.-M. Barnes, M.R. Adams, J.F. Watts, P.A. Zhdan and A.H.L. Chamberlain, *Biofouling* **17**, 1–22 (2001).
250. J. Verran, R.D. Boyd, K. Hall and R.H. West, *J. Food Prot.* **64**, 1377–1387 (2001).
251. A. Vernhet, J.Y. Leveau, O. Cerf and M.N. Bellon-Fontaine, *Biofouling* **5**, 323–334 (1992).
252. M. Kowshik, S. Ashtaputre, S. Kharrazi, W. Vogel, J. Urban, S.K. Kulkarni and K.M. Paknikar, *Nanotechnology* **14**, 95–100 (2003).

253. J. Eriksson, M. Malmsten, F. Tiberg, T.H. Callisen, T. Damhus, K.S. Johansen, *J. Colloid Interface Sci.* **284**, 99–106 (2005).
254. H.L. Buss, S.L. Brantley and L.J. Liermann, *Geomicrobiol. J.* **20**, 25–42 (2003).
255. D.L. Stoner, J.E. Wey, K.B. Barrett, J.G. Jolley, R.B. Wright and P.R. Dugan, *Appl. Environ. Microbiol.* **56**, 2667–2676 (1990).
256. B.E. Kalinowski, L.J. Liermann, S.L. Brantley, A. Barnes and C.G. Pantano, *Geochim. Cosmochim. Acta* **64**, 1331–1343 (2000).
257. T.S. Magnuson, A.L. Neal and G.G. Geesey, *Microb. Ecol.* **48**, 578–588 (2004).
258. A.L. Neal, L.K. Clough, T.D. Perkins, B.J. Little and T.S. Magnuson, *FEMS Microbiol. Ecol.* **49**, 163–169 (2004).
259. V. Di Castro, G. Polzonetti, G. Contini, C. Cozza and B. Paponetti, *Surf. Interface Anal.* **16**, 571–574 (1990).
260. A.L. Neal, J.E. Amonette, B.M. Peyton and G.G. Geesey, *Environ. Sci. Technol.* **38**, 3019–3027 (2004).
261. H.Yang, E. Gong, L. Yang, G. Chen, Y. Fan, Y. Zhao and J. Lue, *Trans. Nonferrous Met. Soc. China* **14**, 1187–1191 (2004).
262. J.M. Mcintosh and L.A. Groat, *Short Course Series – Mineralogical Association of Canada* **25** (Biological–Mineralogical Interactions), 25–62 (1997).
263. G. Chen and C.R. Clayton, *J. Electrochem. Soc.* **144**, 3140–3146 (1997).
264. G. Chen and C.R. Clayton, *J. Electrochem. Soc.* **145**, 1914–1922 (1998).
265. G. Chen and C.R. Clayton, *Surf. Interface Anal.* **27**, 230–235 (1999).
266. L.-S. Johansson and T. Saastamoinen, *Appl. Surf. Sci.* **144–145**, 244–248 (1999).
267. I.B. Beech, V. Zinkevich, R. Tapper, R. Gubner and R. Avci, *J. Microbiol. Methods* **36**, 3–10 (1999).
268. B.H. Olesen, R. Avci and Z. Lewandowski, *Corrosion Sci.* **42**, 211–227 (2000).
269. M. Yokota, F. Sugaya, H. Mifune, Y. Kobori, K. Shimizu, K. Nakai, S.-I. Miyahara and Y. Shimizu, *Mater. Trans.* **44**, 268–276 (2003).
270. H.A. Videla, *Rev. Metal. (Madrid)*, Vol. Extr., 256–264 (2003).
271. B.D. Ratner, in *Biomaterials Science: An Introduction to Materials in Medicine*, B.D. Ratner, A.S. Hoffman, F.J. Schoen and J.E. Lemons (eds.), Academic Press, San Diego, 1996, pp. 21–35.
272. J.P. Lucchini, J.L. Aurelle, M. Therin, K. Donath and W. Becker, *Clin. Oral Implant. Res.* **7**, 397–404 (1996).
273. M. Morra, C. Cassinelli, G. Bruzzone, A. Carpi, G. Di Santi, R. Giardino and M. Fini, *Int. J. Oral Maxillofac. Implants* **18**, 40–45 (2003).
274. M.C. Davies, R.D. Short, M.A. Khan, J.F. Watts, A. Brown, A.J. Eccles, P. Humphrey, J.C. Vickerman and M. Vert, *Surf. Interface Anal.* **14**, 115–120 (1989).
275. B.D. Ratner and R.W. Paynter, in *Polyurethanes in Biomedical Engineering*, H. Plank, G. Egbers, I. Syre (eds.), Elsevier, Amsterdam, 1984, pp. 41–68.
276. V. Sa Da Costa, D. Brier-Russell, E.W. Salzman and E.W. Merrill, *J. Colloid Interface Sci.* **80**, 445–452 (1981).
277. B.D. Ratner, *Prog. Biomed. Eng.* **5**, 87–98 (1988).
278. T.G. Grasel and S.L. Cooper, *Biomaterials* **7**, 315–328 (1986).
279. A.D. Cook, J.S. Hrkach, N.N. Gao, I.M. Johnson, U.B. Pajavni, S.M. Cannizzaro and R. Langer, *J. Biomed. Mater. Res.* **35**, 513–523 (1997).
280. M. Okubo and H. Hattori, *Colloid Polym. Sci.* **271**, 1157–1164 (1993).
281. J. Ji, L. Feng, J. Shen and M.A. Barbosa, *J. Biomed. Mater. Res.* **61**, 252–259 (2002).
282. S. Guo, L. Shen and L. Feng, *Polymer* **42**, 1017–1022 (2001).
283. A. Yamasaki, Y. Imamura, K. Kurita, Y. Iwasaki, N. Nakabayashi and K. Ishihara, *Colloids Surf. B: Biointerfaces* **28**, 53–62 (2003).
284. K. Ishihara, T. Hasegawa, J. Watanabe and Y. Iwasaki, *Artif. Organs* **26**, 1014–1019 (2002).
285. X.H. Wang, D.P. Li, W.J. Wang, Q.L. Feng, F.Z. Cui, Y.X. Xu, X.H. Song and M. van der Werf, *Biomaterials* **24**, 3213–3220 (2003).
286. N.P. Huang, R. Michel, J. Voros, M. Textor, R. Hofer, A. Rossi, D.L. Elbert, J.A. Hubbell and N.D. Spencer, *Langmuir* **17**, 489–498 (2001).

287. M. Morra and C. Cassinelli, *Surf. Interface Anal.* **26**, 742–747 (1998).
288. K. Merrett, C.M. Griffith, Y. Deslandes, G. Pleizier, M.A. Dube and H. Sheardown, *J. Biomed. Mater. Res. A* **67A**, 981–993 (2003).
289. P. Kingshott, J. Wei, D. Bagge-Ravn, N. Gadegaard and L. Gram, *Langmuir* **19**, 6912–6921 (2003).
290. T.W. Chung, Y.F. Lu, H.Y. Wang, W.P. Chen, S.S. Wang, Y.S. Lin and S.H. Chu, *Artif. Organs* **27**, 155–161 (2003).
291. Y.S. Lin, S.S. Wang, T.W. Chung, Y.H. Wang, S.H. Chiou, J.J. Hsu, N.K. Chou, K.H. Hsieh and S.H. Chu, *Artif. Organs* **25**, 617–621 (2001).
292. J. Marchand-Brynaert, E. Detrait, O. Noiset, T. Boxus, Y.J. Schneider and C. Remacle, *Biomaterials* **20**, 1773–1782 (1999).
293. D. Klee, Z. Ademovic, A. Bosserhoff, H. Hoecker, G. Maziolis and H.J. Erli, *Biomaterials* **24**, 3663–3670 (2003).
294. M. Morra, C. Cassinelli, G. Cascardo, P. Cahalan, L. Cahalan, M. Fini and R. Giardino, *Biomaterials* **24**, 4639–4654 (2003).
295. Y. Chevolot, J. Martins, N. Milosevic, D. Leonard, S. Zeng, M. Malissard, E.G. Berger, P. Maier, H.J. Mathieu, D.H.G. Crout and H. Sigrist, *Bioorg. Med. Chem.* **9**, 2943–2953 (2001).
296. Y. Li, K.G. Neoh, L. Cen and E.T. Kang, *Biotechnol. Bioeng.* **84**, 305–313 (2003).
297. B. Lindberg, R. Maripuu, K. Siegbahn, R. Larsson, C.G. Goelander and J.C. Eriksson, *J. Colloid Interface Sci.* **95**, 308–321 (1983).
298. H.J. Salacinski, G. Hamilton and A.M. Seifalian, *J. Biomed. Mater. Res. A* **66A**, 688–697 (2003).
299. Z. Ding, J. Chen, S. Gao, J. Chang, J. Zhang and E.T. Kang, *Biomaterials* **25**, 1059–1067 (2004).
300. L. Dai, H.A.W. StJohn, J. Bi, P. Zientek, R.C. Chatelier and H.J. Griesser, *Surf. Interface Anal.* **29**, 46–55 (2000).
301. S.P. Massia, M.M. Holecko and G.R. Ehteshami, *J. Biomed. Mater. Res. A* **68A**, 177–186 (2004).
302. M. Shen, Y.V. Pan, M.S. Wagner, K.D. Hauch, D.G. Castner, B.D. Ratner and T.A. Horbett, *J. Biomater. Sci.: Polym. Ed.* **12**, 961–978 (2001).
303. Z.F. Li and E. Ruckenstein, *J. Colloid Interface Sci.* **269**, 62–71 (2004).
304. J.G. Archambault and J.L. Brash, *Colloids Surf. B: Biointerfaces* **33**, 111–120 (2004).
305. B. Gupta, C. Plummer, I. Bisson, P. Frey and J. Hilborn, *Biomaterials* **23**, 863–871 (2002).
306. J.A. Hayward, A.A. Durrani, Y. Lu, C.R. Clayton and D. Chapman, *Biomaterials* **7**, 252–258 (1986).
307. K.C. Popat, R.W. Johnson and T.A. Desai, *Surf. Coat. Technol.* **154**, 253–261 (2002).
308. M. Chen, P.O. Zamora, P. Som, L.A. Pena and S. Osaki, *J. Biomater. Sci.: Polym. Ed.* **14**, 917–935 (2003).
309. N.L. Hernandez de Gatica, G.L. Jones and A. Joseph Jr., *Appl. Surf. Sci.* **68**, 107–121 (1993).
310. S. Lerouge, C. Guignot, M. Tabrizian, D. Ferrier, N. Yagoubi and L.H. Yahia, *J. Biomed. Mater. Res.* **52**, 774–782 (2000).
311. J. Hyun, Y. Zhu, A. Liebmann-Vinson, T.P. Beebe Jr. and A. Chilkoti, *Langmuir* **17**, 6358–6367 (2001).
312. J.L. Dewez, J.-B. Lhoest, E. Detrait, V. Berger, C.C. Dupont-Gillain, L.M. Vincent, Y.-J. Schneider, P. Bertrand and P.G. Rouxhet, *Biomaterials* **19**, 1441–1445 (1998).
313. M. Winkelmann, J. Gold, R. Hauert, B. Kasemo, N.D. Spencer, D.M. Brunette and M. Textor, *Biomaterials* **24**, 1133–1145 (2003).
314. C.R. Wu, J.O. Nilsson and W.R. Salaneck, *Phys. Scr.* **35**, 586–589 (1987).
315. K. Uvdal, P. Bodoe, A. Ihs, B. Liedberg and W.R. Salaneck, *J. Colloid Interface Sci.* **140**, 207–216 (1990).
316. A. Ihs, B. Liedberg, K. Uvdal, C. Toernkvist, P. Bodoe and I. Lundstroem, *J. Colloid Interface Sci.* **140**, 192–206 (1990).
317. K. Uvdal, P. Bodoe and B. Liedberg, *J. Colloid Interface Sci.* **149**, 162–173 (1992).
318. J.M. Gold, M. Schmidt and S.G. Steinemann, *Helvetica Phys. Acta* **62**, 246–249 (1989).

319. R.W. Paynter and B.D. Ratner, in *Surface and Interfacial Aspects of Biomedical Polymers*, Vol. 2, J.D. Andrade (ed.), Plenum Press, New York, 1985, Chap. 5, pp. 189–216.
320. H. Fitzpatrick, P.F. Luckham, S. Eriksen and K. Hammond, *J. Colloid Interface Sci.* **149**, 1–9 (1992).
321. J.E. Sundgren, P. Bodo, B. Ivarsson and I. Lundstroem, *J. Colloid Interface Sci.* **113**, 530–543 (1986).
322. J.-B. Lhoest, E. Detrait, Ph. Van den Bosch de Aguilar and P. Bertrand, *J. Biomed. Mater. Res.* **41**, 95–103 (1998).
323. R.W. Paynter, B.D. Ratner, T.A. Horbett and H.R. Thomas, *J. Colloid Interface Sci.* **101**, 233–245 (1984).
324. Q. Garrett, R.C. Chatelier, H.J. Griesser and B.K. Milthorpe, *Biomaterials* **19**, 2175–2186 (1998).
325. R. Margalit and R.P. Vasquez, *J. Protein Chem.* **9**, 105–108 (1990).
326. C.G. Gölander, S. Joensson, T. Vladkova, P. Stenius and J.C. Eriksson, *Colloids Surf.* **21**, 149–165 (1986).
327. M.S. Wagner, S.L. McArthur, M. Shen, T.A. Horbett and D.G. Castner, *J. Biomater. Sci.: Polym. Ed.* **13**, 407–428 (2002).
328. P. Kingshott, S. McArthur, H. Thissen, D.G. Castner and H.J. Griesser, *Biomaterials* **23**, 4775–4785 (2002).
329. M.S. Wagner, T.A. Horbett and D.G. Castner, *Langmuir* **19**, 1708–1715 (2003).
330. M.S. Wagner, T.A. Horbett and D.G. Castner, *Biomaterials* **24**, 1897–1908 (2003).
331. V. Bartzoka, M.A. Brook and M.R. McDermott, *Langmuir* **14**, 1887–1891 (1998).
332. J.L. Dewez, Y.J. Schneider and P.G. Rouxhet, *J. Biomed. Mater. Res.* **30**, 373–383 (1996).
333. E. Detrait, J.B. Lhoest, P. Bertrand and Ph. van den Bosch de Aguilar, *J. Biomed. Mater. Res.* **45**, 404–413 (1999).
334. J.G. Jolley, G.G. Geesey, M.R. Hankins, R.B. Wright and P.L. Wichlacz, *Surf. Interface Anal.* **11**, 371–376 (1988).
335. Z. Zhang, M. King, R. Guidoin, M. Therrien, C. Doillon, W.L. Diehl-Jones and E. Huebner, *Biomaterials* **15**, 1129–1144 (1994).
336. Y.W. Tang, R.S. Labow, I. Revenko and J.P. Santerre, *J. Biomater. Sci.: Polym. Ed.* **13**, 463–483 (2002).
337. H.H. Huang, *J. Biomed. Mater. Res.* **60**, 458–465 (2002).
338. H. Lin and J.D. Bumgardner, *Biomaterials* **25**, 1233–1238 (2004).
339. B. Feng, J. Weng, B.C. Yang, S.X. Qu and X.D. Zhang, *Biomaterials* **24**, 4663–4670 (2003).
340. K. Mustafa, J. Pan, J. Wroblewski, C. Leygraf and K. Arvidson, *J. Biomed. Mater. Res.* **59**, 655–664 (2002).
341. S. Hiromoto, T. Hanawa and K. Asami, *Biomaterials* **25**, 979–986 (2004).
342. S. Seal, T.L. Barr, S. Krezoski, D.H. Petering and W. Antholine, *J. Vac. Sci. Technol. A* **15**, 1235–1245 (1997).
343. S. Seal, T.L. Barr, S. Krezoski and D. Petering, *Appl. Surf. Sci.* **173**, 339–351 (2001).
344. S. Seal, S. Krezoski, D. Petering, T.L. Barr, J. Klinowski and P. Evans, *J. Vac. Sci. Technol. A* **14**, 1770–1778 (1996).
345. S. Seal, S. Krezoski, T.L. Barr, D.H. Petering, J. Klinowski and P.H. Evans, *Proc. R. Soc. London Ser. B* **263**, 943–951 (1996).
346. J. Hofmann, R. Michel, R. Holm and J. Zilkens, *Surf. Interface Anal.* **3**, 110–117 (1981).
347. S. Kamalesh, P. Tan, J. Wang, T. Lee, E.T. Kang and C.H. Wang, *J. Biomed. Mater. Res.* **52**, 467–478 (2000).
348. R.W. Paynter, H. Martz and R.G. Guidoin, *Biomaterials* **8**, 94–99 (1987).
349. Z. Zhang, Y. Marois, R.G. Guidoin, P. Bull, M. Marois, T. How, G. Laroche and M.W. King, *Biomaterials* **18**, 113–124 (1997).
350. M. Richard, Th. Le Mogne, A. Perret-Liaudet, G. Rauwel, J. Criquelion, M.I. De Barros, J.C. Cetre and J.M. Martin, *Appl. Surf. Sci.* **240**, 204–213 (2005).
351. T.A. Wollin, T.C. Tieszer, J.V. Riddell, J.D. Denstedt and G. Reid, *J. Endourol.* **12**, 101–111 (1998).

352. G. Reid, R. Davidson and J.D. Denstedt, *Surf. Interface Anal.* **21**, 581–586 (1994).
353. S.L. McArthur, K.M. McLean, H.A.W. St.John and H.J. Griesser, *Biomaterials* **22**, 3295–3304 (2001).
354. G.M. Bruinsma, J. Devries, H.C. van der Mei and H.C. Busscher, *J. Adhesion Sci. Technol.* **15**, 1453–1462 (2001).
355. G.M. Bruinsma, M. Rustema-Abbing, J. De Vries, B. Stegenga, H.C. van der Mei, M.L. Van der Linden, J.M.M. Hooymans and H.J. Busscher, *Invest. Ophthalmol. Vis. Sci.* **43**, 3646–3653 (2002).
356. L. Pickart, M.M. Millard, B. Beiderman and M.M. Thaler, *Biochim. Biophys. Acta* **544**, 138–143 (1978).
357. R.G. Meisenheimer, J.W. Fischer and S.J. Rehfeld, *Biochem. Biophys. Res. Commn.* **68**, 994–999 (1976).
358. J.M. Schakenraad, H.C. van der Mei, P.G. Rouxhet and H.J. Busscher, *Cell Biophys.* **20**, 57–67 (1992).
359. J.M. Solletti, M. Botreau, F. Sommer, W.L. Brunat, S. Kasas, T.M. Duc and M.R. Celio, *Langmuir* **12**, 5379–5386 (1996).
360. A. Ulman, *Chem. Rev.* **96**, 1533–1554 (1996).
361. M. Mrksich and G.M. Whitesides, *Annu. Rev. Biophys. Biomolec. Struct.* **25**, 55–78 (1996).
362. E. Ostuni, L. Yan and G.M. Whitesides, *Colloids Surf. B: Biointerfaces* **15**, 3–30 (1999).
363. M. Zharnikov and M. Grunze, *J. Phys.: Condens. Matter* **13**, 11333–11365 (2001).
364. A.S. Duwez, *J. Electron Spectrosc. Relat. Phenom.* **134**, 97–138 (2004).
365. M.C. Martins, B.D. Ratner and M.A. Barbosa, *J. Biomed. Mater. Res. A* **67A**, 158–171 (2003).
366. V.A. Tegoulia and S.L. Cooper, *J. Biomed. Mater. Res.* **50**, 291–301 (2000).
367. H. Elwing, S. Welin, A. Askendal, U. Nilsson and I. Lundstroem, *J. Colloid Interface Sci.* **119**, 203–210 (1987).
368. T.G. Ruardy, J.M. Schakenraad, H.C. Van der Mei and H.J. Busscher, *Surf. Sci. Rep.* **29**, 1–30 (1997).
369. M. Henry, C. Dupont-Gillain and P. Bertrand, *Langmuir* **19**, 6271–6276 (2003).
370. T. Kinoshita, *J. Electron Spectrosc. Relat. Phenom.* **124**, 175–194 (2002).
371. P.G. Rouxhet, A.M. Misselyn-Bauduin, F. Ahimou, M.J. Genet, Y. Adriaensen, T. Desille, P. Bodson and C. Deroanne, *Surf. Interface. Anal.* **40**, 718–724 (2008).
372. J. Walton and N. Fairley, *Surf. Interface Anal.* **36**, 89–91 (2004).
373. C.C. Dupont-Gillain and P.G. Rouxhet, *Nano Lett.* **1**, 245–251 (2001).
374. R.S. Pembrey, K.C. Marshall and R.P. Schneider, *Appl. Environ. Microbiol.* **65**, 2877–2894 (1999).
375. F. Ahimou, C.J.P. Boonaert, Y. Adriaensen, P. Jacques, Ph. Thonart, M. Paquot and P.G. Rouxhet. *J. Colloid Interface Sci.* **309**, 49–55 (2007).

Appendix

Symbols and acronyms used in the text are defined below. To present a consistent picture some symbols have been changed with respect to the original papers.

Lowercase Letters

ac	Acetal
ad	Adsorbed
bdl	Below detection limit
com	Computed
cont	Contamination
d	Distance traveled by a photoelectron
exp	Experimental
$h\nu$	Energy of X-ray photons
i	Sensitivity factor
k, k^*, k''	Constants
l	Attenuation length (AL)
m	Exponent in the equation giving the dependency of λ on E_k
n	Exponent in the equation giving the dependency of T on E_k
su	Substrate
t, t^*	Thickness
x	Cartesian coordinate in the analyzed solid
y	Cartesian coordinate in the analyzed solid
z	Cartesian coordinate in the analyzed solid; depth from which an electron is photoemitted

Capital Letters

A	Element
A^{+*}	Ion in an excited state
Aa	Peak a of element A
B	Element
Bb	Peak b of element B
C	Concentration; carbon; element
C_{ox}	Peak components due to carbon bound to oxygen or nitrogen
D_{Aa}	Efficiency of the detector
E'_k	Kinetic energy measured by the spectrometer
E_b	Binding energy
E_{bl}	Binding energy at which the low binding energy side of a peak meets the baseline
E_c	Charging term

E_k	Kinetic energy
E_p	Pass energy
G	Particular direction of photoelectron collection
H	Path traveled by photoelectrons
HC	Hydrocarbonlike compounds
I	Peak intensity
J	Average X-ray flux
J'	X-ray flux along direction X
$K_{\alpha 1,2}$	Main X-ray line doublet
$K_{\alpha 3,4}$	Secondary X-ray line
K_β	Secondary X-ray line
L'_{Aa}	Angular asymmetry factor
$L_1, L_{2,3}$	Notation used for 2s and 2p subshells in Auger spectroscopy
L_{Aa}	Average angular asymmetry factor
N_{nonpr}	Nonprotonated nitrogen
N_{pr}	Protonated nitrogen
P_A	Constant for given element A and recording conditions
Pr	Proteins
PS	Polysaccharides
Q	Probability that a photoelectron leaves a solid; amount adsorbed per unit area
R_0	Median radius of the analyzer
R_1	Radius of analyzer internal hemisphere
R_2	Radius of analyzer external hemisphere
R_{Aa}	Factor accounting for surface roughness
S	Irradiated area; surface area
T'_{Aa}	Transmission function
T_{Aa}	Average transmission function
V_1	Electrical potential of the analyzer internal hemisphere
V_2	Electrical potential of the analyzer external hemisphere
W	Slit width; displacement of energy levels upon the formation of a doubly ionized atom
X	Cartesian coordinate for the sample stage; particular direction of incident X-rays; element
Y	Cartesian coordinate for the sample stage; element
Z	Cartesian coordinate for the sample stage; element

Greek Letters

ΔE	Absolute resolution of the HSA
ΔE_n	Intrinsic line width
ΔE_p	Width given by the kinetic energy analyzer
ΔE_x	Width of the excitation source

Φ_{sp}	Work function
Γ_X	Surface concentration of element X
α	Angle of injection of electrons to the tangential direction in the analyzer; Auger parameter
α'	Modified Auger parameter
β	Factor used to calculate the angular asymmetry factor L
δ	Angle between the direction of incident X-rays and the normal to the sample-holder plane
γ	Angle between the photoemission direction and the incident X-ray direction
φ	Degree of surface coverage
λ	Inelastic mean free path
θ	Angle between the direction of electron photoemission and the normal to the surface of the solid
ρ	Density
σ	Photoemission cross-section
τ	Life time of the core hole created by the photoemission

Acronyms

AFM	Atomic force microscopy
AL	Attenuation length
ARXPS	Angle resolved X-ray photoelectron spectroscopy
CAE	Constant analyzer energy
CHA	Concentric hemispherical analyzer
CHO	Polymers containing C, H, and O
DLD	Delay line detector
DPPC	Dipalmitoyl phosphatidylcholine
DPPE	Dipalmitoyl phosphatidylethanolamine
ELP	Energy loss peak
EPS	Extracellular polymeric substances
ESCA	Electron spectroscopy for chemical analysis
FEP	Perfluorinated ethylene-propylene copolymer
FWHM	Full width at half maximum
GRGD	Glycine-arginine-glycine-aspartic acid
HOPG	Highly oriented pyrolytic graphite
HSA	Electrostatic hemispherical analyzer
IEP, iep	Isoelectric point
IERF	Intensity/energy response function
IgG	Immunoglobulin G
IKVAV	Isoleucine-lysine-valine-alanine-valine
IMFP	Inelastic mean free path
IR	Infrared
LB	Langmuir–Blodgett

LCS	Lens care solution
LDPE	Low-density polyethylene
LLDPE	Linear low-density polyethylene
LTA-TA	Lipoteichoic and teichoic acids
MIC	Microbiologically influenced corrosion
NC+CA	Nitrocellulose + cellulose acetate
PA-6	Polyamide 6 (Nylon ™)
PAA	Poly(acrylic acid)
PBS	Phosphate buffer saline
PC	Poly(bisphenol A carbonate)
PC	Phosphatidylcholine
PDMS	Poly(dimethyl siloxane)
PE	polyethylene
PEEK	Poly(aryl ether ether ketone)
PEG	Poly(ethylene glycol)
PEI	Poly(ethylene imine)
PEO	Poly(ethylene oxide)
PET	Poly(ethylene terephthalate)
PET-COOH	Carboxylated poly(ethylene terephthalate)
PET-OH	Hydroxylated poly(ethylene terephthalate)
PETox	Oxidized poly(ethylene terephthalate)
PEU	Polyetherurethane
PHEMA	Poly(2-hydroxyethyl methacrylate)
PLA	Poly(lactic acid)
PLL	Poly(L-lysine)
PMMA	Poly(methyl methacrylate)
PP	Polypropylene
PPO	Poly(propylene oxide)
PS	Polystyrene
PSox	Oxidized polystyrene
PTFE	Poly(tetrafluoroethylene) (Teflon™)
PTMSP	Poly(1-trimethylsilyl-1-propyne)
PU	Polyurethane
PVA	Poly(vinyl alcohol)
PVC	Poly(vinyl chloride)
PVdF	Poly(vinylidene fluoride)
PVdFox	Oxidized PVdF
PVMK	Poly(vinyl methyl ketone)
PVTFA	Poly(vinyl trifluoroacetate)
RGD	Arginine-glycine-aspartic acid
S/N	Signal/noise ratio
SAC	Sample analysis chamber
SAM	Self-assembled monolayer
SEM	Scanning electron microscopy

SIMS	Secondary ion mass spectroscopy
SMA	Spherical mirror analyzer
SRB	Sulfate-reducing bacteria
SS	Stainless steel
TCPS	Tissue-culture polystyrene
TGF-β2	Transforming growth factor β2
ToF-SIMS	Time-of-flight secondary ion mass spectroscopy
TPP-2	Tanuma, Powell and Penn relation
TPP-2M	Improved Tanuma, Powell and Penn relation
UHV	Ultrahigh vacuum
UPS	Ultraviolet photoelectron spectroscopy
XPS	X-ray photoelectron spectroscopy

Index

A
Abelcet®, 187
Abou-Arab, T.W., 204
Abraxane®, 187
Acrivos, A., 219
Adsorbed biomolecules, 99
Aerodynamic diameter, of inhaled particle, 192
Aerosols
 definition of, 206
 particle motion of, 213, 214
 as respiratory therapeutic agent, 191, 192
AFM examination
 basidiospores, 128
 bilayer, 153
 monolayers, 153, 154
 patterned collagen layers, 145–147
 PS surface, 99, 100
 surface coverage-thickness, adsorbed phases, 96, 97, 147
Agarwal, J.K., 202
Ahmadi, G., 201–204, 224, 239, 241, 243, 245, 256–258, 263
Aidun, C.K., 269
Air-jet milling process, for pharmaceutical manufacturing, 176
Alberts, B., 258
Albumin adsorption, 147, 155
Alpha-lactalbumin, 131
Alveolar cavities, particle transport and deposition in, 256–258
Ambisome®, 187
Amine functions, 142
Amorphous gate (bitmap), 2
Analyzer transmission function, 72, 73
Angle-resolved XPS
 sample charging
  and charge stabilization, 58–60
 flood gun energy, 59
 monochromatized X-ray beam, 58
 secondary electrons and photoelectrons, 57–58
 semi-quantitative use of background shape, 89
 PP, 87
Angular asymmetry factor
 for photoelectron emission, 70
 for spectrometers, 71
Anjilvel, S., 239
Antara™. *See* Fenofibrate drug
Antibody-conjugated beads, and light scatter shifts of, 16
Arnason, G., 201, 202
Artefacts, 104, 105
Artoli, A.M., 263
ARXPS. *See* Angle-Resolved XPS
Asgharian, B., 204, 239
Asher, S.A., 289
Atherosclerosis, 264–267
AUC. *See* Extents of absorption
Auger electron kinetic energy, 64
Auger peaks
 and Auger parameter, 64, 65
 photoionization and, 63, 64
Azospirillum brasilense
 proteins, 112
 surface composition of, 122, 123

B
Bacillus subtilis
 C_{ox}/C for, 108
 N 1s peak of, 106, 107
 peak decomposition of, 110, 111
 wide scan XPS spectra of, 49
Background subtraction, 77, 78

Bacterial adhesion, 131, 139
Balashazy, I., 239
Ball milling process, for pharmaceutical manufacturing, 176
Baron, P.A., 206
Basset-Boussinesq-Oseen equation, 201
BBO. *See* Basset-Boussinesq-Oseen equation
Bead–blood assays
  flow cytometry advantages in, 19
  using CD3 antibody–beads
    antibody lot dependence, 6, 7
    bead titer dependence, 7, 8
    factors affecting performance of, 4, 5
Bead–blood mixing time, 5, 7
Bead chemical sensors
  application for, 19
  Flow-Cal 575 fluorospheres, 18
Bead probes
  biological cell assays, problems in, 16
  CD3 antibody–beads, 4–9
  for flow cytometric analysis, 1
  light scatter nanoparticle
    fluorescence emission, 17–27
    luminescence emission, 27–29
    Raman scattering, 32–37
    surface plasmon resonance, 15–17
  light scatter shifts, 16
Bead suspension array technology, advantages of, 21
BEM. *See* Boundary element method
Berger, S.A., 258, 261, 262
Binding energy, 46
Biochemical compounds
  chemical functions of, 81–83
  peak components of, 80, 83
Biocorrosion, thin copper films, 150
Biodegradable polymers, 136
Biofilms, material surface conditioning with, 129, 131
Biomedical applications, 129
  materials, titanium implants, 135
  model systems, 153
  schematic representation of, 134
Biomolecules, adsorption
  binary and ternary mixtures, 148
  from single protein solutions, 144
Bio-Plex^{TM} suspension array system, 21
Biosorption process, 130
Biosystems, 43
Bisphosphonate ester, usage, 178
Blood flow
  atherosclerosis, 264–267
  composition of, 259, 260
  DNS methods of, 267–270
  red cells flow behavior, 261–264
  rheology of, 260, 261
Bogush, G. H., 284
Bone formation, 135
Boundary element method, 267
Bradley (initial not found), 291
Brady, J.F., 290
Bremsstrahlung radiation, 53
Brenner, H., 208, 261
Brewing yeast. *See Saccharomyces cerevisiae*
Britter, R.E., 202
Brooke, J.W., 204
Brownian motion process, 221–224
Büchel, G., 294
Budesonide particles, 177, 178, 191, 195
Buerk, D.E., 258
Buoyancy effect, of particle, 216, 217
Burshteyn, A., 4, 25

C
Cai, H., 265
Calabrese, R.V., 201
Cao, J., 204, 297
Caro, C.G., 265
Caruso, F., 297
Caruso, R.A., 297
CD3 antibody–beads
  bead–blood assays
    antibody lot dependence, 6, 7
    bead titer dependence, 7, 8
    factors affecting performance of, 4, 5
  bead-to-cell ratios for, 8, 9
CD45 antibody–magnetic beads, 9
CD3/CD3-PE competitive binding experiments, 7
CD4+ lymphocytes
  antibody activation, and bead–blood mixing time, effects of, 5, 6
  micrographs of, 3, 4
CD8+ lymphocytes, selective depletion of, 14
CD38-PE biological assay, 19
CdSe q-dots^{TM}, 29, 30
CdS particles
  emission colors of, 28
  energy transfer, 30–32
Cell assays
  needs for, 29, 30
  problems in, 16
  receptor sites and light scatter pattern, 16
Cell membrane, 153
Cells adhering, 131
CFCs. *See* Chorofluorocarbons
Chaffey, C., 261
Chaffey, C.E., 261

Index 311

Chang, I-S., 256–258
Channel plates, 56
Channeltrons, 56
Chan, T.L., 239
Charge stabilization devices, 60
Charnay, C., 297
Chemical analysis, 43, 44
Chemical functions
   of biochemical compounds, 83
   components of, 79, 80
   peak decomposition and assignment
      oxygen and nitrogen peak, 111
   of solids, 81, 82
Chemical shifts, 135
   biochemical compounds, 80–87
   on oxygen and carbon, 80
   polymers, 79, 80
Chen, B.T., 206
Chen, C.P., 204
Cheng, Y.S., 246, 253
Chen (initial not found), 269
Chen, J.F., 297
Chen, Q., 204
Cherukat, P., 218
Chorofluorocarbons, 191
Ciftcioglu, M., 286
Circulaire™ Aerosol Drug Delivery System
   jet nebulizer, 196
Cleaver, J.W., 202
Clickhaler®, 193
Cohen, B.S., 239
Cokelet, G.R., 262
Collins, L.R., 205
Colloidal crystals, 287
Comer, J.K., 239
Concentric hemispherical analyzer (CHA). *See*
   Electrostatic hemispherical analyzer
Contact lenses, 158
Continuous adlayer
   heterogeneity of zone, 91
   organic contamination influence on, 92, 93
   silicon wafers, 90
Controlled precipitation, of pharmaceutical
   products, 177
Corrsin, S., 201
Corynebacterium glutamicum, phosphate
   concentration of, 130
Coulter cellular therapies, 14
Coulter ® EPICS™ EliteTM series flow
   cytometers, 1
Coulter ® STKS2B™ hematology analyzer
   assay

   for CD3+ lymphocytes
      antibody bead lots, 6, 7
      DF 8 (median angle light scatter) *vs.*
         Coulter volume, 5, 6
      vs. bead titers with whole blood, 7, 8
Cremaphor® EL, 187
Crowe, C.T., 204
Cryofixation, 118
Csanady, G.T., 201
Cundy, K. C., 179
Cunningham correction factor, 208, 209
Cyclosporine, 177

D
Danazol drug, 177, 179
Dandy, D.S., 218
Danocrine®. *See* Danazol
1D3 antibody–ferrite bead titers
   effect on neutrophils, 12
Darquenne (initial not found), 256
Davies, C.N., 202
Davies, P.F., 265
De Kruif, C.G., 290
Deleu, M., 153
Dental implants, 135
Derjaguin, B.V., 235
Diatrizoic acid, usage, 189, 190
Ding, E.-J., 269
Direct numerical simulations, 204, 205
   for human blood flow, 267–270
Discontinuous adlayer
   film organization, 96
   substratum
      peak intensity, 94
      plot of surface coverage, 95
Diskhaler® inhalation device, 193
Dispersed solids, 43, 44
DMT model, for particle adhesion, 235, 237, 238
DNA synthesis, of hepatoma cells, 153
DNS. *See* Direct numerical simulations
Donnelly, J. P., 187, 188
Doolen, G.D., 268
Doxil®, 187
DPI. *See* Dry powder inhaler device
DPPC monolayers, 154
Drag force and drag coefficient
   for liquid droplets, 209
   models for, 207
   wall effects on, 208
Drolet, F., 205
Drug dissolution time, 179
Drug particles size, reduction of, 175, 176
Drug residence time, 178, 179
Drust, F., 204

Druzhinin, O.A., 205
Dry powder aerosols delivery, to lung, 192–194
Dry powder inhaler device, 192
Dulling, B.R., 266
DuPin, M.M., 269
Durst, F., 202
Dwyer, H.A., 218
Dye-embedded core-shell nanoparticles, 33
Dynamic shape factor, of particle, 213

E

Eaton, J.K., 204, 205
Edwards, D. A., 193
Electroactive aniline polymer plates, 151
Electron
 photoejection process, 47
 trajectories of, 52
Electron microprobe, 44
Electron transmission function, 72
Electrostatic hemispherical analyzer
 components of, 54
 electron, trajectories of
  pass energy, 55
  tangential, 54
Electrostatic lens, electrons
 acceleration/retardation by, 55
 collection by, 54
Elghobashi, S.E., 202, 204, 205
Ellipsoidal particles, drag force for, 210–213
Ellison, J., 205
Emend®, 176
Emmett, S., 284, 286, 290
Entocort® EC drug, 185
Enzyme adsorption, 132
Extents of absorption, 180, 183
Extracellular polymer substances (EPS), 131

F

Fåhraeus–Lindqvist effect, 262
Fan, F.G., 202–204
Faxen, H., 201, 208
FEM. *See* Finite element method
Fenofibrate drug, 182
Fernandez de la Mora, J., 202
Ferrite-nanoparticle–monoclonal-antibody
 conjugates, 11, 12
Fibrinogen adsorption, 137
Fichman, M., 202
Fine-particle active agents, in pharmaceutical
 products methods of preparing, 175–177
 oral delivery of
  food effects on, 180–183
  oral bioavailability, 177–180

 rapid absorption, 183–185
 topical administration to gastrointestinal
  tract, 185, 186
 parenteral administration of, 186–191
 particle transport and deposition process
  adhesion of, 229–238
  aerosol particle motion, 213, 214
  aerosols, 206, 207
  atherosclerosis, 264–267
  blood composition, 259, 260
  Brownian motion process, 221–223
  computer simulation for, 224, 225
  DNS of blood flow, 267–270
  droplets, 209
  ellipsoidal and nonspherical particles,
   210–213
  Fokker–Planck equation, 223
  hydrodynamic forces, 207–209
  lift force of, 217–219
  in lung airways, 239, 240
  in multibifurcation airways, 240–245
  in nasal passages, 239, 240, 245–253
  oral airways, 253–258
  particle motion, equation of, 221
  particle shape factor, 213
  particles on plane, lift force, 219, 220
  red cells, flow behavior, 261–264
  rheology of blood flow, 260, 261
  terminal velocity of, 214–217
  in viscous sublayer, 225–229
  works on, 201–206
 pulmonary delivery of, 191–197
Finite element method, 267
First-principles approach, 74
Fisher, D.M., 239
Flood gun energy, 59
Flovent® Rotadisk®, 193
Flow-Cal 575 fluorospheres
 for CD38 assay, 19
 flow cytometric histogram of, 18
Flow chamber
 particle/cell suspension, 2
 sheath fluid, 1
Flow cytometry
 bead chemical sensors for
  application for, 19
  Flow-Cal 575 fluorospheres, 18
 blood analysis by
  CD3 antibody–beads, 4–9
  PS latex beads, 3–4
 CD3/CD3-PE competitive binding
  experiments, 7
 flow chamber, 1
 particle-based immunoassays for, 21

Index

particle/cell suspension, flow of, 2
photomultiplier tubes (PMTs), 2
suspension arrays, 21, 22
Flunisolide, 191
Fluorescence-labeled antibody, 25
Fluorescence resonance energy transfer
  CdS nanoparticle and PE acceptors, 30
  nanoparticle donors and tetramethylrhodamine (TMR) acceptors, 32
  tryptophan residue donors and nanoparticle acceptors, 30, 31
Fluorescent beads, 17, 18
Fluorescent dyes, 2
Fluticasone, 191
Fokker–Planck equation, 223, 224
Freeze-dried copolymers, 137
FRET. *See* Fluorescence resonance energy transfer
Friedlander, S.K., 202
Fuchs, N.A., 201
Full width at half the maximum intensity (FWHM)
  components of, 105
  peak decomposition, 78, 79
Fulwyler, Mack J., 3, 18, 19
Fung, Y.C., 258, 261

G
Gallily, I., 204
3G8 antibody (IgG1 isotype), 23
fluorescence intensity, 25
Gao, X., 297
Gastrointestinal (GI) tract treatment, topical drug administration, 185–186
3G8/3G8-FITC competitive binding trial
  antibody capture, 27
  competitive binding, 26
  fluorescence intensity, 24, 25
  Langmuir adsorption isotherm, 25
Giesche, H., 284
G-jitter excitation, 204, 205
Goldenberg, M., 204
Gold nanoparticles
  on aminodextran-coated PS beads, 34
  light scattering, 15
  plasmon extinction bandwidths, 17
Goldschmidt, V.W., 201
Goldsmith, H.L., 262
Gradon, L., 204
Gram-negative bacteria, 157
Granulocytes, depletion of, 11

H
Haefeli-Bleuer (initial not found), 256
Hahn, I., 246
Hall, D., 220
Hamaker constants, for dissimilar materials, 232
Hamaker, H.C., 232
Hanratty, T.J., 202
Harrison, D.G., 265
Hydrophobicity recovery, 63, 100
He, C., 202–204
Hedley, A.B., 202
Hematite surfaces, 132
Henglein, A., 28
Hetsroni (initial not found), 261
Heyder, J., 239
HFAs. *See* Hydrofluoroalkanes
Hidy, G. M., 202
High-energy milling process, for pharmaceutical manufacturing, 176
High performance liquid chromatography (HPLC), 290, 291
Hinds, W.C., 201, 245
Hinze, J. O., 201, 202
Histograms, 2
Hofmann, W., 239
Holm, R., 109
Homogenization process, for pharmaceutical manufacturing, 176
Hook effect, 19, 20
HP-β-XD. *See* Hydroxypropyl-β-cyclodextrin
HSA. *See* Electrostatic hemispherical analyzer
Human oral airways, particle transport and deposition in, 253–258
Human serum albumin (HSA)
  adsorbed on polystyrene (PS), 80, 85
  wide scan XPS spectra of, 49
Hydrofluoroalkanes, 192
Hydrogen sulfide gas, microbial production of, 130
Hydrophobic polymers, surface contamination of, 75
Hydroxypropyl-β-cyclodextrin, 180

I
IDD®tP MicroParticle technology, 183
IERF. *See* Intensity/energy response function
Iglesias, A.J., 239
Imaging applications, diagnostic, 189, 190
Inelastic mean free path (IMFP)
  energy band gap for, 98
  vs. photoelectron kinetic energy, 71, 72
Inipenem administration, 188

Intensity/energy response function
  of spectrometer, 72
  and XPS data, 73
Intramuscular drug administration, 188, 189
Intravenous drug administration, 186–188
Ion gun, 53
Itraconzole drug, 188

J
JKR Model, for particle adhesion, 234, 236, 237
Johnson, J.R., 239
Johnson, K.L., 234
Johnstone, H.F., 202
Joseph, R., 263
Jou, L.-D., 258, 261, 262
Jurewicz, J.T., 204

K
Kaazempur-Mofrad, M.R., 262, 267
Kader, B.A., 202
Kaiser, C., 284
Kamm, R.D., 258
Kelly, J.T., 246
Kendall, K., 234
Kerker, M., 32
Keyhani, K., 246
Key–lock recognition groups, 298
Kim, C.S., 239
Kinkel, J.N., 285, 291
Kirkland, J.J., 291
Kleinstreuer, C., 253
Kricka, L. J., 3
Ku, D.N., 258
Kvasnak, W., 202, 204

L
Lactobacillus helveticus strain, surface analysis of, 115
Lactococcus lactis
  surface of, 115
  XPS analysis of, 116
Lane, D.D., 202
langevin equation, 222–224
Lattice Boltzmann method, 267
LBM. *See* Lattice Boltzmann method
Leal, L.G., 261
Lee, S.L., 202
Leighton, D.T., 219
Lens care solutions (LCS), 158
Leuprolide acetate, administration, 188
Levich, V., 201
Li, A., 202, 204, 224, 239, 241

Light scatter bead
  CD3 antibody–beads, 4–9
  PS latex beads, 3, 4
  surface plasmon resonance
    gold colloid-labeled cells, 15
    side scatter intensities, 16
Light scatter shifts
  of antibody-conjugated beads, 16
  of metal nanoparticles, 17
Limit of detection, 109
Limit of quantification, 109
Liu, B.Y.H., 202
Liversidge, G.G., 179, 180
Loratadine, 177
Low Reynolds number (LRN) k-ω *turbulence model*, 253
Lumley, J.L., 201
Lysozyme molecules, 147

M
Magnetic beads
  CD45 antibody-conjugated, 9
  depletion characteristics of
    CD8+ lymphocytes, 14
    granulocytes, 11
    leukocytes, 14
    neutrophils, 11, 12, 26
    RBCs, and platelets, 11, 12
    WBCs, 9
  normal cells, removal of, 14, 15
  size distribution and magnetism in, 9
Ma, J., 295
Mammalian cell surfaces, analysis of, 152, 153
Manganese ferrite nanoparticles
  monoclonal antibodies conjugated, 11
  SEM photomicrograph of, 10
  size distribution of, 9, 10
Maranzano, B.J., 290
Markowitz (initial not found), 285
Marshall, C, 110
Marshall, J.R., 205
Martonen, T.B., 239, 246
Mass median aerodynamic diameter, 192
Matrix metalloproteinases, 265
Maugis, D., 235
Maugis–Pollock model, 235, 236
Maxey, M.R., 201
Ma, Y., 295
Mazaheri, A.R., 239, 241, 243, 245
McCoy, D. D., 202
Mckee, S., 202
McLaughlin, J.B., 203, 204, 218
MCP1. *See* Monocyte chemoattractant peptide-1

Megace®gES. *See* Megestrol acetate drug
Megestrol acetate drug, 181
Mei, R., 218
Meloxicam, 183, 184
Mercer, T.T., 201
Merret, K., 138
Metal ion species, 130
Metal nanostructures, scattering excitation profiles for, 16, 17
Methyprednisolone acetate administration, 188
Microbial sample preparation, XPS
    approaches, 116
    freeze drying, 117
    Saccharomyces cerevisiae
        freeze dried, 117, 118
        with nonmonochromatized radiation, 119
    vitrification, 118
Microbial surfaces, 129
    liquid environment, interactions between, 130
    molecular composition of, 114, 115
Microbiologically influenced corrosion (MIC), 133
Microemulsion technique, 285
Microfluidizer®, 176
Microorganisms
    bioleaching, 132
    material surface conditioning with, 131
    surface analysis of, 109, 134
        Azospirillum brasilense, 122
        Bacillus subtilis, 110, 111
        Lactobacillus helveticus strain, 115
        Ph. crysosporium, 115
        Pseudomonas fluorescent strain, 114
    surface composition of
        Azospirillum brasilense, 122, 123
        brewing yeasts, 121, 122
        gram-positive and gram-negative bacteria, 120, 121
        hydrophobicity, relations with, 126, 127
        surface electrical properties, relations with, 124–126
Middleman, S, 201
Millard, M.M., 110, 153
Milne, S.J., 286
Mineral surface, 132
MMAD. *See* Mass median aerodynamic diameter
MMPs. *See* Matrix metalloproteinases
Mobic®. *See* Meloxicam
Mollinger, A.M., 220
Molybdate, 130
Monoclonal antibody-conjugated magnetic beads, 13, 14

Monoclonal-antibody conjugates
    CD101 and CD160, 22, 23
    ferrite-nanoparticle, 11
Monocyte chemoattractant peptide-1, 265
Mononuclear phagocytic system, 187
Monte Carlo simulation technique, 256
MPS. *See* Mononuclear phagocytic system
Muller, V.M., 235
Multi-analyte detection system, 19
Myoglobin, 147

N
NanoCrystal® colloidal dispersion, of meloxicam, 184
NanoCrystal® technology, 176
NanoCrystal solid dosage, 184
Nanoparticle bead
    fluorescence emission
        PS fluorospheres, 18
        silica-fluorescent dye beads, 20, 21
    magnetic beads, 9–14
    surface plasmon resonance
        gold/silver nanoparticles, 16, 17
        metallic microrods, 15
Naprosyn®, 188
Naproxen drug, 188
Nasal passages, particle transport and deposition in, 245–253
Natan, M., 15
Nerem, R.M., 265
Neutrophils
    1D3 antibody–ferrite bead titers effects on, 12–13
    depletion effi-ciency graph for, 11, 12
Newton regime, 207
Nieuwstadt, F.T.M., 220
Nitrogen signal, 138
Nonspherical particles, drag force for, 210
Nonsteroidal anti-inflammatory drugs (NSAIDS), 183
Noyes–Whitney equation, 179

O
Octreotide acetate, administration, 188
Omron MicroAir®, 195
Oral bioavailability, of drugs, 177–180
Organic emitters, 17
Ostrander, K. D., 195
Ounis, H., 202–204, 223, 225, 226
Ozin, A., 287

P
Paiva (initial not found), 256
Papavergos, P.G., 202
Particle-based immunoassays, 21, 22

Particle/cell suspension, 2
Particle deposition rate, simulation of, 202–204
Particle image velocimetry, 246
Particulate matter (PM) deposition, 239, 240
Patterson, G.S., Jr., 202
Paynter, R.W., 151
Peak decomposition
  potentialities and limitations of, 85, 86
  procedure description, 78, 79
Peak smoothing, 77
Peak width, 105
Peclet number and nose capture efficiency, 251, 252
Pentaamineruthenium, 147
Peroxisome proliferator-activated receptor α, 182
PET, modification scheme of, 140
Pharmaceutical dosage forms, composition of, 175
PHEMA grafts
  C 1s peak of, 101
  surface reorganization of, 100
Phenacetin drug, 179
Phosphate buffer saline (PBS), 131
Photoelectric effect, 46
Photoelectrons
  inelastic collisions, 48, 50
  injection of, sample, 55
Photolithography, 142, 143
Photomultiplier tubes, 2
3-D Photonic-bandgap structures, 287, 288
Phycoerythrin (PE), bilin groups in, 21
Physical adsorption, 139, 145, 149
PIV. See Particle image velocimetry
Plasma discharge, 142
Plasma polymerization, 138, 142
Plasma-sprayed surfaces, 135
Plasmon structures, 67
Platelet-rich plasma, 137
Platelets (PLTs) consumption
  by polyurethanes, 136
  depletion efficiency graph for, 11, 12
Pluronic adsorption, 149
pMDI. See Pressurized metered-dose inhaler
PMTs. See Photomultiplier tubes
Pollock, H.M., 235
Polychromatic X-ray beam, 53
Polyclonal antibody, 130
Poly(ethylene oxide) (PEO), 137
Polystyrene latex beads
  for blood cell analysis
    diameter of, 5
    light scatter patterns of, 3
  for CD16 antigen, 22

Polyurethane vascular grafts, 152
Polyvinylpyrrolidone, 180
Pore structure
  controlled, 297
  octahedral/tetrahedral, 286
Porous solids, 43, 44
Powder sintering, 286, 287
Pozrikidis, C., 258, 267
PPARα. See Peroxisome proliferator-activated receptor α
Prednisone drug, 191
Prescott, L. F., 179
Pressurized metered-dose inhaler, 191
  in fine particle delivery, 194, 195
Proctor, D.F., 245
Prolate ellipsoid of revolution, motion of, 211
Protein adsorption, 131, 132
Pruitt, J. D., 179, 183
Pseudomonas fluorescent strain, 114
PS latex beads. See Polystyrene latex beads
PS–magnetite beads
  light scatter shifts in, 9
Pulmicort® Respules®, 195
Pulmicort® Turbuhaler, 192
Pumping systems
  as contamination source, 75
  used on XPS spectrometers, 52
PVP. See Polyvinylpyrrolidone

Q

Qbead^{TM}, 29
Qi, D., 270
Quantum dotencoded bead system, applications of, 29

R

Raabe, O.G., 240
RAFC/RACS. See Raman-activated flow cytometer/cell sorter
Raman-activated flow cytometer/cell sorter, 32
Rapamune®, 176
Reactive oxide species, 265
Red blood cells (RBCs), depletion effi-ciency graph for, 11, 12
Reeks, M.W., 202
Respirable inhalation powders, preparation of, 193
Respiratory aerosols, in respiratory disease treatment, 191, 192
Reynolds number, 207
Rhone-Poulenc magnetic latex beads, 9, 10
Riley, J.J., 202
Risperidone, administration, 188
Rizk, M.A., 202
Roberts, A.D., 234

Index

ROS. *See* Reactive oxide species
Rouhiainen, P.O., 202
Ruddy, S. B., 185, 186
Ryde, N. P., 179

S

Saccharomyces cerevisiae, 130
　precision of analysis of, 119, 120
　sample preparation
　　freeze dried, 117, 118
　　with nonmonochromatized radiation, 119
　　scanning electron micrographs of, 117
　　XPS analysis of, 118
Saffman, P.G., 217, 218, 221
Sample preparation
　adsorbed biomolecules, 99, 100
　low temperature analysis, 101, 102
　sample alteration during, 102, 103
　surface reorganization, 100, 101
Schamberger, M. R., 204
Scherer, P.W., 247
Schiller, C.F., 239
Schreck, S., 246
Schroter, R.C., 239
Secondary electrons, 56
Sehmel, G.A., 202
Seinfeld, J.H., 201
Semiconductor nanoparticles
　close packing of, 29
　luminescence emission, 27, 28
Sensitivity factors, spectrometer, 74
SERS. *See* Surface enhanced Raman scattering
Shams, M., 225
Shan (initial not found), 269
Shapiro, M., 204
Signal/noise ratio (S/N), XPS instruments, 105–107
Siiman, O., 1–36
Silica-fluorescent dye beads, concentration quenching of, 20, 21
Silicate glass surfaces, 132
Silicon wafers
　for atomic force microscope investigations
　　XPS spectra of, 90
　Auger peaks, background rise, 62
　wide scan spectra of, 61
Silver nanoparticles
　on aminodextran-coated PS beads, 34
　elastic light scattering work on, 34
　light scattering, 15
　plasmon extinction bandwidths, 17
Skalak, R., 258, 261
Smith, D.H., 204
Snyder, W.H., 201

So, J-H, 290
Soltani, M., 203, 204
Spurny, K.R., 201
Squire, J.M., 266
Squires, K.D., 204
SRB. *See* Sulfate-reducing bacteria
Stachiewiz, J.W., 202
Steric repulsion, 130
Stöber silica particles
　medical applications of, 297–299
　preparation of, 283–285
　surface and bulk modifications of, 285
　technological applications of
　　chromatography, 290–295
　　3-D photonic-bandgap structures, 287, 288
　　hard sphere suspensions, rheology of, 290
　　particle packing and pore structure, 286
　　pigments, 295–297
　　powder sintering, 286, 287
　　sensors, 289
Stock, D.E., 202, 204
Stokes drag force, 207
Stopping distance, of particle, 215
Streptococcus sobrinus, 130
Strong J. C., 246
Stukel, J.J., 202
Sulfate-reducing bacteria, 130
Supercritical fluids, usage of, 177
Surface analysis
　vs. bulk analysis
　　depth profiling, 129
　　force–distance curves, 128
　　infrared bands, 127
　methods of, 48–51
　of microorganisms, 109
　　Azospirillum brasilense, 122
　　Bacillus subtilis, 110, 111
　　Lactobacillus helveticus strain, 115
　　Pseudomonas fluorescent strain, 114
　quantitative relationships
　　peak shape and elemental composition, 114
　　polysaccharides, 113, 114
　　proteins and esterphosphates, 113
　　proteins and polysaccharides, 112, 113
Surface chemical composition, 138
Surface coverage, 145
Surface electrical properties, 129
Surface enhanced Raman scattering, 32
　of gold nanoparticles, 32, 33
　red-shifted wavelength, 36
　of R6G
　　on gold nanoparticle/PS beads, 35
　　on silver nanoparticle/PS beads, 36

Surface modifications, 139
  biological exposure, 149
    bacterial culture, 150
    explantation, 151
  of biomaterials, 142
  by immobilization, 141
Surface-modified materials, 138
Surface reorganization
  freeze-drying, 118, 123
  silicone rubber substrate, 100
Surface topography, 135
Suspended particle dispersion, computer simulation techniques for, 202–204
Swift, D.L., 239, 245, 246, 250

T

Tavoularis, S., 202
Taxotere®, 187
Tchen, C.M., 201
Tetraethoxysilane (TEOS), 283
Therapeutic active agent, topical administration of, 185, 186
Tian, L., 239, 240
Tissue-culture polystyrene (TCPS), 142
Toporov, Y.P.T., 235
Triamcinolone acetonide, 191
TriCor®, 176, 182
Triglide^{TM}. See Fenofibrate drug
Trinh, E.H., 205
Trusdell, G.C., 202
Ttriptorelin pamoate, administration, 188
Tunable synchrotron sources, 53
Twomey, S., 201

U

UHV. See Ultra-high vacuum
Ultra-high vacuum
  contamination sources in, 74, 75
  Photoelectron spectroscopy, 46
Ultraviolet Photoelectron Spectroscopy (UPS), 46
Unger, K.K., 290, 291
Uronic acids, 131

V

Vacassy, R., 295
Vacuum-dried copolymers, 137
Van der Waals force, for particle, 230–232
Van Helden, A.K., 284, 290
Vascular cell adhesion molecule-1, 265
VCAM-1. See Vascular cell adhesion molecule-1
Vibrational fine structure, 67
Vinals, J., 205

Vincent, J.H., 201
Vink, H., 266
Vrij, A., 284

W

Wagner, J., 290
Wagner, M. S., 148
Wang, L.-P., 202
Wang, Q., 204
Wang, Z., 206
WBCs
  biological assays for, designing, 3
  depletion of, leuko-rich sample, 9–11
  enumeration of subsets of, 11, 12
  light scatter markers for, 11
Weibel, E.R., 240, 256
Weinbaum (initial not found), 266
Wood, N.B., 202
Wood, P.E., 204
Wright, J., 4

X

xMultianalyte Profile (xMAP^{TM}) technology, 21
XPS applications
  interfacial phenomena, microorganisms, 129
    biomedical applications, 133
    marine bacterial strains analysis, 130
    surface composition, 129
XPS imaging, image acquisition
  parallel, 57
  serial, 56
XPS peaks
  binding energy of, 48
  homogeneous solid, 49, 50
  human dental enamel, 57, 59
XPS spectra
  analyzed zone as homogeneous
    adsorbed contaminants, influence of, 74, 75
    dipalmitoyl phosphatidylcholine (DPPC), 76
    first-principles approach, 74
    general expression for, 73, 74
    sensitivity factors, 74
  Auger peaks
    and Auger parameter, 64, 65
    photoionization and, 63, 64
  energy scale referencing
    binding energy, 61
    kinetic energy, 60
  photoelectron peaks and background
    background intensity, 62
    energy loss, 61

photoionization cross-section, 70
quantification of
  basic equation for, 69, 70
  sample irradiation, X-rays and emitting photoelectrons, 67–69
secondary lines due to
  multiplet splitting, 66
  plasmon structures, 67
  shake-up satellites, 66, 67
  source satellites, 65, 66
valence band
  factors affecting spectrum in, 63
  spectra of LLDPE and PP, 62, 63
XPS spectrometer
  analyzed zone, heterogeneity of
    aluminum coating, 89
    background shape, 89
    function of tilting, 87
  chemical shifts
    biochemical compounds, 80–87
    polymers, 79, 80
  components of, 51
  data interpretation, by simulation
    continuous adlayer, 90–94
    discontinuous adlayer, 94–97
    mixture of particles, 97
    rough surfaces and porous solids, 97
  data interpretation, general
    background subtraction, 77, 78
    peak decomposition, 78, 79
    peak smoothing, 77
  instrument setting
    accuracy, 107–109
    artifacts, 104, 105
    limit of detection and quantification, 109
    peak width, 105
    signal/noise ratio, 105–107
  intensity/energy response function (IERF)
    of area of collection, 72
    dependence on, 73
  lens system and energy analyzer, 54, 55
  performance evaluation criteria for, 103
  pumping systems, 52
  sample handling
    powder specimens, 52
    specific accessories, 53
  sample preparation
    adsorbed biomolecules, 99, 100
    low temperature analysis, 101, 102
    sample alteration during, 102, 103
    surface reorganization, 100, 101
  schematic representation of, 47

systematic errors, 108
vacuum in, importance of, 52
X-ray sources
  metal target, 52
  tunable synchrotron sources, 52, 53
X-ray fluorescence, 104
X-ray image enhancement, application of, 189, 190
X-ray photoelectron spectroscopy
  analyzed area, 56
  applications, 44 (*see also* XPS applications)
  chemical composition analysis, 44
  constant analyzer energy (CAE) mode, 55
  dental implants, 135
  electron energy levels, 48
  electron multipliers, 56
  image acquisition
    parallel, 57
    serial, 56
  limitations of, 149
  line scans of, 143
  microbial sample preparation
    approaches, 116
    freeze drying, 117
    saccharomyces cerevisiae, 117, 118
    vitrification, 118
  principle of
    electron photoejection process, 47
    kinetic energy, electrons, 46
    photoelectric effect, 46
  surface composition, 138
  thin layer analysis, 50
X-ray sources
  magnesium, 65
  metal target, 53, 54
  monochromatized, 105
Xu, H., 298

Y

Yaglom, A.M., 202
Yates, B., 202
Yin, Y., 286
Yoshinaga, K., 285, 291
Yu, C.P., 204
Yu, G., 247

Z

Zahmatkesh, I., 253, 254
Zamankhan, P., 247, 248, 250
Zhang, H., 203, 204, 239, 245
Zukoski, C.F., 284

Printed in the United States of America